Natural and Enhanced Attenuation of Contaminants in Soils

Second Edition

Raymond N. Yong
Catherine N. Mulligan

CRC Press
Taylor & Francis Group
Boca Raton London New York

CRC Press is an imprint of the
Taylor & Francis Group, an **informa** business

CRC Press
Taylor & Francis Group
6000 Broken Sound Parkway NW, Suite 300
Boca Raton, FL 33487-2742

Printed on acid-free paper

International Standard Book Number-13: 978-1-1380-6637-3 (Hardback)

Library of Congress Cataloging-in-Publication Data

Names: Yong, R. N. (Raymond Nen), author. | Mulligan, Catherine N., author.
Title: Natural and enhanced attenuation of contaminants in soils / Raymond N. Yong and Catherine N. Mulligan.
Other titles: Natural attenuation of contaminants in soil
Description: Second edition. | Boca Raton : Taylor & Francis, a CRC title, part of the Taylor & Francis imprint, a member of the Taylor & Francis Group, the academic division of T&F Informa, plc, 2019. | Previously published: Boca Raton : Lewis Publishers, c2004 under the title Natural attenuation of contaminants in soil.
Identifiers: LCCN 2018057925 | ISBN 9781138066373 (hardback : acid-free paper)
Subjects: LCSH: Soil remediation. | Hazardous wastes--Natural attenuation.
Classification: LCC TD878 .Y657 2019 | DDC 628.5/5--dc23
LC record available at https://lccn.loc.gov/2018057925

Visit the Taylor & Francis Web site at
http://www.taylorandfrancis.com

and the CRC Press Web site at
http://www.crcpress.com

Contents

Preface

Over the past two decades, there has been increased acceptance of the use of the natural attenuation capability of soils as a tool/technique in remediation of contaminated sites. So long as proper and complete knowledge of site properties and characteristics such as underlying soils, hydrogeological settings, etc. are obtained together with knowledge of the nature and distribution of the contaminants in the site, the use of natural attenuation (NA) as a remediation tool can be fruitfully considered. Acceptance of this kind of passive remediation tool has been gained in many jurisdictions around the world, and the experiences garnered from there have directed more attention to the use of NA as a remediation tool. The key requirements accompanying the promotion and use of this passive remediation tool are: (a) competent monitoring of remediation progress, (b) multiple lines of evidence throughout, and (c) complete awareness and acceptance of the knowledge-based caveat.

To a large extent, determination of applicability of NA has relied on bench and pilot tests, history of natural remediation processes in the contaminated site area, previous experience, records of contaminant transport at the candidate site, etc. as input for development of the knowledge-based decision for application of NA as a passive remediation process. The primary aim of most NA treatment of contaminated sites is to eliminate or reduce bioavailability of the contaminants in the contaminated site. For this to happen, a detailed knowledge of the outcome of interactions between contaminants and soil compositional features and activities of microorganism is needed. Monitoring the progress of NA at the contaminated site is one of the key requirements in keeping track of "what is happening in the ground." The complete NA process, which is termed *monitored natural attenuation* (MNA), has been well formalized by several regulatory bodies through the issuance of guidelines and protocols.

The focus of this book has been to develop an appreciation of the interaction mechanisms between contaminants and soil compositional features and the activities of micro-organisms in degrading and transforming organic chemical compounds. In short, the discussions in the various chapters of the book try to answer the questions of "What is happening in the contaminated soil?", "How do these interactions affect the bioavailability of the contaminants?", and "What do we know about the transport and fate of the contaminants?"

Chapter 1 discusses the various kinds of inorganic- and organic-type contaminants that find their way on and into the ground, and application of the knowledge-based passive remediation technique of contaminated sites/ground using the natural attenuation capability of the subsoil known as MNA. Reliance is placed on the natural processes occurring in the soil that serve to reduce the toxicity of the contaminants and/or the concentration of the contaminants. They involve a range of physical actions, chemical and biologically mediated reactions, and combinations of all of these. Detailed protocols and guidelines which can be site-specific or generic, are designed to cover the essential requirements, procedures, standards, criteria, etc. that need to be used in determining the multiple lines of evidence and the effectiveness of the MNA process.

Chapter 2 deals with the nature of soils and pays particular attention to those soil fractions that contribute directly to the assimilative capacity of soils. The main items and issues considered in this chapter include: (a) nature of soil, such as origin, geologic, and regional controls, soil fractions and compositional control on development of pertinent soil features, properties, and characteristics, (b) chemically reactive groups (defined as *surface functional groups*) on the surfaces of the various soil fractions making these surfaces reactive, thus allowing the particles to react chemically with solutes and other dissolved matter in the porewater, and (c) mechanisms associated directly with contaminant assimilation and attenuation (i.e., soil uptake) that can be optimized when all the available soil solids' surfaces interact with the solutes in the porewater.

The discussions in Chapter 3 deal with interparticle actions in the presence of an aqueous phase. The intent of this chapter is to inform the reader that there are many more mechanisms involved in contaminant attenuation than microbial intervention. The nature of the reactive surfaces in the soil-water system and how these surfaces are obtained will give us insight into how the soil conditions in the field will impact directly on the transport and fate of the contaminants under consideration. This requires one to develop a more detailed understanding of the interactions or processes that occur in the soil depend not only on the properties and characteristics of the contaminants, but also on the properties and characteristics of the soils. All too frequently, assessments and evaluations of natural attenuation of contaminants overlook the fact that the abiotic and biotic reactions between contaminants and soils need the same detail in determination of the various attributes of the soil that participate in those reactions.

Chapter 4 discusses the interactions between heavy metal contaminants and soil solids, together with the various reactions in contaminant-particle interactions that contribute to the assimilative capacity of soils. These are prime mechanisms and processes that contribute to the attenuation of heavy metal contaminants during their transport in soils. One needs to have knowledge of (a) the interactions between soil particles and heavy metals, and particularly of (b) processes that lead to partitioning. In addition, there are at least five factors one should consider in evaluating environmental mobility and bioavailability of heavy metal contaminants in the ground. These are: (1) hydrogeological setting, (2) changes in pH-regime of the affected ground, (3) changes in the ionic strength of the contaminated ground or soil-water system, (4) changes in the oxidation-reduction potential of the system, and (5) formation of complexes.

The discussion in Chapter 5 deals with the interactions between organic chemicals and soil fractions—i.e., soil material making up the subsoil. Although some of the mechanisms and processes involved will be like those attributed to interactions between soil fractions and inorganic contaminants, others will be different due to the specific properties of organic chemical contaminants. Chemical mass transfer is responsible for partitioning of contaminants in the fate and transport of contaminants. Oxidation-reduction reactions can also play important roles in the fate of the contaminants. We need to know how partitioning of the organic chemical compounds occur and also the degree to which an organic chemical pollutant is available for biologically mediated transformations.

Chapter 6 is concerned with the role of micro-organisms in soils in respect to the degradation and transformation of organic contaminants in particular, and certain applicable inorganic contaminants. The various types of common micro-organisms found in typical soils are first presented to indicate to the reader the wide variety of micro-organisms available for interaction with the organic chemicals and methods for identifying their presence and diversity. The main emphasis in this chapter is on the biological treatment of contaminants. Various types of organisms involved will be described, and growth and metabolism of bacteria, in particular, will be emphasized. The types of chemicals that can be biodegraded or transformed, and the pathways of conversion are discussed as well as the toxicity and availability of several chemicals. A proper understanding of these aspects is very important because they will serve as the foundation of knowledge required for determining the potential for natural attenuation, and especially those facets dealing with natural bioremediation.

How does one know whether the natural attenuation process is applicable as a remediation tool to the contaminated site in question? Chapter 7 presents the essential items needed for assessment of the potential for NA as a treatment process. Numerous case histories of recent and ongoing MNA treatment of contaminated sites are presented. These are followed by the discussion in Chapter 8 focusing on the use of MNA as a passive remediation and/or intrinsic remediation technique, stating that MNA can work well when evidence- and science-based pieces of information are properly assessed. A prime requirement for application of the natural processes in the soil to remediate contamination by natural attenuation processes is: (a) a strict monitoring program, (b) supporting laboratory tests, (c) studies on partitioning mechanisms, (d) development of mechanistic and analytic transport and fate models, and (e) protocols and acceptance criteria for the multiple lines of evidence.

In Chapter 9, the discussion deals with the two main reasons for utilizing enhanced natural attenuation processes as a remediation tool include: (a) shortening the time required for NA processes to reach remediated standards or criteria, and (b) controlling and directing—i.e., physically managing—the contaminant plume to optimize NA effectiveness. All the enhancement options falling within these two main reasons or categories should result in the improvement of NA as a passive or semi-passive contaminated site remediation tool. If assessment of NA is not working as expeditiously as expected as a remediation tool in the contaminated site under treatment, one should investigate the use of enhanced NA (i.e., ENA) treatments. The two avenues for enhancing natural attenuation processes are: (a) introduction of "aids" to the indigenous micro-organisms to obtain greater effectiveness from these micro-organisms, and (b) improving the substrate soil to obtain more effective partitioning of the contaminants.

As we have stated before, the material presented in this book provides the picture of contaminant attenuation from the soils' point of view. Because of the multi-disciplinary nature of the material covered in this book and the importance of the role of soil solids in the natural attenuation of contaminants, we have had to make some difficult decisions as to the amount and level of "basic theory" needed to be presented. The authors wish to acknowledge the benefit received from interactions and discussions with their colleagues, research students, and from reports of the professionals in this challenging work.

Raymond N. Yong

Catherine N. Mulligan

Authors

Raymond N. Yong is the William Scott Professor Emeritus at McGill University, Canada. He has authored and co-authored 14 other textbooks, over 500 refereed papers in the various journals in the disciplines of Geoenvironmental Engineering and Earth Science and holds 42 patents. He is a fellow of the Royal Society (Canada), and a Chevalier de l'Ordre National du Québec. He and his students were among the early researchers in Geoenvironmental Engineering and Science engaged in research on the physico-chemical properties and behavior of soils, their use in buffer/barriers for HLW (high-level radioactive waste) and HSW (hazardous solid waste) containment and isolation, and restoration/remediation of contaminated sites. He and his colleagues are currently engaged in research on Geoenvironmental sustainability.

Catherine N. Mulligan holds a Concordia Research Chair in Geoenvironmental Sustainability (Tier I) and is a full professor in the Department of Building, Civil and Environmental Engineering, Concordia University, Montreal, Canada. She has authored more than 100 refereed papers in various journals, authored, co-edited, or co-authored 8 other books, holds 3 patents, and has supervised to completion more than 60 graduate students. Her research is related to the decontamination of soils, sediments and mining residues and treatment of surface water and wastewater. She is the director of the Concordia Institute of Water, Energy and Sustainable Systems that performs research into new systems, technologies, and solutions for environmental sustainability. She is a fellow of the Canadian Society for Civil Engineering (CSCE) and the Engineering Institute of Canada (EIC) and the winner of the John B. Sterling Medal of the EIC.

1 Managing Soil Contamination

1.1 SITE AND SOIL CONTAMINATION

The label *contaminated sites* is often used to identify sites that have been found to have underlying soils that contain noxious substances that can cause serious injury to the immediate environment and its biotic receptors. These noxious substances are most often referred to as *contaminants, pollutants* and *toxicants.* They commonly occur in soils in connection with anthropogenic activities associated with the range of industries involved with the production of goods and services—including municipal and domestic services. Inadvertent spills, liquid and solid waste discharges, and waste containment and management strategies are also some of the sources of the contaminants.

For the discussions in this book, we will concentrate on the many aspects and factors that attend the application of natural and enhanced contaminant attenuation capabilities of soils for management and remediation of contaminated soils (in contaminated sites). Soil contamination is a central issue in contaminated ground since it is the mantle of the solid land mass that we call earth. To fully appreciate why soil contamination is a major issue and why managing implementation of remedial measures is also a major concern, two particular points need to be considered. Firstly, soils play a central role in sustaining life-support systems and all the other aspects associated with urbanization and industrialization, development and growth. Simply put, one could ask, "Why are soils so important a factor in our lives?" There are many answers to that one simple question—beginning with a realization of the fact that soil is not only a natural resource, but also a habitat and an essential element in the overall life-support systems of humankind, flora and fauna. In short, one needs to take into account the fundamental role and primary importance of soil in terms of a *soil ecosphere* in the overall context of a geoenvironment (see Figure 1.1). Soil is a natural capital of the geosphere and any degradation or reduction of its quality is a loss in its potential functionality. Two significant attributes of soil impacted by soil contamination are *soil quality* and *site functionality.* The term *soil quality* refers to the value of the soil in its many functional capabilities, i.e. what the soil can be used for, and the term *site functionality* refers to the many uses of site where the soil resides. These attributes, which describe the importance and role of soil in life supporting system, will be discussed in further detail in the latter part of this chapter.

Secondly, soil contamination degrades the value or usefulness of the site wherein the soil resides. The terms *contaminated ground* and *contaminated site* identify sites where soil contamination exists and where usefulness of the site (*site functionality*) is compromised. Further than that aspect is the potential of the contaminants in the contaminated site to become environmental mobile, thus presenting health threats

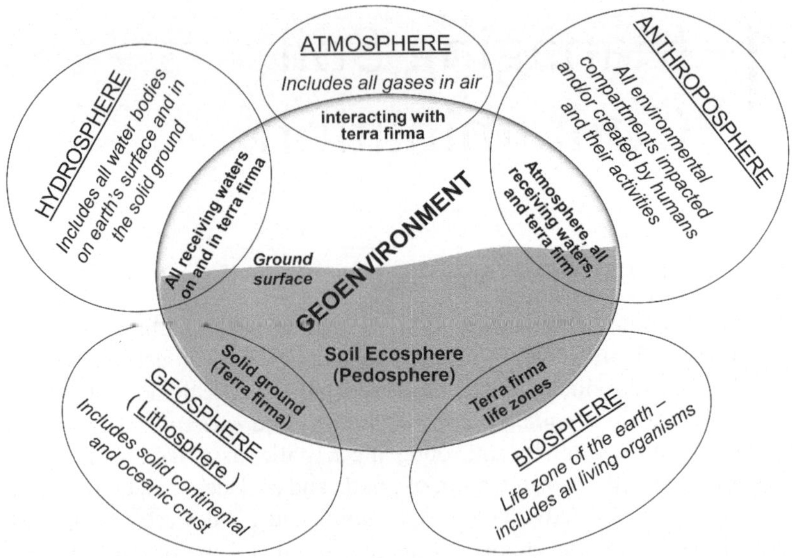

FIGURE 1.1 Venn diagram showing the various elements in the five ecosphere components (atmosphere, hydrosphere, geosphere, biosphere, and anthroposphere) that make up the geoenvironment. (From Yong, R.N. et al., *Sustainable Practices in Geoenvironmental Engineering*, CRC Press, Boca Raton, FL, 535, 2015.)

to biotic receptors. These aspects will be discussed in brief in the latter part of this chapter and will be further discussed in the latter chapters of this book.

1.1.1 Soil and the Geoenvironment

The term *geoenvironment* refers to the *terra firma* of the environment. Included in the geoenvironment are the receiving waters in the landscape, such as ponds, rivers, lakes, wetlands, estuaries, groundwater, and aquifers. The major constituent of terra firma is the *soil ecosphere*. Figure 1.1 shows the geoenvironment as a compartment consisting of various elements of the five ecosphere components (atmosphere, hydrosphere, geosphere, biosphere, and anthroposphere). These include a significant portion of the geosphere and portions of the atmosphere, hydrosphere, biosphere, and anthroposphere. The interactions of the elements from the different ecospheres in the land mass that contains the soil ecosphere create an environment that is specific to the land mass and all that lie within and on the land mass. This is what defines the geoenvironment. The soil ecosphere, which is the *pedosphere*, is the sum total of the interactions of the various compartments or ecospheres. This is the biophysical system wherein all the biotic and abiotic organisms in the various ecospheres interact, and is a system where the various individual elements and organisms interact singly or collectively to the advantage or detriment of the whole. It is easy to see why soil contamination is a condition that bears severe consequences. In essence, soil is an essential primary resource in the production of many of the goods required for life support.

The *geosphere* consists of the inorganic mantle and crust of the earth. These include the land mass, the oceanic crust, and the solid layers (soil and rock mass) stretching downward from the mantle and crust. The *terra firma* component of the geosphere forms the solid mass of the geoenvironment while the *hydrosphere* includes all forms of water on Earth such as the oceans, rivers, lakes, ponds, wetlands, estuaries, inlets, aquifers, groundwater, coastal waters, snow, ice, etc. Not all of these forms of water are in the geoenvironment. The geoenvironment contains all the receiving waters located within the terra firma in the hydrosphere. These include not only the rivers, lakes, ponds, inlets, wetlands, estuaries, coastal marine waters, groundwater, and aquifers, but also the coastal (marine) environment since these waters are impacted by the discharge of contaminants in the coastal regions from land surface runoffs, and discharge of polluted waters from rivers or streams.

The *biosphere* is the ecozone that contains all living organisms. However, only those life zones in or on terra firma from the biosphere are considered to be part of the geoenvironment. In respect to the geoenvironment, the elements of concern in the *atmosphere* are all the gases in air that interact with terra firma. Whilst the *anthroposphere* refers to all the ecospheres and their compartments impacted by humans and their activities, the geoenvironment aspects of the anthroposphere include only those compartments of the various ecospheres that interact with terra firma and its receiving waters.

1.1.2 Contaminants, Pollutants, and Toxicants

The terms contaminants, pollutants, and toxicants have been used in the literature in reference to perceived undesirable (noxious) substances on the ground surface and in the ground, e.g., contaminants in soil, pollutants in soil, contaminated ground, etc. without sufficient clarification as to the real import of the terms. These substances can include hazardous materials or elements, toxic substances, pollutants, and all other substances that are non-threatening to human health and the environment. For this book, the term *contaminants* used in the context of soil contamination refers to substances that are not indigenous to the location under consideration. More specifically, in the case of the soil ecosphere, the term contaminants refers to substances or materials that are not directly associated with soil formation, or simply put, "what soil is," i.e., an accumulation of disintegrated rock fragments with particle sizes ranging from somewhere close to 100 mm to sizes that cannot be seen by the naked eye—less than 0.0001 mm (in the μm range)—obtained from the physical and/or chemical disintegration of rock. The subject of "what soil is" will be discussed in detail in Chapter 2.

Pollutants in soils are contaminants that have been or are classified by regulatory agencies as substances being harmful to the health of humans, other biotic receptors, and the environment. The toxic chemicals or compounds listed in the U.S. Environmental Protection Agency (EPA) Priority Pollutants List are good examples of pollutants associated with industrial activities which have been found in contaminated sites. These include chlorinated benzenes and ethanes, chlorinated phenols and naphthalenes, and pesticides—i.e., PCBs, PAHs, CHCs, VOC, POPs, etc. All pollutants are contaminants, but not all contaminants are pollutants.

Toxicants are contaminants that have been judged or classified by regulatory agencies as being toxic, and hence are harmful to human health and other biotic receptors. All toxicants are contaminants, but not all contaminants are toxicants. The discussions in this book will use the more general encompassing term contaminants in most instances, without necessarily meaning that these contaminants are pollutants or toxicants. When necessary, for emphasis or for clarity, the terms pollutants and/or toxicants will be used.

1.2 SOURCES OF CONTAMINANTS FOUND IN CONTAMINATED SITES

While the greater proportion of contaminants found in soils underlying contaminated sites in the natural landscape of the geoenvironment and in developed environments is to a very large extent directly or indirectly associated with anthropogenic activities, one should be aware that geogenic contaminants do exist as a natural consequence of soil formation and weathering. Natural causes or events can result in contaminating ground surface and/or soil, for example such phenomena as earthquakes, floods, tornadoes, landslides, and dust storms. The December 26, 2004, Banda Aceh and March 11, 2011, Fukushima earthquakes are good examples of not only sources of contaminants from distressed structures and infrastructures, but also sources of contaminants from the results of subsequent events, such as tsunamis and floods triggered by the earthquakes. The release of radioactive materials into the atmosphere from the explosions of Units 1, 2, and 3 of the Fukushima Daiichi Nuclear Power Plant buildings, in the period of March 12–15, 2011, resulted in contamination of ground surfaces by these same materials. Radioactive nuclides such as Kr, Xe, I, Cs, Sr, Ru, and Ce, released into the atmosphere contaminated a wide area of the surrounding land surface (Nakano and Yong, 2013). Radioactive cesium ^{137}Cs, because of its strong photon energy and long half-life, is a pollutant of considerable concern, not only because of its presence on the land surface but also because of the potential for transport (infiltration) into the underlying soil. Anecdotal reports estimate that about 20% of ^{137}Cs released in the explosions were deposited on the land surfaces between 60 and 400 km from the power plant site.

Many of the anthropogenic-related actions responsible for direct and/or indirect contamination of the geoenvironment and soil in particular can be attributed to the operation of life-support industries and activities, and production of goods and services. Figure 1.2 shows some of these activities and how they contribute or result in contaminating the underlying soil. While many of these are not deliberately designed or intended to contaminate the geoenvironment and the soil, they nevertheless are consequences derived as a result of anthropogenic actions.

Human activities in development, growth, and support of industrialization and urbanization produce waste streams and discharges that if not properly treated and contained will contaminate the geoenvironment—meaning that they will ultimately contaminate the underlying soil. The construction and operation of tailings dams to contain discharged tailings from resource mining activities, and the establishment and management of landfills, waste isolation, and barrier systems are activities that require total capture of waste products. Unfortunately, mishaps and long-term

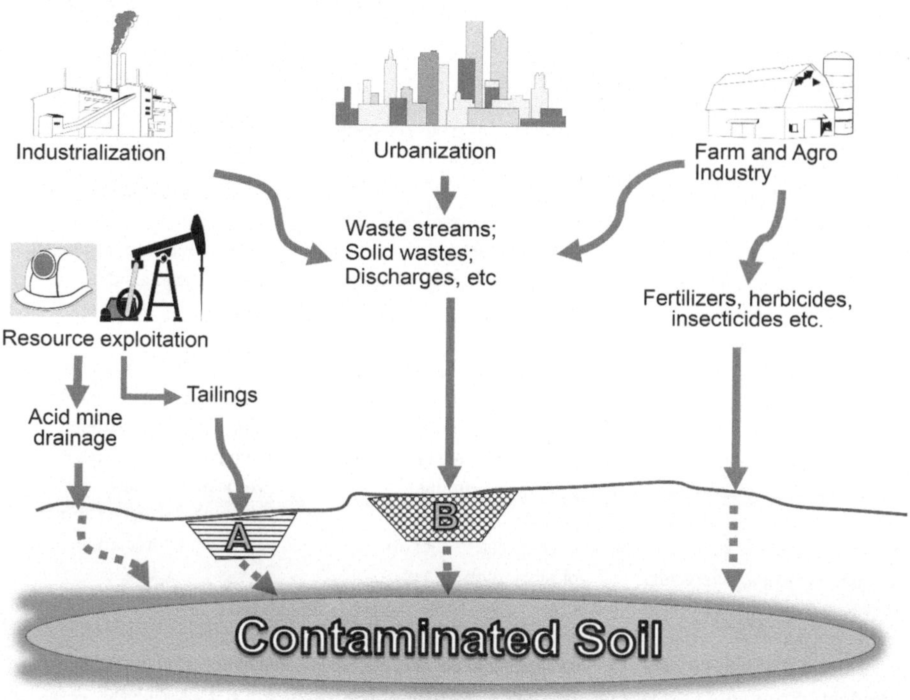

FIGURE 1.2 Sources of contaminants due to anthropogenic activities. Symbols A and B in trapezoids above Contaminated Soil refer to tailings pond (A) and waste landfill (B), respectively.

system competence issues will result in the release of contaminants into the ground and the surrounding soil. Wastewater and solid waste discharge, spills, leaks, and other forms of discharges to the land environment, together with leachates escaping from landfills designed to contain wastes and other hazardous materials in the land environment, are notable sources of contaminants. Other sources of contaminants in soils can be traced to the use of chemical aids in agricultural (agro) industry practices and in pest control such as fertilizers, herbicides, insecticides, pesticides, and fungicides. Agricultural activities in conjunction with agro industry will produce farm wastes, animal wastes together with fertilizers, insecticides, pesticides, etc. Herbicides and pesticides are: (a) persistent in the environment, (b) highly mobile, and (c) can accumulate in the tissues of animals, producing a variety of ill effects. These will serve as non-point source contaminants that find their way into the ground and receiving waters of the geoenvironment.

Table 1.1 gives a short account of some of the typical contaminants found in the waste streams of various industries, and Figure 1.3 gives a more detailed look at some of the kinds of discharges that find their way into the geoenvironment and the underlying soil. Some of the more common contaminants found in many contaminated sites reported in the literature and field reports include: heavy metals (compounds of lead, mercury, copper, cadmium, chromium), petroleum hydrocarbons (PHCs),

TABLE 1.1
Typical Contaminants Obtained from Various Industrial Waste Streams

Typical Industries	Sources for Major Waste Streams	Typical Contaminants
Metal Production, Industries and Users Electronic industry Metal manufacture and goods Metallurgical activities Machinery production and use	Steel pickling; metal finishing; metal cleaning; metal surface treatment; galvanizing; plating; anodizing; metal working and production; smelting	Spent acids (sulphuric, chromic, pickle liquor, etc.); spent alkalis; spent reagents; metal sludges; metal residues; solvents
Chemical Production, Industries, and Users Pharmaceuticals; Production of chemicals and feedstocks, e.g., inks, plastics, petrochemicals, explosives, pesticides, insecticides, fungicides, soaps, detergents, petrochemicals, paints varnishes, glues, etc.	Chemical manufacture and synthesis; pigments; fine chemicals manufacture; pharmaceuticals manufacture; petroleum refining; production of chemical intermediates, phenols, plastics, rubbers, etc.	Spent acids and alkalis; metal sludges; chlorinated solvents; aromatic amines; nitrated phenylamines; chlorinated hydrocarbons; halogenated and non-halogenated solvents; petroleum hydrocarbons; etc.
Agriculture, Pulp and Paper Products, Printing Agriculture and Animal husbandry; Horticulture; Paper and cardboard production; Printing and photography	Application of fertilizers, fungicides, herbicides; food processing; livestock wastes; photographic and printing processes	Phenolics, carbamates, benzoic acids, organochloride compounds, phenoxyacids, triazines, substituted ureas, etc.; non-halogenated solvents, methanol, acetone, various glycols, ketones and acetates, cyclohexane, xylene, etc.
Petroleum, Gas, Consumer Goods—Production Petroleum, oil and gas extraction and production; Textile and clothing industry; Leather and wood working industries	Gas works; oil and petroleum refining; wood preservatives; textiles and leather production	Spent caustic; acid tars; heavy metal sludges; oil sludges; distillation residues; halogenated and non-halogenated solvents; etc.

halogenated and non-halogenated organic chemicals and solvents, and PCBs. The common types of contaminants found in contaminated ground can be grouped into two specific groups. These include:

- *Heavy metals* (*HMs*)—Heavy metals are elements that easily lose electrons to form positive ions, and although there are 39 elements which classify as HMs, the majority of these are not found in significant quantities in contaminated ground. The HMs are generally obtained as waste streams from mining, ore refining, metal producing and electroplating industries. The common ones which have been reported in contaminated ground include: cadmium, chromium, copper, iron, lead, mercury, nickel, silver, tin, and zinc.

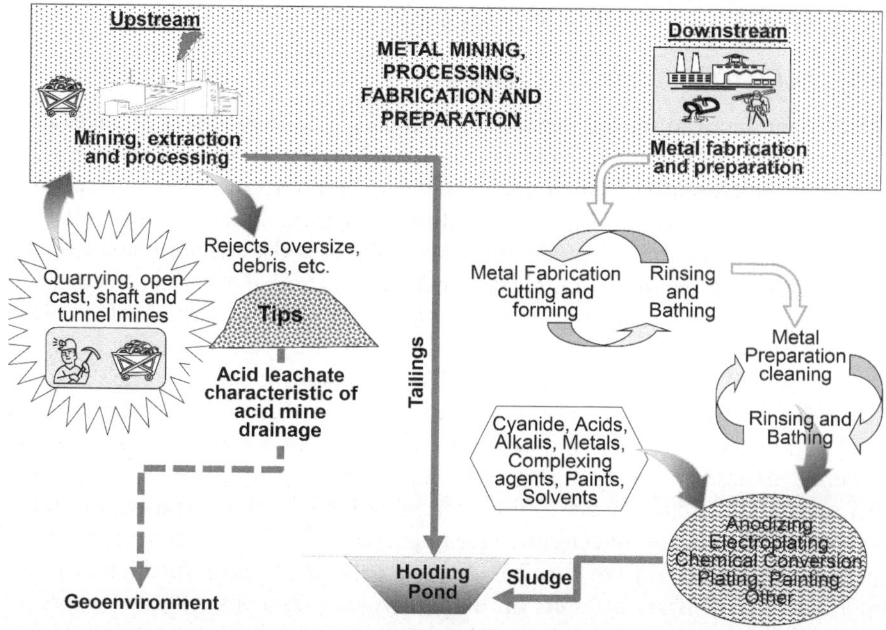

FIGURE 1.3 Example of some of the more prominent types of stressors originating from metal mining, processing, fabrication, and preparation. Not shown in the illustration are the various wastewater discharges at the various stages of extraction-processing and rinsing-bathing sequences. (From Yong, R.N. et al., *Environmental Soil Properties and Behavior*, CRC Press, Boca Raton, FL, 435, 2012.)

They can exist in their elemental forms or as compounds in the porewater or attached (partitioned) to the soil solids. The detailed discussion of the interactions between these contaminants and soils can be found in Chapter 4.

- *Organic chemicals/compounds*: Activities associated with industrial production and use of petroleum products, solvents, pesticides, and explosives are some of the major sources of organic chemicals/compounds found in contaminated sites. Production of industrial intermediates using aliphatic and aromatic compounds can also be included in the above. The chemicals and compounds listed in the *Toxicity Characteristic Leaching Procedure* (TCLP) regulatory levels limit for characterization of a chemical as *toxic* are good examples of many of the chemicals of concern. These include: benzene, carbon tetrachloride, chlordane, chlorobenzene, *o*-cresol, *m*-cresol, *ρ*-cresol, 1,4-dichlorobezene, 1,2-dichloroethane, 1,1 dichloroethylene, 2,4 dinitrotoluene, endrin, heptachlor, hexachlorobenzene, hexachloro-1,3-butadiene, hexachloroethane, lindane, methoxychlor, methyl ethyl ketone, nitrobenzene, pyridine, tetrathcloroethylene, toxaphene, trichloroethylene, 2,4,5-trichlorophenol, 2,4,6-trichlorophenol, silvex and vinyl chloride. Chapter 5 gives a detailed discussion of the interactions between these contaminants and soils.

1.3 IMPACT OF CONTAMINANTS IN SOILS

1.3.1 Soil Status or Functionality

A significant factor in the degradation of land use, i.e., functionality of a particular site, is the presence of contaminants in the soil underlying the site. As we have stated previously, the presence of contaminants in the soil presents threats (perceived, real, or otherwise) to human health, the immediate ecosystem, and the environment. To better appreciate the impact of contaminants found in soils, it is necessary to firstly establish or determine the *ab-initio* state or status (functionality) of the piece of soil/site in question. This is achieved by determining a soil parameter that describes the functionality of a soil, i.e., the state of the soil and its purpose or role in the site wherein the soil resides (*ab-initio* state). The term *soil functionality* (*SF*) is used to describe the functionality of a soil at any one time when human intervention is introduced or when some event has impacted the soil in question. In that case, we can define SF as the capacity of a specific piece of soil to function under designed or other unplanned circumstances. Since contamination of soil is a reality, it is necessary to establish a means to determine the impact of soil contamination on its natural functionality. The quality of a soil is directly dependent on its functionality. It is important to remember that soils are living dynamic systems with characteristics and properties that are the products of their: (a) origin, i.e., source material; (b) type and level of maturation, i.e., evolution; (c) location; (d) climate and environmental envelope; and (e) bio-ecosystem and human activities.

Soil functionality is a vital component of *site functionality*. A contaminated soil results in a contaminated site and is therefore a restricted site—meaning that site access and use by humans is, in most jurisdictions, strictly restricted or even prohibited. In most jurisdictions, regulations require clean-up of the site to pristine uncontaminated condition before human access and use of the site is permitted. A measure of how useful a soil is, i.e., a measure of soil quality, before and after human intervention is the *soil functionality index* (*SFI*). In essence, the SFI is a qualitative or quantitative measure of the ratio of *SF*(*after*) to *SF*(*before*), i.e., a soil functionality degradation (or improvement) index. For example, contamination of an uncontaminated soil in an agricultural setting results in a decrease of the usefulness of the soil for agricultural and food growing purposes. This leads to a decrease in the functionality of the soil in regard to food growing purposes and is reflected in a decrease in the SFI of the soil and of the site itself. $SF(after) = SF(contaminated) < SF(before)$. A detailed development of soil functionality determination and soil functionality indicators can be found in Yong et al. (2012).

1.3.2 Health Threats

Health threats resulting from the presence of contaminants in the ground are constant issues and concern with the general public and those parties charged with ensuring public safety. That being said, we need to determine, "What constitutes a threat situation?" It is not unusual for the general public to be alarmed when confronted with a situation where contaminants have been discovered on the ground surface or in the ground nearby or underneath. While this is a normal reaction, there needs to be careful examination of how the health threat will materialize. In other words, how will the contaminant (pollutant or toxicant) in the soil come into contact with humans and other biotic receptors? Would it be through movement of the contaminants to the biotic receptors? Will the contaminants remain dormant or fixed in position in the soil, and if so, will they still pose a threat to human health and other biota? To address the questions and many other related concerns, it is necessary to determine how the threat is delivered to the biotic receptors—i.e., how the threat becomes a reality. This is achieved by examining the different available pathways from the source to the receptors. This process or technique is called the *source-pathway-receptor* (SPR) approach, shown in Figure 1.4. The pathway realized from contaminant source is as much dependent on the type of contaminant as on the specific site conditions—e.g., hydrogeological setting, topography, soil type, etc.

The schematic illustration of the SPR approach depicted in Figure 1.4 shows that materialization of threats to human health and the environment can occur through direct and indirect pathways. This SPR approach is a useful method of analysis for evaluation of the level of threat to the biotic receptors. Actual physical contact by the biotic receptors of contaminants from the source occurs via direct exposure, ingestion and inhalation. The primary means for direct exposure and contact by humans, for example, are for the contaminants to be transported in the ground (a) to a drinking water resource thereby polluting the drinking water resource; (b) to receiving waters such as lakes, ponds, etc. resulting in polluting the resident aquatic species;

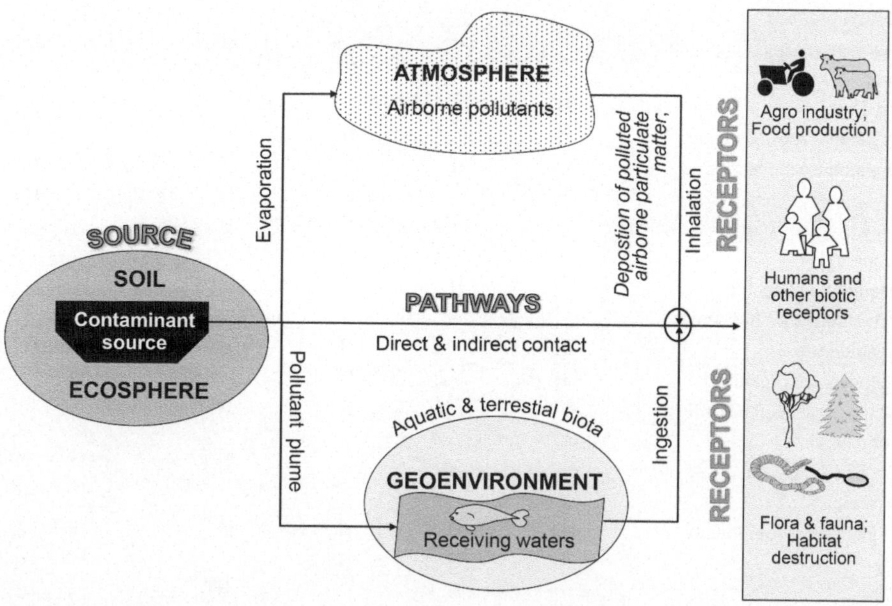

FIGURE 1.4 Schematic showing contaminant source in the ground as the potential source of health and environmental threats. Contaminants carried into the atmosphere by evaporation and deposited on land and receiving waters are potential threats via air, land, and water pathways to the various receptors shown at the right-hand side of the illustration.

and (c) to an agricultural region where the polluted soil serves as a food production resource. While drinking polluted water is classified as *direct contact*, ingestion of polluted aquatic species and polluted agricultural products are generally classified as *indirect contact* with contaminants. Evaporation of volatiles such as the *volatile organic compounds* (VOCs) can lead to diminished air quality and inhalation problems for humans and other creatures. If atmospheric particulate matter becomes contaminated with toxic substances, and if these airborne particulates are deposited onto ground surfaces, they will pose health and environmental threats.

Type and extent of exposure and contact with contaminants are issues that require careful consideration when contaminants are found in the ground. A very limited exposure to a high concentration of a contaminant classified as a pollutant or toxicant may not be as critical in comparison to a long-term exposure at a much lower concentration, to the same substance. The information from Yong et al. (1992) shown in Table 1.2 highlights the acute and chronic effects of some selected contaminants in waste streams on human health, using statistically verifiable information—i.e., obtained from studies

TABLE 1.2
Acute Effects of Some Contaminants on Human Health

Contaminant	Nervous System	Gastro System	Neuro System	Respiratory	Skin	Death
Halogenated organic pesticide	H		H	H		H
Methyl bromide			H			
Halogenated organic phenoxy herbicide					H	
2-4-D	H					
Organophorous pesticide	H		H	H		H
Organonitrogen herbicide		H				H
Carbamate insecticide	H		H	H		H
Dimethyldithiocarbamate fungicide				H		
Aluminum phosphide		H				
Polychlorinated byphenyls		A			H	
Cyanide wastes	H		H	H		H
Halogenated organics	H		H		H	
Non-halogenated volatile organics			H	H		
Zn, Cu, Se, Cr, Ni		H		H	H	
Arsenic		H			H	H
Organic lead compounds	H	H	H			
Hg	H	H	H			H
Cd		H		H		H

Source: Yong, R.N. et al., *Principles of Contaminant Transport in Soils*, Elsevier Science Publishers, Amsterdam, the Netherlands, 327, 1992.

Note: H, A = statistically verifiable effects on humans (H) and animals (A), respectively.

on animals (A) or humans (H). The difficulties in extrapolation from studies on animals relate to the fundamental problem of dosage and basis for extrapolation. The problem of determination of what constitutes a health threat is not easily resolved since determinations of lethal hazards require considerable amounts of biostatistics that may not be full available. Regulatory agencies in most jurisdictions provide toxicant listings and limiting exposure/intake concentrations or values as guidelines or code requirements or restrictions.

1.3.3 Impact on Water and Water Resources in the Geoenvironment

Soil contamination results in contamination of the porewater in soils and by extension (transport of contaminants in the soil), contamination of existent groundwater and the aquifers. The importance in protection of available freshwater—and especially groundwater—is shown in Figure 1.5 which shows that non-saline (non-seawater) is just less than 5% of global water, and more significantly, the largest contribution to the available non-saline water is groundwater. Whilst porewater in compact clays is not easily or readily extractable, the same cannot be said for granular materials such as silts and sands. These materials are more porous than clays and water contained in these kinds of materials are readily extractable. Aquifers, in general, are seams or layers of soils that are more granular and permeable than clays. The term *groundwater* is used to mean both soil porewater (above and below the water table) and water in the aquifers. Contaminants on the land surfaces are transported into the

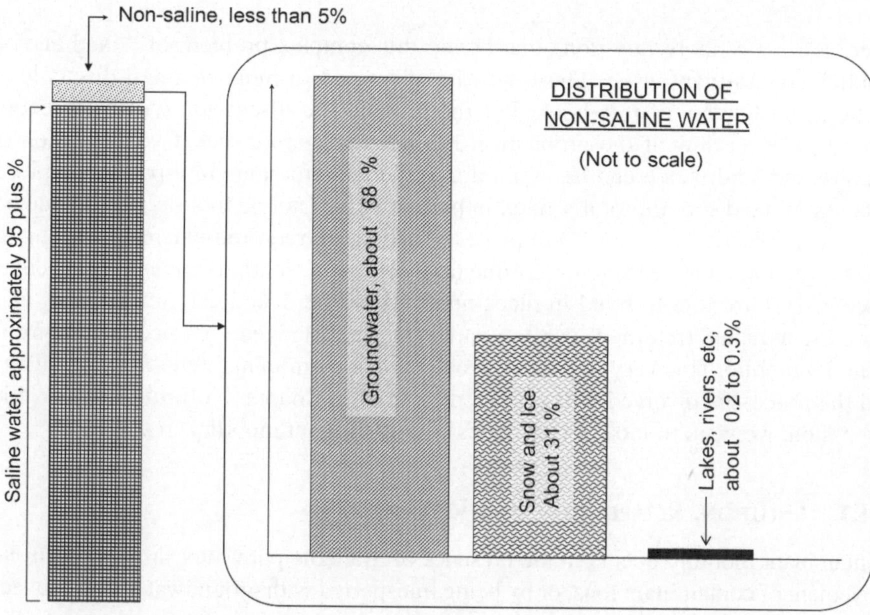

FIGURE 1.5 Sketch of distribution of global water. Both fresh and saline water are shown. (From Yong, R.N. et al., *Sustainable Practices in Geoenvironmental Engineering*, CRC Press, Boca Raton, FL, 535, 2015.)

subsoil via infiltration and other transport mechanisms such as dispersion and diffusion. Pesticides, herbicides, insecticides, organic and animal wastes, ground spills, other discharges of noxious substances on surfaces, etc., when transported into the underlying soil and groundwater will eventually find their way into receiving waters such as aquifers, rivers, lakes, etc. through advective forces and diffusive means. Impairment of groundwater quality means a reduction in the functionality of the host soil and ingestion of polluted groundwater can be detrimental to the health of humans and other biotic receptors.

1.4 MOBILITY AND ATTENUATION OF CONTAMINANTS

The detailed discussion of the transport and fate of contaminants in soils can be found in Chapter 3. For this chapter, the discussion is directed towards the earlier statements regarding the mobility of contaminants in the SPR (source-path-receptor) context. There are several questions to consider when one is confronted with the presence of contaminants in the soil underlying the site of concern.

- Do the contaminants in the contaminated site pose a threat to human health and the environment if they remained immobile—assuming that volatilization is not an issue?
- What will make or allow the contaminants to become mobile?
- What are the mechanisms or processes involved in contaminant mobility?
- And would not contaminant mobility result in attenuation of the contaminants?

There are many more questions that attend this complex problem of "fixed and/or mobile" contaminant issue. These will be addressed in more detailed discussions in the later chapters of this book. For the present, the discussion will be directed towards an overview of the problem. It has been suggested that if volatilization is not an issue, and if one can be assured that the contaminants in a particular piece of contaminated soil are totally fixed in place and will not be mobile, should one be overly concerned? Two significant pieces of action are required when responding to this suggestion. One needs to determine (a) whether the contaminants in a particular piece of soil are totally fixed in-place and will not be dislodged; and (b) how and when the attached (retained) contaminants can be dislodged, i.e., become environmentally mobile. This very important problem of contaminant retention mechanisms and the processes involved will also be discussed in Chapters 3 through 5. For this discussion, we want to look at the modes of contaminant mobility in soils.

1.4.1 Dilution, Retardation, and Retention

Contaminant mobility occurs in the presence of available porewater through diffusive movement of contaminant ions, or by being transported with groundwater flow (advective means). In the situation where the amount or concentration of contaminants in the soil of interest is fixed, i.e., no more contaminants are added to the soil or to the site where the soil resides, any movement of contaminants away from the affected region would in essence reduce (attenuate) the amount or concentration of contaminants left in the affected region. The attenuation of contaminants refers to the reduction of concentrations of contaminants and can be accomplished by: (a) dilution because of mixing with uncontaminated groundwater, (b) interactions between contaminants and soil solids which can lead to partitioning of the contaminants between the soil solids and porewater, and (c) transformation of the contaminants that reduce the toxicity threat posed by the original contaminants. The term *partitioning* refers to the apportioning of contaminants in the porewater between those that are retained (sorbed) by the soil solids and those that remain in the porewater. It is important to note that while sorption processes serve to produce retention of contaminants by soil solids, total or absolute retention of contaminants is not assured since not all the sorption mechanisms provide for permanent retention of contaminants by the soil solids. As will be discussed in Chapters 3 through 5, much depends on the nature of the contaminants and the kinds of soil solids. The surface properties of both these interacting participants are the key to whether retention of contaminants by the soil solids is permanent or temporary.

The schematic illustration shown as Figure 1.6 portrays the results of movement of a pulse (load) of contaminants from a source region. The total amount of contaminants in the pulse is shown by the area framing the vertical axis. The three modes of movement of the original contaminant pulse, dilution, retardation, and retention are

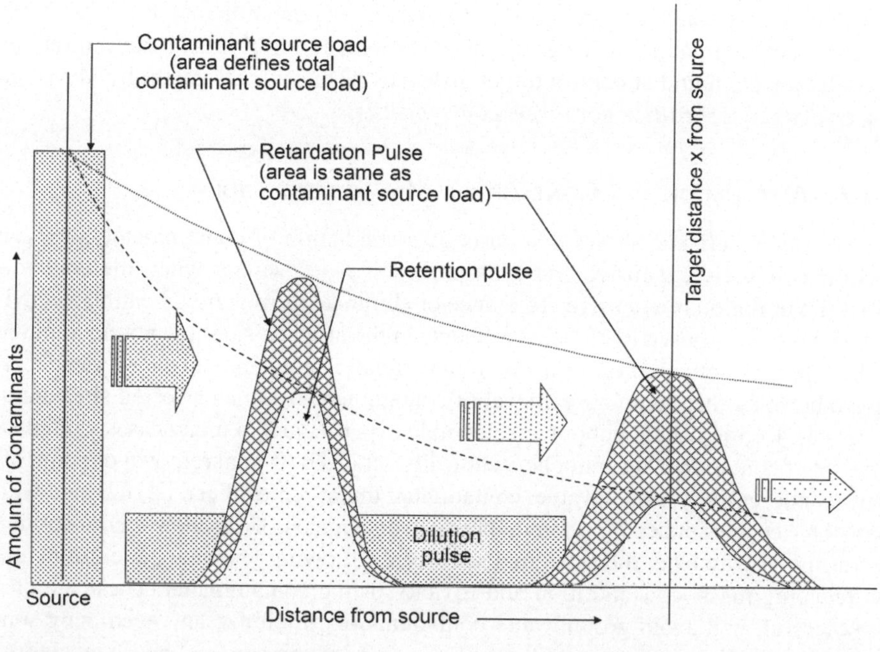

FIGURE 1.6 Illustration of progress of various contaminant pulses. Areas of dilution and retardation pulses are constant and are equal to the original contaminant pulse load. Because of retention mechanisms, the retention pulses will decrease in size as one progresses further away from the contaminant source. Note that the bell-shaped retardation and retention pulses will most likely have different arrival times at the target distance—i.e., the horizontal axis represents distance from source and not elapsed time.

all assumed to be solely responsible for contaminant attenuation. The rectangular-shaped dilution pulse aligned atop the horizontal axis portrays a homogeneous and uniform dilution of the contaminants in the original contaminant pulse. What this shows is the total amount of contaminants from the source attenuated by dilution, and that the concentration of contaminants in the dilution pulse is a function of the amount of water involved in the dilution process. For the other two modes of contaminant mobility, the difference between retardation and retention of contaminants is the delivery of contaminants at distances away from the source. Consider the target distance x away from the contaminant source (Figure 1.6). and the areas under the two bell-shaped curves. The area under the *retardation pulse* is equal to the area of the rectangle framing the vertical axis—i.e., the total amount of contaminants at the source point. Retardation of contaminant mobility is due to the tortuosity of movement of the contaminants presented by the soil solids. While attenuation of contaminants during movement of the contaminants results from mobility retardation, the total load of contaminants will sooner or later be delivered to the target point.

In the case of the retention pulse illustrating the movement and partitioning of contaminants, attenuation of contaminants occurs as a result of retention of contaminants by the soil solids. This partitioning process reduces the total amount of contaminants delivered to the target x some distance away from the contaminant source. This is important because attenuation of contaminants at target x reduces the level of health threat posed by the reduced concentration of contaminants. It is quite common for many regulatory agencies to impose criteria and/or limits on concentrations of contaminants found at control target distances from source—generally identified as a *critical concentrations* or *target concentrations*.

1.4.2 Attenuation as a Contaminant Management Tool

Present-day experience shows that costs for remediation of contaminated sites can become extremely expensive and even prohibitive—especially when the contaminated site or region is extensive. To reduce or eliminate the potential health and environmental threats posed by the contaminants in the affected site, it is not uncommon to think in terms of complete removal of the offending contaminants. Experience has shown however that complete removal of contaminants from an affected site cannot always be accomplished without considerable expense, and in many cases, complete removal of contaminants cannot be technically achieved. If total removal of contaminants is not feasible or achievable, contaminant threat removal can be obtained with these three courses of action: (a) immobilizing contaminants to disallow their migration or transport away from their original site, (b) disallowing leachate generation, and (b) reducing mass, concentration and toxicity of the contaminants. These actions, if successful, will avoid groundwater contamination, minimize and even deny contaminant migration and communication between contaminants and biotic receptors.

1.4.2.1 Natural Attenuation

Natural attenuation of contaminants in a soil as a passive remediation treatment process is generally thought to mean the reduction of health threats to biotic receptors and reduction of impacts on the environment through natural processes.

In most cases, this means reduction in the toxicity, concentration, and mass of the contaminants in the soil. The natural attenuation of immobile contaminants most often occurs with processes that include abiotic and biotic reactions, transformations and degradations. However, the more common and general cases for application of natural attenuation as a remedial process are those that involve transport of the contaminants away from their source—as shown in the source-path-receptor illustration of Figure 1.4.

We have seen, from the earlier part of this section and in Figure 1.6 that if partitioning or retention of contaminants by soil solids and soil structure occur during transport of contaminants from source to receptors, a reduction in mass and concentration of contaminants occurs as transport distance increases. This should also result in a reduction in toxicity of the contaminants—assuming that toxicity is as much associated with type of contaminant as with concentration of contaminants. The ability of the soil solids and soil structure to retain contaminants and the various mechanisms and processes involved are discussed in detail in the next three chapters. For this present discussion, the subject of natural and enhanced attenuation will be introduced as a means to manage soil and site contamination. The full discussion of natural and enhanced attenuation can be found in Chapter 3.

The *partitioning* of contaminants is the term that describes the removal of contaminants from soil porewater through various mechanisms of interaction between contaminants and soil solids. From the soil viewpoint, partitioning occurs as a result of the *assimilative capability* of soils, and the capability for natural attenuation of contaminants is referred to as the *natural attenuation capacity* of soils. The assimilative capability or capacity of soils derives its properties from physical and physico-chemical partitioning mechanisms, and chemical and biologically mediated reactions. The assimilative capacity of a soil is a direct function of the surface properties of both soil solids and contaminants and contaminant-soil interactions—i.e., a combination of physical, chemical, and biologically mediated reactions in the soil. The nature and extent of the reactions are dependent on the physical and chemical properties of both contaminants and soil-water systems. In short, contaminant attenuation achieved through the use of the natural attenuation capacity of soils is the net result of the sum of processes occurring in the soil-water system that reduce the toxicity and/or concentration of contaminants in transport through or interaction with these soils. These attenuating processes include: (a) abiotic and biotic reactions, transformations, and degradations; (b) partitioning via sorption, volatilization, and accumulation; and (c) dispersion and dilution. The first two processes occur as a direct result of interactions between soil solids, porewater, pore space, and contaminants, and also within the porewater itself.

We can take advantage of the assimilative capacity of soil by using it as a set of *contaminant sieves*, i.e., sieves or barriers that will remove contaminants as they move through the barrier. These contaminant attenuating soil-engineered barriers can be designed to restrict the rate of flow through the barrier so as to allow maximum reaction between the soil solids and the contaminants. In the illustration shown in Figure 1.7, attenuation of the contaminants in the leachate emanating from the contaminated soil in the affected site occurs in the soil in front of the first soil-engineered barrier. With sufficient information on the surface properties of the

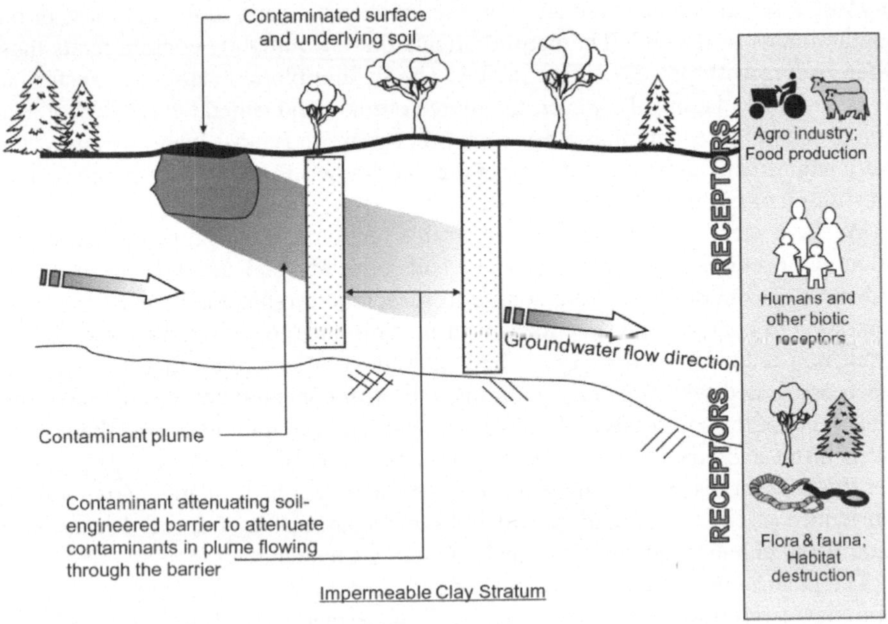

FIGURE 1.7 Schematic illustration of management of contaminants in the contaminated site using an intercepting contaminant attenuating soil-engineered barrier placed so that the advance of the contaminant plume into and through the barrier will attenuate the contaminants.

contaminants and both the subsoil and soil used in the soil-engineered barriers, it is possible to design the system so that by the time the contaminant leachate plume reaches the biotic receptors, little harm will occur. This subject will be discussed in greater detail in Chapter 9.

The use of the natural attenuation capacity of soils as a contaminant management tool is in essence a passive remediation technique. Monitoring of the progress of contaminant threat reduction is required and its effectiveness is assessed through the use of guidelines and criteria. The complete process is called *monitored natural attenuation* (MNA) and is a vital part of the management and remediation of contaminated soils. The protocols and procedures will be discussed in detail in Chapter 8.

1.5 REGULATORY ATTITUDES

The processes and mechanisms of interactions between contaminants and soils are site specific, soil specific, contaminant(s) specific, situation specific, etc. In other words, because the kinds of contaminants are highly variable in amount, type, species, concentration, etc. and because soils are highly variable in composition, etc. it would be difficult for any regulatory agency to set forward exact and detailed specifications for application of natural attenuation as a treatment procedure a contaminated site. Instead, most jurisdictions that accept and promote the use of natural attenuation as a passive treatment procedure have (a) issued guidelines and recommendations on its use; (b) required detailed and comprehensive knowledge of the contaminated site itself, the contaminants and the soils involved; (c) required robust monitoring programs and reporting; and (d) issued guidelines regarding progress of efficacy of natural attenuation through the use of protocols defined as *lines of evidence.* Central to these guidelines and protocol is the requirement for a comprehensive knowledge of the nature of the two "parties" (contaminants and soils) involved. The adoption of guidelines and protocols constitute a passive remediation process generally called MNA. These protocols, which are generally issued by on oversight body (e.g., regulatory commission or body) are designed to ensure *de minimus* practice requirements for protection of "the public good and the environment" at all times. They are especially essential if one chooses to implement remediation of contaminated sites using the MNA procedures. The protocols are designed to require detailed information on the nature and properties of the contaminants—in addition to the development of a good understanding of the mechanisms contributing to the natural attenuation capacity of the soil.

1.5.1 Monitored Natural Attenuation

Monitored natural attenuation (MNA) has been defined in the U.S. Environmental Protection Agency—Science Advisory Board (USEPA-SAB 2001) to mean *a remediation approach based on understanding and quantitatively documenting naturally occurring processes at a contaminated site that protect humans and ecological receptors from unacceptable risks of exposure to hazardous contaminants.* The report further indicates that this is *a knowledge-based remedy because instead of imposing active controls, as in engineered remedies, scientific and engineering knowledge is used to understand and document* ***naturally*** *occurring processes to clearly establish a causal link.*

The use of natural attenuation (NA) as a treatment procedure is "remediation by natural processes." The important point to be noted in almost all guidelines issued by regulatory agencies is that NA is a knowledge-based natural treatment process. This means that one needs a proper understanding of the many principles involved in the natural processes of contaminant-soil interaction that contribute to the end result. These interactions will be discussed in detail in Chapters 4 and 5. Monitoring of the contaminant plume at various positions away from the source is a key element of the use of MNA. Remembering that this (MNA) is a contaminant and soil-specific phenomenon, one generally tracks a very limited number of contaminants, and specifically the ones considered to be the most noxious. Historically, more attention has been paid to documenting the properties and characteristics of the contaminants. By and large, the contaminants tracked have primarily been the organic chemicals including, for example, chlorinated solvents [perchloroethylene (PCE), trichloroethylene (TCE), and dichloroethylene (DCE)], and hydrocarbons such as benzene, toluene, ethylbenzene, xylene (BTEX). However, more attention is now being paid to the nature of the soils involved—not only from the viewpoint of initial soil composition and potential interactions, but also from the perspective of the role of soil fractions. We need to learn about the role of the various soil fractions in relation to processes involving electron transfer and partitioning of contaminants. The outline protocols shown in Figure 1.8 summarizes the various decision steps needed when contemplating application of NA.

The importance for requirement of "knowledge-based" information when considering application of NA as a remediation treatment option cannot be overstated. Figure 1.8 shows that unless proper site conditions exist, there will be no opportunity for NA to work properly. Favorable consideration of NA as a treatment process is based on three separate groups of historical and ongoing evidence, namely (a) historical ground water and/or soil chemistry data that demonstrates a trend of declining contaminant concentration, (b) hydrogeologic and geochemical data that demonstrate NA processes and rates, and (c) field or microcosm studies. The basic pieces of knowledge required vis-à-vis site conditions are topography, hydrogeological setting, hydrogeochemistry, soil type and composition, available micro-organisms, etc. In addition to the information related to the natural setting, detailed information on the nature of the contaminants in the affected site is needed (see Chapters 3 through 5).

Because of site, soil, and contaminant specificities, two other sets of requirements need to be addressed. These are (a) laboratory or table-top contaminant partitioning and release experiments, and (b) analytical studies (modelling) directed towards establishing the fate and transport of released contaminants in the affected soil. Modelling of the performance of the NA process at the specific site of application provides one with some indication of the effectiveness of the process, and also

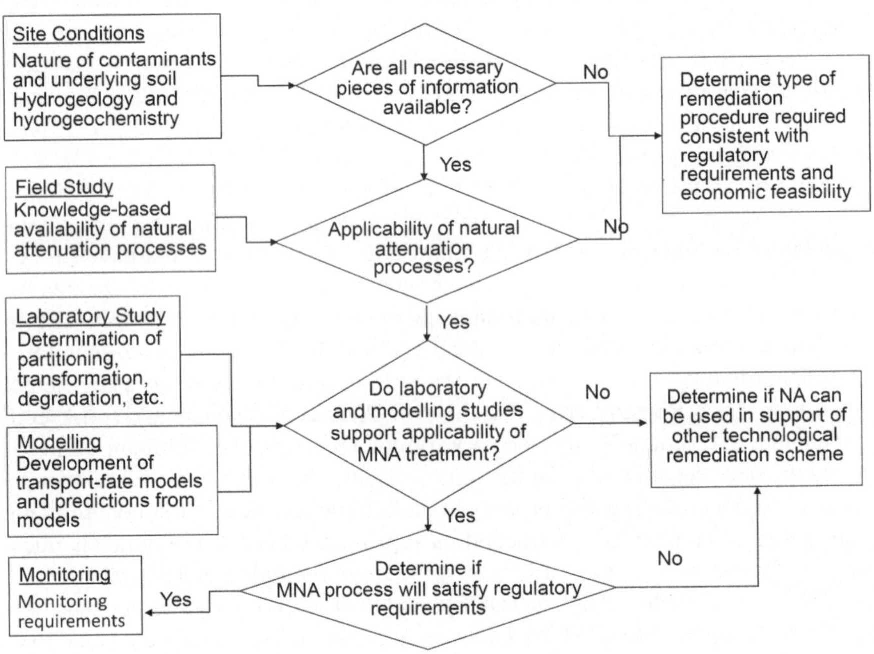

FIGURE 1.8 General protocol for considering natural attenuation (NA) as a remediation tool. MNA = monitored natural attenuation.

informs one of the likely changes in the performance characteristics of the process. These models provide support for the multiple lines of evidence (see Chapter 8 for full details) required as part of the regulatory approval/acceptance of MNA as a treatment procedure. The information obtained from experiments and analyses will provide both guidance and also checks on ongoing performance of NA phenomena—i.e., the information obtained will be in support of the monitoring program as to fulfill the regulatory requirement for the overall natural attenuation remediation treatment. In short, the tests and analyses in combination with the monitoring program constitute the regulatory requirements of most jurisdictions for application of NA—generally known as the MNA process. Monitored natural attenuation is appropriate as a remedial approach only where it can be demonstrated to achieve remedial objectives within a reasonable time frame, and will meet the applicable remedy selection criteria for the particular regulatory program. Detailed discussions of specific MNA applications can be found in Chapter 8.

The keys to successful use of the natural attenuation potential or capacity of soils are:

1. Compatibility between contaminants and soil composition: Proper matching of the contaminants and soils is required to obtain the necessary reactions to obtain: (a) effective and optimum partitioning of the contaminants with the soil solids; (b) transformations and alterations that will reduce toxicity of contaminants.
2. Specification of target concentrations or critical concentrations: Without proper knowledge of target or critical concentrations, i.e., limits beyond which serious threats to environment and biotic receptors could occur, it would not be possible to determine whether attenuation would be a viable tool for pollution management. These limits are obviously contaminant or contaminant-specific and location-specific.
3. Proper monitoring system: The judicious placement of monitoring wells to monitor leachate quality and capture the necessary samples for testing is obvious. Wells upstream within the contaminated zone and downstream of the contaminant plume will be necessary.
4. Continuous analysis of samples retrieved from monitoring stations in a well-developed monitoring protocol: This includes quality testing of both porewater and soil solids to determine partitioning and mobile concentration of contaminants, and also for determination of natural attenuation rates and avoidance of impact on potential receptors.
5. Realistic viable forecasting (analytical modelling) of transport and fate of contaminants.

1.5.2 Enhanced Natural Attenuation

When active controls or agents are introduced into the soil to render attenuation more effective, this is called *enhanced natural attenuation* (ENA). Enhanced natural attenuation refers to the situation where, for example, nutrient packages and other micro-organisms are added to the soil system to permit enhanced biodegradation to occur, or where catalysts are added to the soil to permit chemical reactions to occur more effectively. The use of MNA and ENA for management and control of leachate plumes is now a reality in treatment of contaminated sites (USEPA-SAB, 2001; NRC, 2000; NATO-CCMS, 1999). A full discussion of the necessary guidelines and protocols to be followed in use of these procedures is given in Chapter 9.

It is useful to observe that the attenuation process requires two partners: (a) contaminants, and (b) soil-water system. How they interact to successfully promote attenuation depends on the actions of BOTH partners. Paying attention only to one partner alone (contaminants) denies one the opportunity or capability to fully exploit the process. The intent of the discussions in the succeeding chapters of this book is to provide in-depth discussions of both partners vis-à-vis interactions and resultant attenuation.

1.6 CONCLUDING REMARKS

Natural attenuation refers to the situation when attenuation of contaminants results because of the processes that contribute to the natural assimilative capacity of soil. Reliance is placed on the natural processes occurring in the soil that serve to reduce the toxicity of the contaminants and/or the concentration of the contaminants. The term contaminants is used instead of pollutants and/or toxicants because the attenuation processes apply to the broad range of contaminants and not solely to pollutants and/or toxicants. These natural processes of contaminant attenuation include dilution, partitioning of contaminants, and transformations. They involve a range of physical actions, chemical and biologically mediated reactions, and combinations of all of these.

Managing contaminated sites so that they do not pose any threat to biotic receptors and the environment can take several forms. Active management techniques include aggressive remedial procedures involving the use of such techniques ranging from the traditional "dig and dump" to the more complex techniques involving chemical, biological, hydraulic, and mechanical means in various combinations. These may be well suited to contaminated sites of limited extent, but they may become greatly expensive and complicated when the sites are extensive. The use of a passive approach to management of contaminated sites is the use of the assimilative capacity of soils. With such a soil property, it is possible to obtain attenuation of contaminants as they are transported in the soil. With proper design procedures and knowledge of the interactions that can be favorably achieved to produce partitioning of contaminants, it is possible to structure a contaminated site management program that would reduce or even eliminate health threats to biotic receptors and damage to the environment. In short, the assimilative capacity of soil for contaminants is a fundamental element in the natural attenuation of contaminants in soil. The various contributing factors and interactions in the natural attenuation process include dilution, and reactions in the soil which are physically, chemically, and biologically mediated.

Detailed protocols and guidelines are generally issued by the regulatory or oversight agencies to cover general and specific contaminants. These detailed protocols, which can be site-specific or generic, are designed to cover the essential requirements, procedures, standards, criteria, etc. that need to be used in determining the lines of evidence and the effectiveness of the MNA process.

REFERENCES

Nakano, M., and Yong, R.N. (2013), Overview of rehabilitation schemes for farmlands contaminated with radioactive cesium released from Fukushima power plant, *Eng. Geol.*, 155, 87–93.

NRC (National Research Council). (2000). *Natural Attenuation for Groundwater Remediation, Committee on Intrinsic Remediation*, Water Science and Technology Board and Board on Radioactive Waste Management, Commission on Geoscience, Environment and Resources, National Academy Press, Washington, DC, 274p.

USEPA. (1999), NATO/CCMS Pilot study: Evaluation of demonstrated and emerging technologies for the treatment and clean-up of contaminated land and groundwater (Phase III), EPA/542/R-99/008.

USEPA-SAB. (2001), Monitored natural attenuation: USEPA research program—An EPA science advisory board review, EPA-SAB-EEC-01-004.

Yong, R.N., Mohamed, A.M.O., and Warkentin, B.P. (1992), *Principles of Contaminant Transport in Soils*, Elsevier Science Publishers, Amsterdam, the Netherlands, 327p.

Yong, R.N., Mulligan, C., and Fukue, M. (2015), *Sustainable Practices in Geoenvironmental Engineering*, CRC Press, Boca Raton, FL, 535p.

Yong, R.N., Nakano, M., and Pusch, R. (2012), *Environmental Soil Properties and Behavior*, CRC Press, Boca Raton, FL, 435p.

2 Nature of Soils

2.1 INTRODUCTION

Chapter 1 introduced the subject of natural attenuation (NA) of contaminated soils as a knowledge-based passive remedial treatment process for contaminated sites. It has been pointed out that utilization of this natural remediation process should be based on a good understanding of the two interacting partners—namely, contaminants and soil—and that there should be careful monitoring of the progress of remediation by natural attenuation processes occurring in the affected soil. It has also been pointed out that because of the highly variable nature and properties of the two interacting partners, it is not possible to set exact specifications on technique, tools to be used, etc. In place of regulatory specifications, most agencies have issued guidelines and protocols to guide users on the use of NA as a remediation process with specific attention to the process as a site-specific and contaminant-specific "knowledge-based" passive remediation technique.

The discussion in this chapter is directed towards developing an understanding of the nature of the underlying soil, its various attributes, and features that provide the basis for assimilation of contaminants. The interest in regard to soil composition is how soil composition features in the development of its various attributes. In particular, together with the discussions in Chapter 3, we would be interested in developing a better understanding of how the various physical and physico-chemical properties of soils impact on the development of the assimilative capacity of soils. These properties control the interactions between the soil solids and contaminants which result in the attenuation of contaminants.

2.2 PARENT MATERIAL AND SOIL FORMATION

We discuss the source of soil material to point out the reasons for the high degree of variability of soils found in the landscape, and especially to show the reason why it is not feasible to formulate a "one size fits all" set of regulations to govern the use of natural attenuation processes as a remediation tool. The parent material of most soils is rocks, commonly referred to as *source rocks*. The three general classes are: igneous, sedimentary and metamorphic. Igneous rocks, which are the product of magma include such rocks as andesite, felsite, durite, gabbro, basalt, etc., with basalts and granites being the greatest proportion of these kinds of rocks. Sedimentary rocks, which are highly compressed (under geologic pressure and heat) soil material deposited by wind and water agencies include such rocks as sandstone, siltstone, dolomite, shale, and limestone. Metamorphic rocks, formed from metamorphosing of igneous and/or sedimentary rocks include quartzite, schist, slate, gneiss, and marble.

The natural processes involved in transformation of source rocks into soils are *weathering processes*—often simply referred to as *weathering*. The intensity of weathering is highest where interfaces between the atmosphere, hydrosphere, biosphere, and lithosphere meet and overlap, and the rate of weathering is influenced or controlled by several factors such as composition and size of the source rock (rock mass) or rock fragments, size and type of exposed surface areas of rocks, relative solubility of rock mass, fragments and weathered material, permeability of the rock, source and availability of water, chemical composition of water, temperature, topography, oxygen content or availability, organic matter, cyclical freezing and thawing and wetting-drying cycles, climatic envelope, etc. The types of actions and agents responsible for soil-forming forces can be physical, chemical, biological, and various combinations of these as, for example, biologically-mediated chemical actions. *Physical weathering* of rocks involves forces and processes that fragment rocks such as cyclic freezing and thawing, heating and cooling, expansion of water in rock fissures as a result of freezing, abrasive forces from movement of overlying glaciers, etc. Physical weathering does not, in general, produce or result in alteration of the chemical composition of the rock fragments. *Chemical weathering* of source rock results in transformation and in some cases even loss of some of the rock constituents. The kinds of chemical processes involved include hydrolysis, hydration, carbonation, oxidation, reduction, and solution, and the rates of chemical reactions are determined by source material composition, temperature, amount and availability of moisture, leaching capability and rate, and composition of the leaching solution. A detailed discussion of soil formation from source rocks can be found in Yong et al. (2012). The schematic shown in Figure 2.1 provides a capsule view of many of the weathering processes involved in the production of soil from source rocks.

Whilst the composition and texture of source rocks are important factors in the outcome of weathering processes, their influence is relatively short-lived in regions with humid conditions and high temperatures. However, in arid and arctic regions where the opposite conditions exist, the influence of parent material remains indefinitely. Climate and vegetation and the products produced by decay of the vegetative (organic) material are significant factors of weathering. Absence of vegetation means absence of organic acids as weathering agents. Temperature and amount of rainfall are important climatic factors. Rapid weathering occurs in warm and humid climates. Percolation and infiltration of water through weathering products, in the natural landscape are determined by topography. High rainfall in low flat areas that are completely water-saturated will produce conditions that will retard weathering.

The soils formed from source rocks through the various weathering processes can either remain in-place (residual) or be transported to other places via several means such as wind, water, and anthropogenic activities. *Residual soils* may or may not retain most of the characteristics of the constituents in the source rock—depending on in-situ weathering processes after formation of these soils. If they retain the characteristics of the source rocks (parent material) they are called saprolitic soils. Transported soils include *glacial* soils derived as a result of transport and deposition of rock grinding and erosion products, and soil formations obtained from the retreat of glaciers include drifts, moraines, till, drumlins etc.

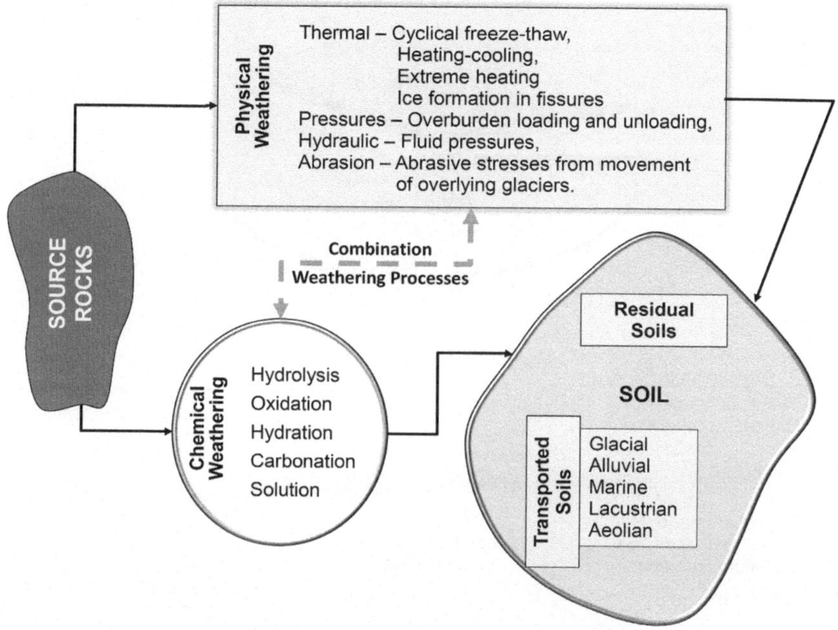

FIGURE 2.1 Soil formation processes of unaltered source-rock mass. Note that the two weathering processes (physical and chemical) can interact—i.e., one can occur first, to be followed by the other, or vice versa. Resultant soil formed can remain in place (residual) or be transported.

Lacustrian deposits are soils obtained from former lake sediments while *fluvial* deposits refer to soils obtained as a result of deposition of soil material transported by streams and rivers. *Alluvial* deposits refer to alluvial fans and deltas and other water-borne deposits not associated with fluvial deposits, and *marine* soils are the result of sediments formed from deposition of suspended material in a marine environment. *Aeolian* deposits are formed from wind transport and the soils contained in the dunes, beaches, loess and drifts include the range from fine particulates to sands.

2.3 SOIL COMPOSITION

The composition of a soil is determined by the various pedogenic processes consistent with the particular region, climate, and anthropogenic activities. As discussed briefly in the previous section, the combination of different types of source materials, various factors and conditions governing weathering and transport processes, regional controls, and anthropogenic activities, result in the high variability of soil composition from location to location. Common terminology refers to soils as three-phase systems consisting of a *solid*, *fluid*, and *gaseous* phase. In general, the gaseous phase is air, and the fluid phase is water with dissolved solutes. The solid phase consists of various types and forms of soil solids commonly referred to as *soil fractions*.

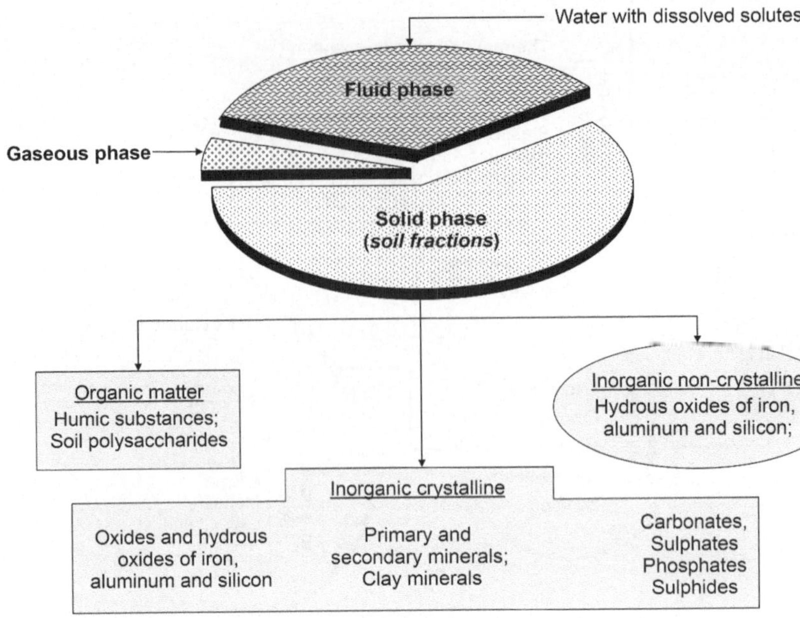

FIGURE 2.2 Pictorial diagram showing the three phases of a soil unit together with the various types of soil fractions in the solid phase. (From Yong, R.N. et al., *Environmental Soil Properties and Behavior,* CRC Press, Boca Raton, FL, 435, 2012.)

While these three phases combine to form the composition of a soil, the key elements in soil composition are (a) the different types of soil fractions that constitute the solid phase, i.e., soil solids; (b) the chemistry and amount of water in the soil mass, i.e., fluid phase; and (c) the structure of the soil, i.e., the manner in which the soil fractions are "packed together" to form the soil mass. As shown in Figure 2.2, these soil fractions consist of (a) *inorganic crystalline* material such as primary minerals, clay minerals, oxides and hydrous oxides of iron, aluminum and silicon, carbonates, sulfates, phosphates, and sulfides, (b) *inorganic non-crystalline* material consisting of hydrous oxides of iron, aluminum and silicon, and (c) *organic matter* consisting of humic substances such as humic and fulvic acids, humins, and polysaccharides.

2.3.1 Size Characterization—Coarse-Grained Soils

The two main groups of soil solids are (a) coarse-grained soils consisting of gravels and sands, and particle sizes larger than 0.06 mm; and (b) fine-grained soils consisting primarily of soils classified as silts and clays with particle sizes less than 0.06 mm, but mainly in the range of sizes less than 2 μm (see Figure 2.3). The behavior of coarse-grained soils, because of their particle size and mass, is governed by gravitational forces. They are *cohesionless* soils—meaning that they lack cohesion between individual and discrete particles. A good example of this is dry sand. Without any water, dry sand cannot be formed into free-standing columns because of lack of cohesion between sand grains, whereas semi-wet sand can be formed

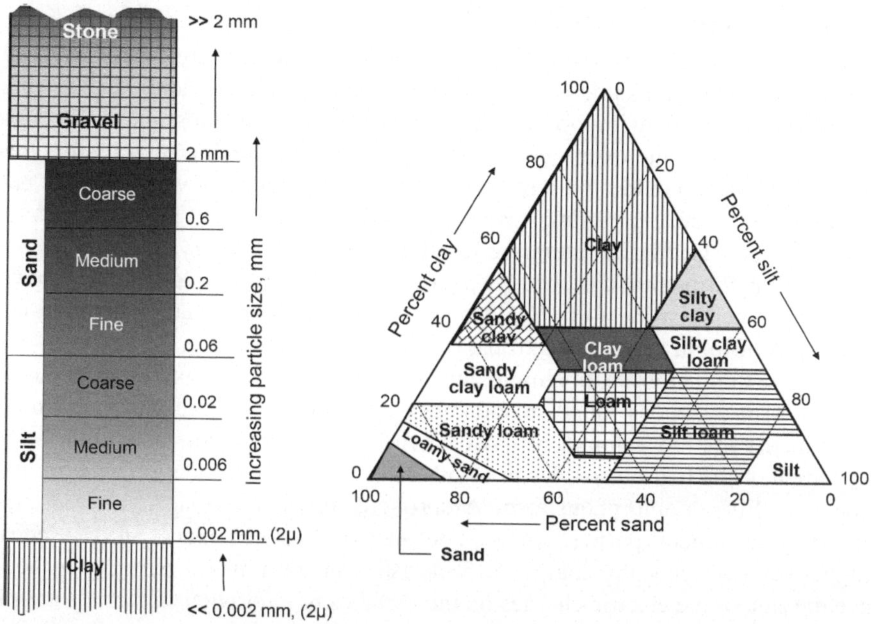

FIGURE 2.3 The left-hand bar chart shows the grain-size (particle-size) differentiation for classification of soils. The right-hand chart shows the textural classification of soils based on proportions of sand, silt and clay in a soil. (From Yong, R.N. et al., *Environmental Soil Properties and Behavior*, CRC Press, Boca Raton, FL, 435, 2012.)

into various shapes because of the capillary forces that serve to hold sand grains together—as witnessed in the phenomenon of sand castles and sculptures on ocean beaches. In general, the cohesionless soil fractions (sands and gravels) consist primarily of inorganic crystalline material—mainly primary minerals such as quartz, feldspar, and primary mica.

2.3.2 Size Characterization—Fine-Grained Soils

Fine-grained soils consist of particles with sizes finer than 0.02 mm. As shown in Figure 2.3, these soils range from medium silts to clays. Strictly speaking, the term *clay* used in most literature refers to soils with particle sizes finer than 0.002 mm, i.e., finer than 2 μm. It is not uncommon to find references to clays without distinguishing between clay-sized soils and the clay minerals that may or may not be present in the clay-sized soil. It is important to distinguish between clay-sized soils and clay minerals because of the role of molecular forces in the development of the properties and response behavior of clay minerals. It is recognized that clay-sized soils as a category or classification will include some clay mineral particles. For the sake of clarity in the discussions to follow, we will refer to clay-sized soils as *clays* and to *clay minerals* when we mean to talk about the minerals as such. These will be discussed later in this chapter and in Chapter 3.

2.3.2.1 Non-clay Minerals

The common non-clay minerals in fine-grained soils (clays) are oxides, hydrous oxides, carbonates, and sulfates. Depending on their proportions and their distribution in the soil, they play important roles in control of soil transmission and strength properties, and to some extent contribute greatly to partitioning of contaminants. The various kinds of oxides present in soil, which are products of weathering of parent rock material include *oxides*, *hydroxides*, and *oxyhydroxides*. The term *hydrous oxides* has been used in the literature to refer to both the non-crystalline form and the crystalline form, although strictly speaking, one should consider the term *oxides* to refer to the crystalline form. Oxides and hydrous oxide minerals are the main constituents of highly weathered tropical soils such as laterites and bauxites—e.g., hematite, goethite, gibbsite, boehmite, anatase, and quartz. Iron oxides are the most common form of oxides found in soils. The oxides of iron, aluminum, and manganese in the pH range found in most natural soils possess low solubilities. The distribution of the hydrous oxides in a soil is very much dependent on whether it is in the crystalline or amorphous form. Amorphous forms of hydrous oxides have the ability to coat mineral particle surfaces because of their differences in net electric charges—thus altering the charge characteristics of the mineral particle surfaces. Since the sign of the electric charges on the surfaces of amorphous materials are pH dependent, it stands to reason that the pH of the soil containing these materials is a factor that needs to be considered, as will be discussed in Chapter 3.

Carbonates and sulfates, because of their relatively high solubilities are more often found in arid and semi-arid regions where limited leaching and high evaporation occur—although smaller amounts of the carbonate mineral can be found in other regions. The carbonates group include calcite ($CaCO_3$), magnesite ($MgCO_3$), siderite ($FeCO_3$), dolomite ($CaMg(CO_3)_2$), trona ($Na_3CO_3HCO_3 \cdot H_2O$), nahcolite ($NaHCO_3$), and soda ($Na_2CO_3 \cdot 10H_2O$), with calcite and dolomite as probably the most common. The presence of carbonates in soils is a significant issue because of its influence on the pH—bearing in mind that many of the chemical processes and reactions occurring in soils are pH sensitive. The sulfate mineral group includes gypsum ($CaSO_4 \cdot 2H_2O$), hemihydrate ($CaSO_4 \cdot 0.5H_2O$), thenardite (Na_2SO_4), and mirabilite ($Na_2SO_4 \cdot 10H_2O$), with gypsum being the most common of the sulfate minerals found in soils—primarily in arid and semi-arid region soils.

2.4 CLAY MINERALS

Clay minerals are referred to as secondary minerals. They are obtained from chemical weathering of primary minerals in metamorphic and sedimentary rocks. The clay minerals formed from dissolution of primary minerals depend not only on the composition of the solution, but also on which ions are lost by leaching in the dissolution process and subsequent to the process. For example, feldspar can weather first to a montmorillonite and subsequently to a kaolinite, or directly to kaolinite depending upon the conditions during weathering. Clay minerals such as montmorillonite or illite are formed from conditions where minimal or restricted leaching exists, i.e., where cations tend to remain in solution. Excessive leaching of silica will produce iron and aluminum oxides.

The influential role of clay minerals in the transmission properties and partitioning of contaminants is derived from their particle size and shape, their large specific surface area, and their surface chemistry. The types of electric charges, the source of charges, and the surface functional groups associated with the type of clay mineral are key to the type and level of physico-chemical activities, and to whether adsorbed contaminants will be weakly or strongly held in a clay. The following discussion of the basic structure and features of clay minerals follows closely the treatment given by Yong et al. (2012).

2.4.1 Unit Cell, Layer, and Mineral Structure

Clay minerals are phyllosilicates (layer silicates)—i.e., they are aluminosilicates with sheets of silicon-oxygen tetrahedral and aluminum-oxygen octahedral units containing small amounts of metal ions substituted within the crystal structure of the minerals—as shown in Figure 2.4. The illustrations in the figure show the aluminum-oxygen and silicon-oxygen combinations forming the basic molecular structural units. These units combine in various configurations to make up the different types of clay minerals found in clay soils. The main groups, grouped according to their mineral structure, include (a) the uncharged 1:1 layer-lattice minerals with no interlayer water such as kaolinites, and (b) the 2:1 layer-lattice minerals with interlayer hydroxide sheets or cations. The latter are the most common and abundant types of clay minerals found in clays.

The basic silica tetrahedral and the Al^- or Fe^- or Mg^- octahedral structural units shown in Figure 2.4 are stacked together to form unit cells as shown in the example

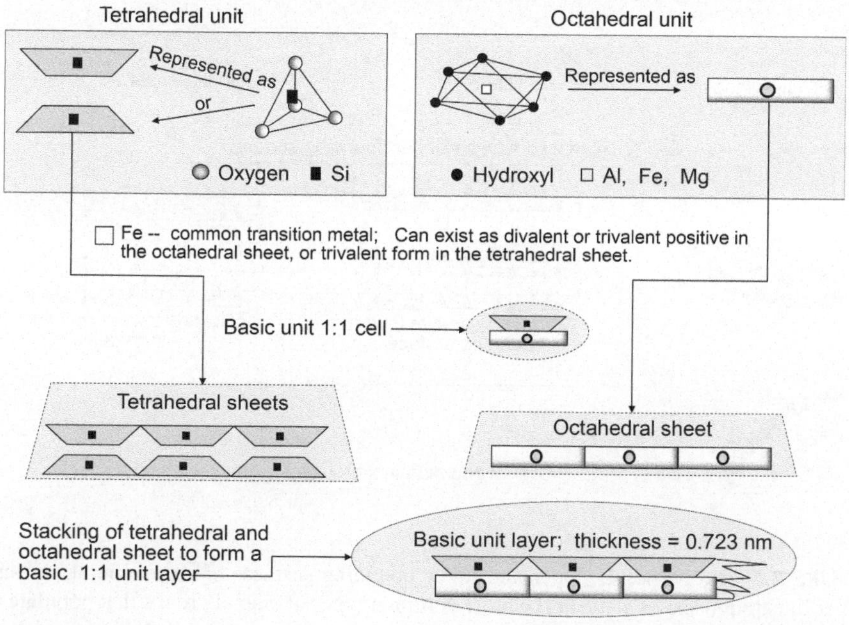

FIGURE 2.4 Basic molecular structural units in clay minerals. (From Yong, R.N. et al., *Environmental Soil Properties and Behavior*, CRC Press, Boca Raton, FL, 435, 2012.)

of a 1:1 unit cell (one tetrahedral unit and one octahedral unit) in the center of the figure. The 2:1 unit cell (not shown in the figure) consists of a top and bottom tetrahedral layer with an octahedral unit sandwiched between the tetrahedral units. The basic unit cells are linked laterally to form unit layers—as shown for example in the bottom portion of the figure. Clay mineral particles are formed from a stack of vertically linked unit layers. The type of links and the nature and distribution of the ions populating the tetrahedral and octahedral units will determine the attributes of the clay mineral obtained.

The important features of these kinds of clay minerals, insofar as interactions with contaminants are concerned are the planar surfaces and the edges of 1:1 and/or 2:1 layer-lattice structures. These surfaces react chemically with porewater and the contaminants and other solutes contained therein. They are often referred to as *reactive surfaces*. The edges of these minerals are also reactive when disruption of the unit layers produces broken bonds at the edges. The details shown in Figure 2.5 in respect to the planar bounding surfaces of 1:1 layer-lattice mineral particles are known as siloxane and gibbsite surfaces respectively for the top and bottom surfaces of the mineral particles. The tetrahedral sheet at the top has an exposed siloxane surface while the lower surface has an exposed gibbsite

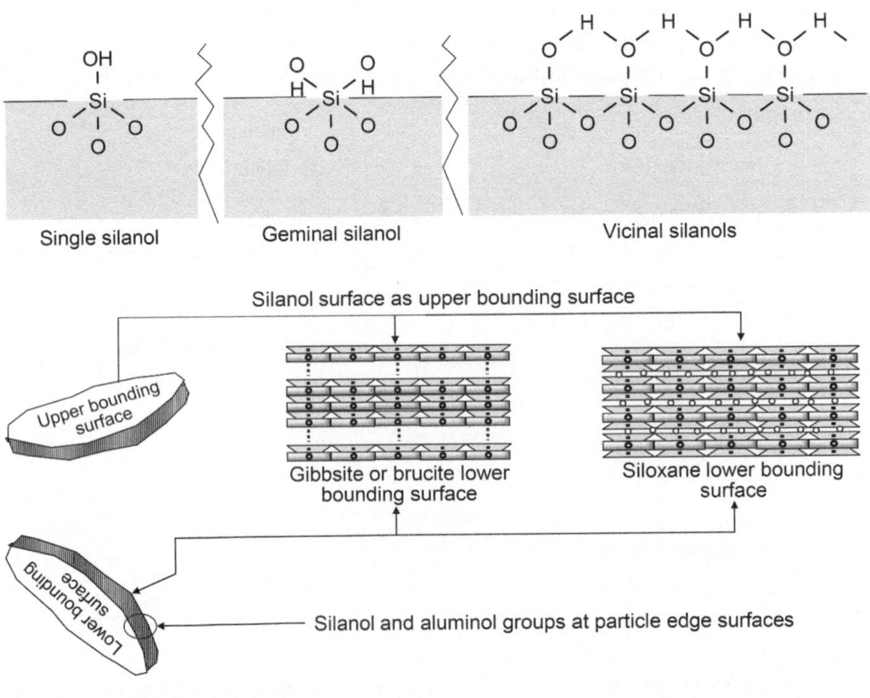

FIGURE 2.5 Properties of upper and lower bounding surfaces of clay mineral particles. Upper illustration shows some of the more common types of silanol groups that populate the siloxane upper bounding surface. The lower bounding surface is a hydroxide sheet which can be a brucite layer or a gibbsite layer. (From Yong, R.N. et al., *Environmental Soil Properties and Behavior*, CRC Press, Boca Raton, FL, 435, 2012.)

surface (see Section 2.5). In the case of the 2:1 layer-lattice structures, siloxane surfaces bound both the upper and lower faces of the mineral particle.

2.4.2 Kaolin Group

The kaolin group of clay minerals is commonly found in warm and moist climates. They are the result of acid leaching of the feldspars and micas in parent rocks. The minerals in the kaolin group are 1:1 layer silicates—i.e., one uncharged tetrahedral sheet linked to one uncharged octahedral sheet as shown in the bottom sketch of Figure 2.4 and in the top sketch of Figure 2.6. The clay mineral, which is commonly known as *kaolinite* belongs to the group of dioctahedral 1:1 layer silicates. The kaolinite mineral is obtained from a stacking of basic unit layers of uncharged tetrahedral and octahedral sheets with thickness of about 0.7 nm. Kaolinites are platy in morphology and their particle sizes are by and large determined by the amount of disorder in the unit layers. The other kaolin minerals in this group include dickites and nacrites. These are variants of the basic kaolinite structure, with dickite being more crystalline than kaolinite.

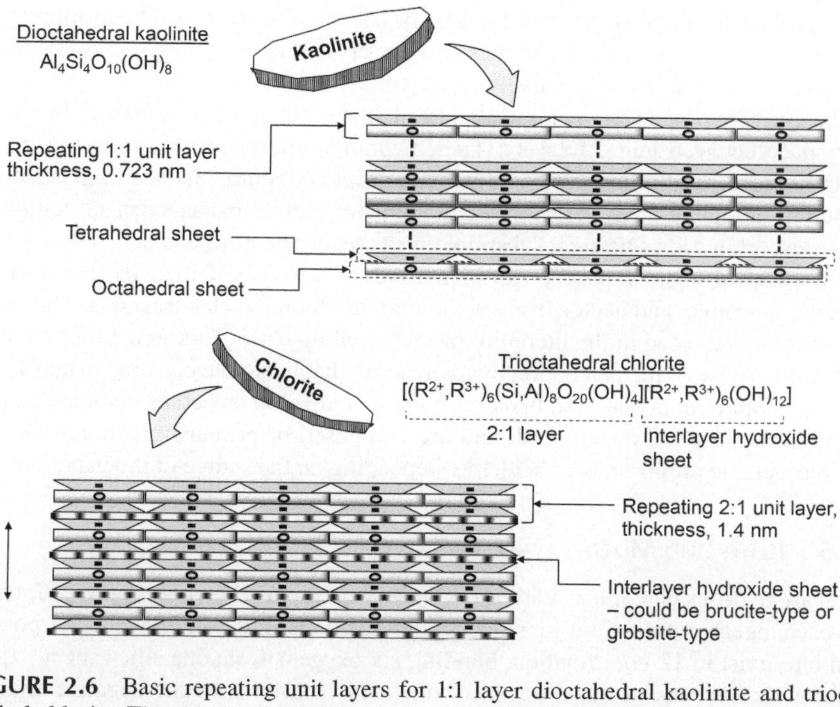

FIGURE 2.6 Basic repeating unit layers for 1:1 layer dioctahedral kaolinite and trioctahedral chlorite. The chlorite shown in the bottom of the figure has a 2:1 repeating unit layer with interlayer hydroxide sheets. R^{2+} and R^{3+} in ideal structural formula represent the sum of divalent and sum of trivalent cations respectively. (From Yong, R.N. et al., *Environmental Soil Properties and Behavior*, CRC Press, Boca Raton, FL, 435, 2012; Structural formula information from Newman, A.C.D. and Brown, G., *The Chemical Constitution of Clays*, Longman Scientific and Technical, Harlow, UK, 1–128, 1987.)

2.4.3 Chlorites

The basic structure of the chlorite mineral shown in Figure 2.6 is basically a unit 2:1 layer structure that has a positively charged interlayer consisting of a coordinated octahedral hydroxide sheet that balances the unit 2:1 layer negative charge. This interlayer sheet differs from the octahedral sheet in the unit layer in that it does not have a plane of atoms which are shared with the adjacent tetrahedral sheet. The coordinated octahedral hydroxide sheets can be trioctahedral (brucite-type) or dioctahedral (gibbsite-type). While cations such as Fe, Mn, Cr, and Cu are sometimes found in the hydroxy sheets, the more common hydroxy sheets are brucite and gibbsite sheets. Typical repeat spacing for the unit layers is 1.4 nm.

2.4.4 Smectites

Smectites are hydrous aluminum silicate clay minerals that possess dioctahedral and trioctahedral structural configurations. The dioctahedral group includes montmorillonite, beidellite and nontronite, and the trioctahedral group includes saponite, sauconite and hectorite. The dioctahedral smectites are 2:1 layer lattice minerals with repeating unit layer consisting of two sheets of SiO_4 tetrahedrons confining a central octahedral layer of hydroxyls with Fe, Mg, or Li ions (Figure 2.7). The terminology in the literature sometimes refer to these unit layers and interlayer spaces as *lamellae* and *interlamellar space*, respectively.

The attribute of note for montmorillonites is the exchangeable cations in the interlayer space between unit 2:1 layers. Their hydration and resultant swelling characteristics, coupled with their various microstructural configurations and large specific surface areas, make them very important elements in the management of contaminant transport in soils. Different substitutions in the tetrahedrons and octahedrons that form the basic sheets will produce different kinds of smectites. Trioctahedral minerals saponite, sauconite, and hectorite are not commonly found as clay fractions. The term *bentonite* is often used in the literature mean "swelling clay." This sometimes creates problems because of the perception by many users that bentonite consists primarily of montmorillonite. In actual fact, bentonites are dioctahedral smectites obtained as the alteration products of volcanic ash and are composed of primarily montmorillonite with measurable proportions of beidellite depending on the source of the bentonite.

2.4.5 Illites and Mixed-Layer Clays

Illites are hydrous clay micas with a 2:1 layer structure as shown in Figure 2.7, with non-exchangeable potassium ions occupying positions in the interlayer. The potassium ions exist in 12-coordination, bonding six oxygens from one silica sheet to the adjacent six oxygens of the silica sheet of the next layer, with the negative charge balancing the potassium cations coming from the substitution of aluminum for silicon in the silica sheet. Illitic clays are non-swelling clays since they do not contain any expanding layer in their structure. On the other hand, illitic clays that are interlayered with montmorillonite, called *interlayered* illites or *mixed-layer*, do exhibit some degree of swelling.

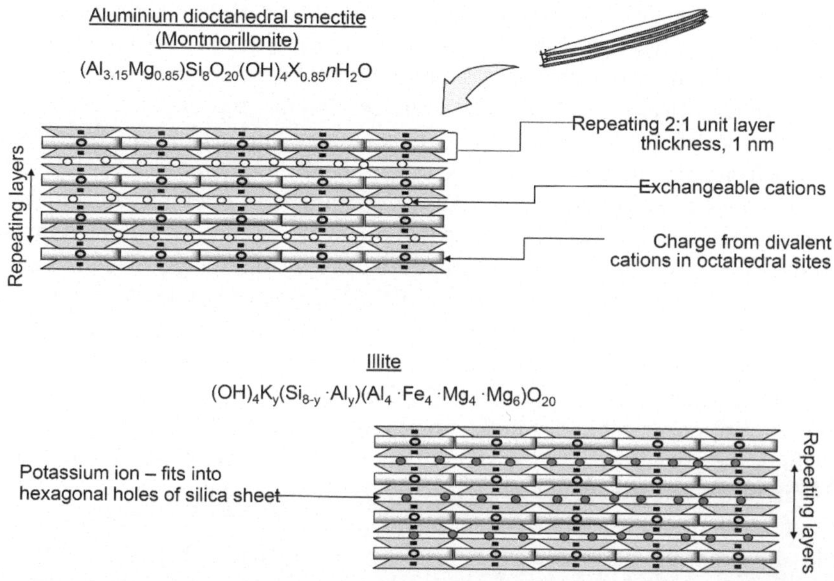

FIGURE 2.7 Basic repeating 2:1 unit layers for montmorillonite and illite. The repeating layers are stacked vertically and stretched spatially to form the respective mineral particles. (From Yong, R.N. et al., *Environmental Soil Properties and Behavior*, CRC Press, Boca Raton, FL, 435, 2012; Structural formula information for smectite from Newman, A.C.D. and Brown, G., *The Chemical Constitution of Clays*, Longman Scientific and Technical, Harlow, UK, 1–128, 1987, and From Grim, R.E., *Applied Clay Mineralogy*, McGraw-Hill, New York, 1962. for illite.)

2.4.6 Soil Organics

Soil organic matter is an important constituent in soils insofar as soil stability and contaminant and solute interactions are concerned. In general, they constitutes anywhere from 0.5% to 5% by weight in a typical soil mass. While peat can be technically considered to be a soil organic matter, because it constitutes 100% of a surface "soil" mass, it is generally considered as a separate entity. The soil organic matter considered as a part of soil originates from vegetative and/or animal sources. Based upon their degree of degradation, they are generally classified as *unaltered organics* and *transformed organics.* As the name implies, unaltered organic materials are fresh and relatively fresh organic materials that have yet to show the results of transformation processes associated with degradation.

Transformed organic materials are obtained as a result of degradation of vegetative and animal sources, with a high degree of degradation—to the point where the organic materials show no morphological resemblance to the parent material, and do not exhibit properties and characteristics of the parent material. The transformations that occur in the degradative process are the result of chemical and biologically mediated reactions that include demethylation and oxidative, reductive, and other electron transfer reactions catalyzed by enzymes (Flaig et al., 1975). *Amorphous soil organics* are highly aromatic polymers that are rich with functional groups (see next section for treatment of soil functional groups). They consist of humic acids, fulvic acids, humins and decayed material such as polysaccharides, lignins, and polypeptides. We should note that the term *amorphous*, in organic chemistry, bears a different meaning from that used in this text. Amorphous materials in soils refer to oxides and hydrous oxides of iron, aluminum, and oxides of silicon which do not exhibit any crystalline structure. In relation to transformed organics, the term *amorphous* is specifically used to indicate that the transformed organics no longer exhibit properties and characteristics of the parent material.

Characterization of these amorphous soil organics is generally obtained through alkali treatment procedures on soil organics extracted from the parent soil as shown in Figure 2.8. The standard procedure for determination of humic and fulvic acids, humins and non-humic fractions follows the protocols shown in the left-hand branch of the schematic diagram shown in the figure. The protocols shown as the right-hand branch of the schematic have been used by Yong and Mourato (1988) to determine the presence of polysaccharides in the non-humic fraction. As seen from the figure, one defines the constituents of the soil organic matter as follows: (a) humic acids—soluble in bases, but precipitate in acids; (b) fulvic acids—soluble in both bases and acids; and (c) humins—insoluble in acids and bases.

Classification of polysaccharides as humic substances sometimes poses a dilemma. Polysaccharides are long molecules that are obtained as by-products of microbial metabolism, generally synthesized as a by-product from the breakdown of animal or vegetal-derived organics. Microbial polysaccharides are synthesized within soil by active microbial populations. Depending on whether polysaccharides

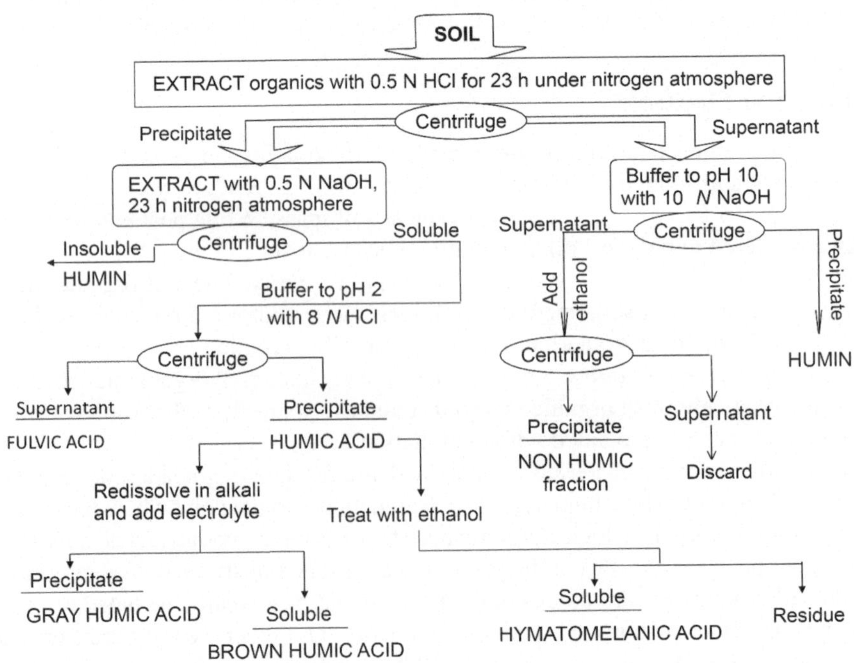

FIGURE 2.8 Extraction and treatment technique for classification of soil organic matter.

are extracted with the fulvic fraction, these carbohydrates are often incorporated as a fulvic-related material. Thus, according to the basic definition of a humic or amorphous substance, polysaccharides do not fall in the class of non-identifiable degraded organics since these polymeric carbohydrates have a definitive structure.

2.5 SURFACE PROPERTIES OF SOILS

Partitioning of contaminants in a soil occurs when contaminants coming into contact with soil solids are held onto the soil solids thus removing them from the transporting liquid (groundwater and soil porewater). The act of partitioning, or the partitioning process is a direct function of the sorption or assimilative capability of soils. The greater the amount of contact between the contaminants and soil solids, the greater is the potential for partitioning—provided that the proper assimilative or sorption capabilities are available. When the soil fractions in a soil mass have maximum surficial contact with the contaminants carried in the soil water (porewater), the optimum condition for uptake of the contaminants by the soil is obtained. This constitutes the partitioning process. Contaminant uptake by soils, defined as *uptake*, refers to the chemical and physical mass transfer of contaminants due to mechanisms, processes and/or reactions between the contaminants in the porewater and the soil solids and their surfaces. For maximum uptake of contaminants, all the available soil solids and their surfaces need to interact with the contaminants in the soil water. Soil structure and the distribution of soil solids/particles (single particles and microstructures) are important factors in soil uptake of contaminants.

2.5.1 Specific Surface Area

The discussion in the preceding section concerning optimum contaminant contact with soil particles and their surfaces leads to the question of how one determines optimum contact. A common procedure is to determine the specific surface area (m^2/g) of the soil in question. By this, we mean the total available solid particle surfaces of the soil. The presence of microstructures—i.e., packets of soil particles, clusters, groups, etc.—will obviously diminish the amount of particle surfaces available for interaction with contaminants in the porewater (see Figure 2.9).

While we can theoretically calculate the surface area of individual particles assuming ideal shapes such as planar surfaces, cubes, spheres, and cylinders or conical shapes, the measurement of surface areas of soil particles in any representative elementary volume (REV) requires the use of an adsorbate—some kind of fluid or gas that would form a monolayer coating around available surfaces. Since soil particles with planar shapes will exhibit the greatest surface area per unit volume or per unit mass, it will be seen from the illustrations in Figures 2.5 through 2.7 that clay minerals will have the largest surface areas in comparison to other soil fractions. Using the *a* and *b* dimensions of a unit cell for a dioctahedral 2:1 mineral such as a smectite, and Avogadro's number of such unit cells, Greenland and Mott (1985) calculated the specific surface area of such minerals to be 757 m^2/g. This refers to all the surfaces available for interaction with water or other fluids. Figure 2.9 gives a highly simplified schematic of available surfaces in relation to some clay minerals

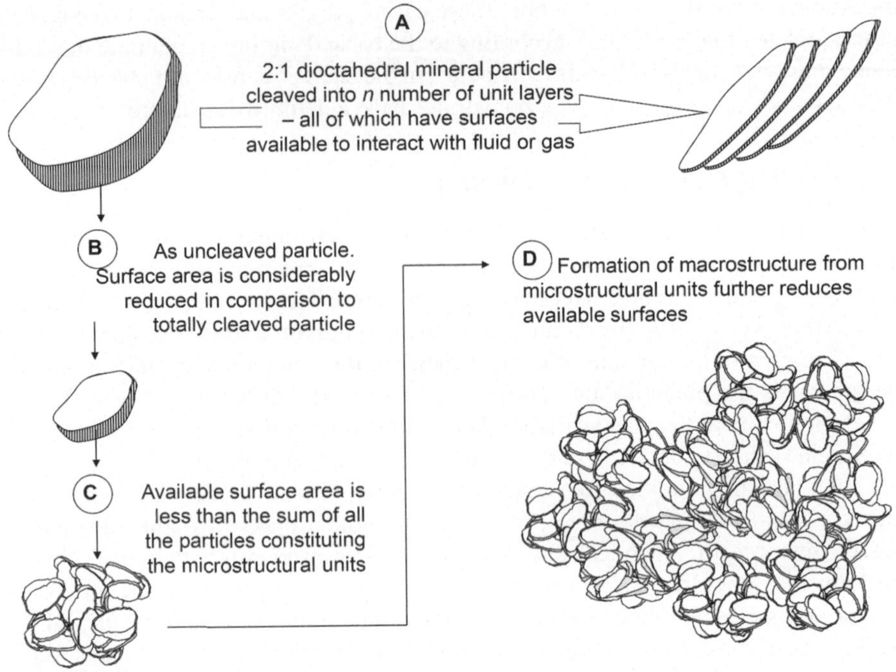

FIGURE 2.9 Surface areas of mineral particles and reduction of available surface areas due to microstructural units and aggregation into macrostructure.

and the role of microstructure in modification of available surfaces. Cleavage of the dioctahedral 2:1 mineral into the component unit layers allows all the surfaces of the unit layers to interact with porewater (top portion of Figure 2.9, "A"). The intact mineral particle denoted as "B" in the figure (e.g., a kaolinite mineral particle) will have considerably less surface area. When combined with other particles in a microstructural unit, the total available surface area becomes less than the sum of all the particles involved because of inaccessibility to fluid conductivity. Wetting of the pore channels defined between microstructural units is confined to those surfaces that define the pore channels used for fluid movement.

Laboratory measurements are often made to determine the specific surface area of soils—since theoretical calculations are not only tedious, but also unrealistic if the soil contains different soil fractions of varying proportions and sizes. The general procedure is to determine the amount of gas or liquid (adsorbate) that forms a monolayer coat on the surfaces of the particles. The choice of the adsorbate and the availability of soil particles in a totally dispersed state are the two most important factors in any laboratory measurement of the specific surface area (SSA) of a soil sample. Accordingly, this makes a laboratory determination of the SSA an *operationally defined* property, i.e., dependent on adsorbate used and degree to which the soil has been properly dispersed, one could easily obtain different SSA values for the same soil sample.

When the adsorbate is a gas, the amount of gas (e.g., number of molecules of nitrogen) sorbed by the soil particles will be dependent on the partial pressure of the gas and also on the temperature. In the test procedure, more than one layer of gas is will sorbed by the soil particles. To determine the amount (volume) of gas equivalent to a sorbed monomolecular layer of gas, the relationship developed by Brunauer et al. (1938) known as the Brunauer Emmett and Teller (BET) equation for multilayer sorption, is probably the best-known relationship used. For adsorbates that are polar fluids, the techniques described by Mortland and Kemper (1965) for ethylene glycol and Carter et al. (1986) for ethylene glycol-monoethyl ether are commonly used.

Since all the techniques require one to obtain a uniform monolayer coating by the adsorbate of all the soil particles, it is easy to see that the values of SSA determined are directly dependent on the procedure used and the analytical technique used to reduce the data obtained. These can all lead to differences in results obtained. Failure to provide for a totally dispersed state of soil particles and failure to obtain uniform monolayer distribution or coating of the dispersed particles are at least two of the more significant factors. A further point of consideration is the presumption that the interactions between porewater and the soil particles' surfaces will mimic those exhibited by the adsorbates.

2.5.2 Reactive Surfaces of Soil Fractions

We define *reactive surfaces* to mean those surfaces which by virtue of their properties are capable of reacting physically and chemically with solutes and other dissolved matter in the porewater. Chemically reactive groups (molecular units) associated with the surfaces of the various soil fractions render the surfaces of these fractions *reactive*. These reactive groups are defined as *surface functional groups*. The literature often refers to these simply as *functional groups*. The soil fractions with reactive surfaces include layer silicates (clay minerals), soil organics, hydrous oxides, carbonates and sulfates. The surface hydroxyls (OH group) are the most common surface functional group in inorganic soil fractions (soil solids) such as clay minerals with disrupted layers (e.g., broken crystallites), hydrous oxides, and amorphous silicate minerals. These surfaces are often referred to as *hydroxylated surfaces*—to acknowledge the presence of these surface hydroxyl groups.

The nature of the reactive surfaces of the clay minerals can traced directly to the structure of the layer lattice structures shown in Figure 2.4. Taking the group of clay minerals with the 1:1 structure such as the kaolinite shown in Figure 2.6, we have siloxane and gibbsite surfaces on opposite basal surfaces of the particles. The siloxane surface is characterized by the basal plane of oxygen atoms which bound the tetrahedral silica sheet. These basal planar surfaces are typical of minerals whose structures have bounding tetrahedral sheets. Thus, whilst kaolinites possesses one siloxane surface by virtue of the 1:1 structure, the illites, montmorillonites, and vermiculites show siloxane-type surfaces on both bounding surfaces. These siloxane-type surfaces are reactive surfaces because of the structural arrangement of the silica tetrahedra and the nature of the substitutions in the layers. The regular structural arrangement of interlinked SiO_4 tetrahedra, with the silicon ions underlying the surface oxygen ions, produces cavities which are bounded by six oxygen ions in a ditrigonal formation. Where no substitution of the silica in the tetrahedral layer and lower valence ions in the octahedral layers occurs, the surface may be considered to be free of any resultant charge. However, if replacement of the ions in the tetrahedral and octahedral layers by lower valence ions occurs through isomorphous substitution, we will obtain resultant charges on the siloxane surface and hence a reactive surface. The resultant charges on these surfaces due to isomorphous substitution are generally positive in nature. At the edges of the kaolinite particles, the broken edges have Al and Si centers that have hydroxyl terminals, i.e., they are terminated by hydroxyls. The aluminols can accept or donate protons. According to Sposito (1984) adsorption of water onto these aluminol sites will produce Lewis acid sites.

2.6 SURFACE FUNCTIONAL GROUPS

Surface functional groups are important features in the development of soil properties and resultant behavior. They are present in all of the solid phase soil fractions. *Surface functional groups* are chemically reactive groups (molecular units) associated with the surfaces of the solid phase soil fractions. They are the reason why the surfaces of the soil solids (particles) are reactive. Consider the illustration of the upper and lower bounding surfaces of clay mineral particle shown in Figure 2.5. The lower bounding surface of the 1:1 layer lattice mineral particle is an octahedral sheet. When two thirds of the available positions in this sheet are filled with Al^{3+} ions, the mineral structure in this sheet is known as a *gibbsite structure* with chemical formula $Al_2(OH)_6$. Layer lattice minerals with two thirds of possible positions in the octahedral sheet filled are known as dioctahedral minerals. When all the positions in this sheet are filled with Mg^{2+} instead of Al^{3+} ions, the resultant structure is known as a *brucite structure* with chemical formula $Mg_3(OH)_6$, and the associated layer lattice mineral is known as a trioctahedral mineral. The structural arrangement of interlinked SiO_4 tetrahedra shown in the figure, with silicon ions underlying the surface oxygen ions, produces cavities bounded by six oxygen ions in ditrigonal formation. Isomorphous substitution in the octahedral and tetrahedral layers results in development of resultant charges on the siloxane surface. By this process, the siloxane surface becomes a reactive surface. The amount of silanol groups associated with the siloxane bounding surface is a function of the crystallinity of the interlinked SiO_4 tetrahedra. Of the various types of silanol groups, isolated (single silanols), geminal (silanediol), and vicinal groups are more common, and when surface silanol groups dominate, the surface will be hydrophilic. *Siloxane bridges* are formed from the condensation of combined surface and internal silanol groups. The hydrophobicity of the siloxane surface is due to the presence of siloxane groups. Siloxanes tend to be unreactive because of the strong bonds established between the Si and O atoms and the partial π interactions.

2.6.1 SURFACE FUNCTIONAL GROUPS—SMECTITES

Siloxane surfaces bound the opposite planar faces of each unit 2:1 layer of montmorillonites. The source of the charges is primarily from isomorphous substitution in the octahedral sheet of the unit layers, with magnesium or iron substituting for aluminum in this sheet. Table 2.1 gives a summary of the sources of charges for the minerals discussed in this section.

TABLE 2.1
Charge Characteristics, SSA and CEC for Some Clay Minerals

Soil Fraction	Cation Exchange Capacity (CEC), meq/100 g	Surface Area, m^2/g	Range of Charge meq/100 g	Reciprocal of Charge Density nm^2/charge	Isomorphous Substitution	Source of Charges
Kaolinite	5–15	10–15	5–15	0.25	Dioctahedral; 2/3 of positions filled with Al.	Surface silanol and edge silanol and aluminol groups (ionization of hydroxyls and broken bonds)
Clay micas and chlorite	10–40	70–90	20–40	0.5	Dioctahedral: Al for Si Trioctahedral or mixed Al for Mg	Silanol groups, plus isomorphous substitution and some broken bonds at edges
Illite	20–30	80–120	20–40	0.5	Usually octahedral substitution Al for Si	Isomorphous substitution, silanol groups and some edge contribution
Mont morillonite[a]	80–100	800	80–100	1.0	Dioctahedral; Mg for Al	Primarily from isomorphous substitution, with very little edge contribution
Vermiculite[b]	100–150	700	100–150	1.0	Usually trioctahedral substitution Al for Si	Primarily from isomorphous substitution, with very little edge contribution

Source: Yong, R.N., *Geoenvironmental Engineering: Contaminated Soils, Pollutant Fate and Mitigation*, CRC Press, Boca Raton, FL, 307, 2001.

[a] Surface area includes both external and intra-layer surfaces. Ratio of external particle surface area to internal (intra-layer) surface area is approximately 5:80.

[b] Surface area includes both external and internal surfaces. Ratio of external to internal surface area is approximately 1:120.

Note that ratios of external:internal surface areas are highly approximate since surface area measurements are operationally defined—i.e., they depend on the technique used to determine the measurement.

2.6.2 Surface Functional Groups—Kaolinites

Kaolinites and chlorites belong to the group of clay minerals with the 1:1 structure that show siloxane and gibbsite surfaces on opposite bounding surfaces of the particles (Figure 2.5). As stated previously, the amount of silanol groups on the siloxane bounding surface depends upon the crystallinity of the interlinked SiO_4 tetrahedra. The broken edges of the kaolinite particle expose hydrous oxide-types of edge surfaces. These surfaces contain both silanol and aluminol groups. The broken octahedral sheets provide for Lewis acid sites [Al(III) · H_2O] that can bind OH groups in single coordination. Note that the gibbsite sheet which acts as the lower bounding surface will also have aluminol groups.

2.6.3 Surface Functional Groups—Chlorites

Substitution of other cations for silicon and aluminum in chlorite results in development of net negative charge in the mineral. Substitution of Al^{3+} for Si^{4+} in the tetrahedral layer results in a negative charge. Substitution of Al^{3+} by Mg^{2+} in the brucite layer gives a positive charge. The result of these substitution is a net positive charge which is the cation exchange capacity of the mineral.

2.6.4 Surface Functional Groups—Oxides

Since oxide surfaces are partially hydrolyzed, their surface charges will be dependent on the pH of the contacting porewater, i.e., the charge developed on the surface of the oxide is dependent on the concentration of protons in the contacting porewater—with the charge developed in relation to the association or dissociation of protons (Bolt and Bruggenwert, 1978). At some point, the surface may be uncharged, and when this occurs, this point is called the *point of zero charge* (*pzc*). The pH at which this occurs is identified as pH_0. Parks (1965) provides a relationship between cation valency and the pzc of its oxide as follows:

Monovalent ions giving an oxide M_2O pzc $<$ pH 11.5
Divalent ions giving an oxide MO $8.5 < pzc < 12.5$
Trivalent ions giving and oxide M_2O_3 $6.5 < pzc < 10.4$
Quadrivalent ions giving and oxide $MO_2 0 < pzc < 7.5$

2.6.5 Surface Functional Groups—Soil Organic Matter

Unlike the clay mineral particles which have mainly hydroxylated surfaces, we have a greater variety of surface functional groups in soil organic matter since carbon constitutes the backbone structure of all soil organics. Carbon can be combined to form saturated or non-saturated rings or chains, and carbon and nitrogen can combine with oxygen and/or hydrogen to form surface functional groups. These functional groups essentially control the properties and reactions of the organic molecules. The illustrative diagram in Figure 2.10 shows common soil organic matter functional

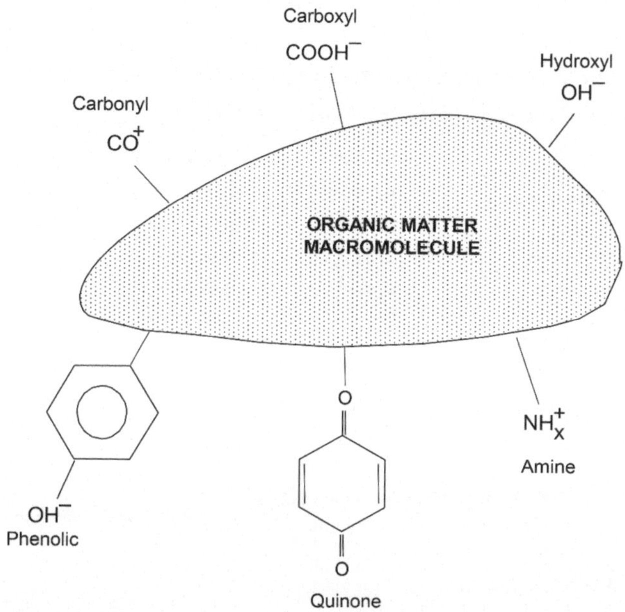

FIGURE 2.10 Some typical functional groups associated with soil organic matter. (Adapted from Yong, R.N., *Geoenvironmental Engineering: Contaminated Soils, Pollutant Fate and Mitigation*, CRC Press, Boca Raton, FL, 307, 2001.)

groups such as hydroxyls, carboxyls, phenolic and amines, which can be considered to be part of the organic matter itself. Because of the large differences in the composition of soil organic matter, as a result of differences in source material, degree of degradation, procedures involved in extraction and testing, it is not uncommon to see wide ranges and values for the proportions of each of the kinds of functional groups. The values shown in Table 2.2 provide a good example of the range of values for some of the typical functional groups. Carboxyls and phenolic groups have generally been considered as the more important functional groups in soil organics, particularly in respect to the development of their Cation exchange capacity (CEC) and chelating capability. Functional groups have the ability to protonate or deprotonate depending on the pH of the surrounding medium. They will develop positive or negative charges depending on the pH of the soil and their respective pK_a and pK_b (i.e., their respective log acidity and log basicity constants).

2.6.6 Surface Charge Density

We have seen from the previous section that silanol surfaces associated with clay particles exhibit resultant positive charges. Additionally, we can see from Table 2.2, and especially Figure 2.10, that the various kinds of functional groups associated

TABLE 2.2
Typical Composition and Common Functional Groups for Fulvic Acid, Humic Acid, and Humin

	Fulvic Acid	Humic Acid	Humin
% Carbon content	40–50	50–60	50–60
% Oxygen	40–50	30–40	30–35
% Hydrogen	4–7	3–6	NA
Carbonyl, %	Up to 5	Up to about 4	NA
Carboxyl, %	1–6	3–10	NA
Quinone, %	2"	1–2	NA
Ketones, %	2"	1–4	NA
Alcoholic OH, %	2.5–4	Up to 2	NA
Phenolic OH, %	2–6	Up to about 4	NA

Source: Yong, R.N., *Geoenvironmental Engineering: Contaminated Soils, Pollutant Fate and Mitigation*, CRC Press, Boca Raton, FL, 307, 2001.

NA = Not applicable.

with soil organic matter also exhibit negative or positive charges as, for example, the carboxyl, phenolic, and hydroxyl groups possess negative charges. Hydrous oxides, oxides, and hydroxide sheets such as brucite and gibbsite will possess charges as a result of association and dissociation of protons. The charges developed on the surfaces of the oxides or hydroxide sheets are due to amphoteric dissociation of the surface hydroxyl groups or from adsorption of H^+ and OH^- ions (Gast, 1997). The nature of these charges will be dependent on both the pH and concentration of electrolytes in the porewater. A more complete discussion of this subject will be found in the next chapter.

The total number of electrostatic charges on the clay particles' surfaces divided by the total surface area of the particles involved provides us with the quantitative determination of the surface charge density. The common procedure is to express this surface charge density in terms of its reciprocal—as shown in Table 2.1. The hydrous oxides such as goethite [–FeOOH] and gibbsite [–$Al(OH)_3$] have been omitted from the table because the range of values for these types of soil fractions are dependent upon: (a) their structure, (b) the specifically adsorbed potential-determining ions, and (c) the pH of the porewater.

2.7 PHYSICAL ATTRIBUTES AND SOIL STRUCTURE

The physical characteristics and properties of soils, i.e., the soil attributes, range from the basic physical description of the nature of the soil such as soil porosity to its performance characteristics such as strength and integrity. In this section, we will be concerned with those physical attributes that relate directly to contaminant attenuation. These are (a) porosity n or voids expressed as void ratio e, (b) pore space continuity, (c) soil macro- and microstructure, and (d) amount of "wetted" surface area. All of these soil properties, which are directly related to the density of the soil mass are dependent on particle packing and arrangement, i.e., soil structure. Whilst deterministic considerations and analyses of soil as a continuum can be used successfully for many soil engineering applications, the behavior of soils that respond to both gravitational and molecular forces require analyses that incorporate gravitational and intermolecular forces. This is particularly true for clay soils with active clay minerals, i.e., clay minerals with active and reactive particle surfaces such as smectites. By and large, clays are generally composed of soil-particle structural units grouped together in some coherent fashion to make up the macrostructure of the clay, as shown previously in Figure 2.9. The microstructural units range from individual mineral particles to what is known as domains, flocs, clusters, peds, and aggregate groups. Figure 2.11 is a sketch that shows how these physical properties impact directly on contaminant transport in soil. The top portion of the sketch shows a contaminated plume (arrow) entering a unit soil volume and leaving the soil as a "cleaned" fluid plume. The bottom portion of the sketch gives an "exploded" view of a portion of the soil mass, showing the physical attributes that participate in the attenuation of contaminants as the contaminated influent passes through the unit soil mass. In the next chapter, we will be making the point that all of these physical attributes are, in fact, central to the control of permeability of the soil for water and air.

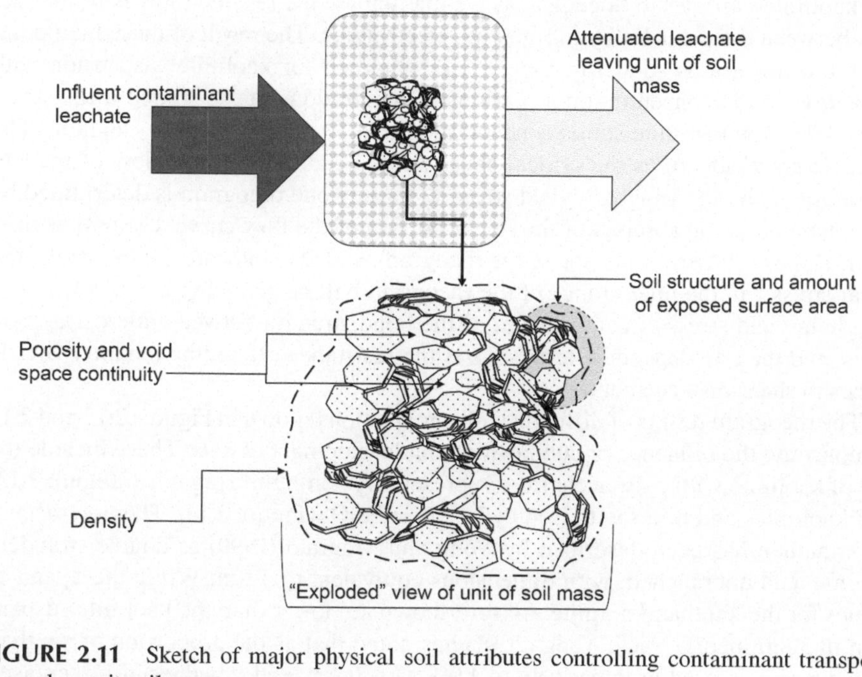

FIGURE 2.11 Sketch of major physical soil attributes controlling contaminant transport through a unit soil mass.

2.7.1 Soil Composition and Soil Structure

Soil structure consists of a combination of individual coarse and fine-grained particles and microstructures in various arrays. Clays are typically made up of a whole host of microstructures clumped together into macrostructures and finally a clay mass. The microstructures, which are packets of clay particles grouped tightly together through bonding forces such as forces of attraction between particles, have been referred to as *peds*, *flocs*, and *clusters*. The types or properties of these microstructures, together with their arrangement in a soil mass, are factors which contribute significantly to the physical and interactive (chemically reactive) properties of the clay soil. In particular, these microstructures have the capability to influence not only hydraulic conductivity, but also partitioning of contaminants in the transport through the soil. This is because of the amount of surface areas exposed to the contaminants, and also because the size and arrangement of the microstructural units or aggregate groups, will control the nature of the pore channels through which the contaminant in the porewater will be transported.

Quite often, composition and formational features such as origin and regional controls are significant influences on the development of the type of microstructures. The various microstructures, which are often referred to as microstructural units—*mus* (i.e., *mus* for plural and *mu* for a single unit), form the building blocks for the macrostructure of most clays. Figure 2.12 shows the results of viscosity experiments on kaolinite soil suspensions conducted to demonstrate the development of *mus* as one changes the local environment of soil particle deposition. In this case, we allow the kaolinite particles to flocculate as we manipulate the resultant forces of interaction between the particles by introducing electrolytes. The result of these manipulations are rheograms shown in Figure 2.12 obtained for kaolinite suspension with 9.1% w/w solids concentration at various concentrations of NaCl, and a rheogram for a 4.9% w/w kaolinite solids concentration in a 10 meq/L $NaHCO_3$ solution. The shear stress τ_o identifies the critical shear stress needed to initiate flow of the soil suspension, and the Bingham yield stress τ_β for any one rheogram is determined by the intercept on the abscissa of the linear portion of the flow curve. Using η to refer to the slope of the linear portion of the rheogram, and D to indicate the shear rate, the shear stress τ in the linear range of the rheogram will be given by: $\tau = \tau_\beta + \eta D$. The Bingham yield stress τ_β is a measure of the bond strength between microstructural units, and the rate dependent term, ηD, is the resistance offered by the interparticle forces to shear displacement.

The rheogram results of different soil compositions shown in Figures 2.13 and 2.14 demonstrate the influence of composition on development of *mus*. These include the use of kaolinite with polysaccharides representing non-humic organics (Figure 2.13) and kaolinites and illites with iron hydroxides, Fe_2O_3 (Figure 2.14). The polysaccharide xanthan has been described by Yong and Mourato (1990) as double stranded, anionic, and unbranched, with dimensions equivalent to 4 μm. While the τ_o and τ_β values for the xanthan-kaolinite mixture shown are lower than the kaolinite suspension in 3000 meq/L NaCl, Yong (2001) has noted that as the proportion of xanthan added was increased progressively to 1.0% w/w, the τ_o and τ_β values also increased progressively. What is significant about the addition of xanthan is that even with very

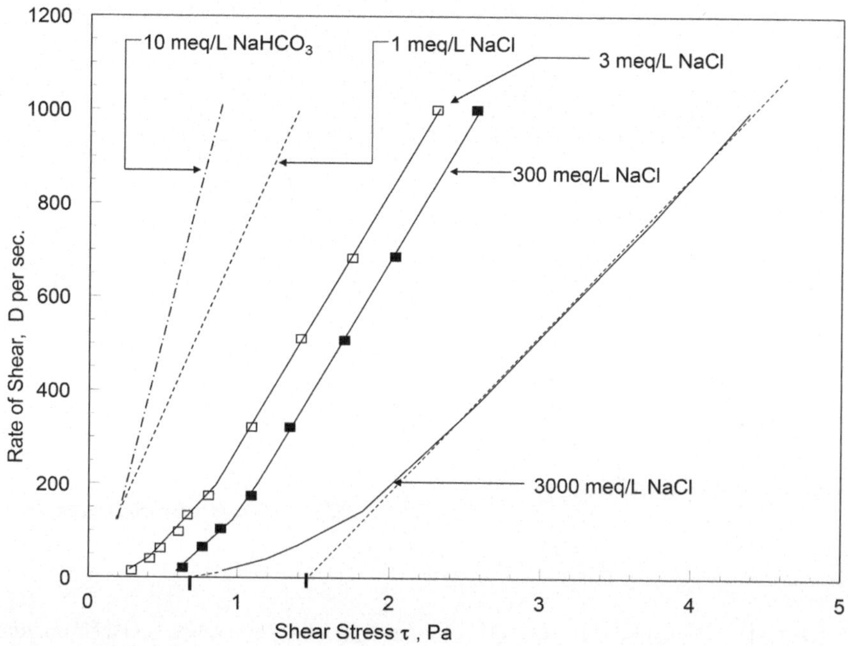

FIGURE 2.12 Rheograms for kaolinite soil suspension showing development of microstructural units as salt concentration increases—as might be deduced by the development of shear stress at very low rates of shear.

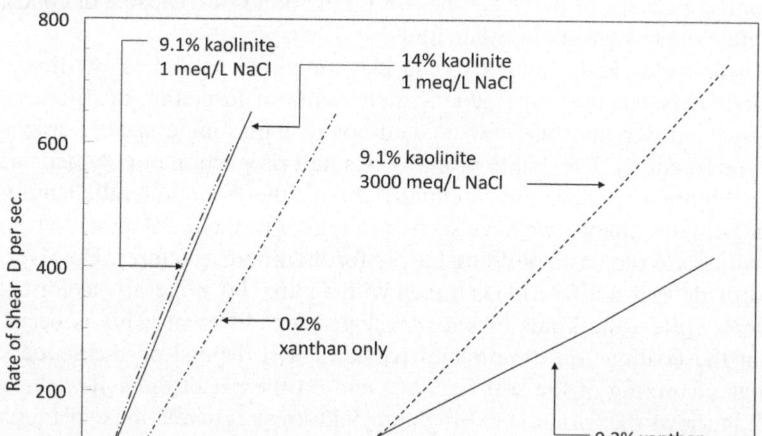

FIGURE 2.13 Rheograms for kaolinite in 1 meq/L NaCl and 3000 meq/L with addition of xanthan (polysaccharide).

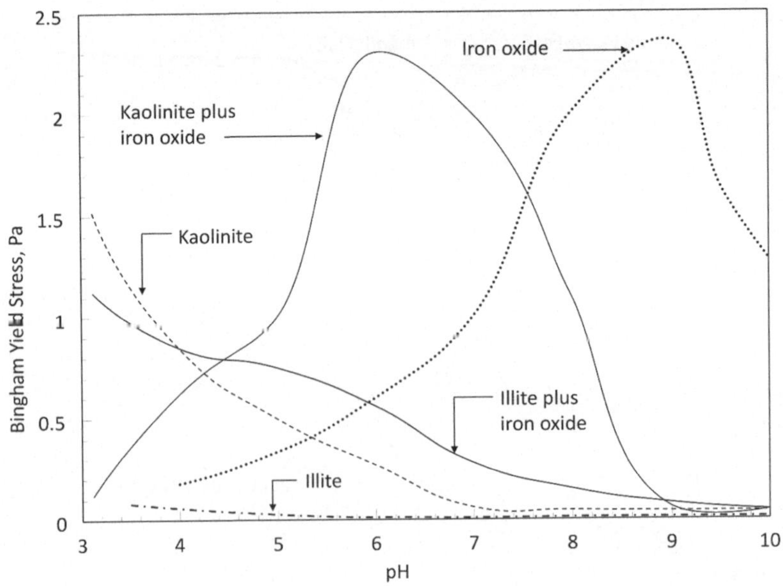

FIGURE 2.14 Bingham yield stress curves for kaolinite, illite, iron oxide, and their mixtures.

small proportions of the polysaccharide, the anionic character of the polysaccharide provides a bridging bond between the positive edge charges of the kaolinite particles and domains, packets, or tactoids. These bridging bonds are capable of contributing greater attachment strength between units.

Iron hydroxides Fe_2O_3 added to the clay minerals kaolinites or illites create interactions between the soil fractions and results in formation of different kinds of *mus*—i.e., different from *mus* created solely with single species minerals or single-type fractions. The addition of a very small proportion of polysaccharide for example (Figure 2.13) does not essentially change the monomineralic nature of the soil. The primary change we have seen is in regard to the development of bridging bonds, which are the basic building blocks for the microstructures. However, when iron hydroxide as a soil fraction is mixed with a pure clay mineral—as for example, kaolinite or illite—the kinds of *mus* developed are different. This is because the nature of the coatings on the mineral particles will depend on the sequence and conditions of mixing of the soil fractions and on the pH of the soil-water system. Figure 2.14 shows the variation in Bingham yield stress τ_β in relation to pH for kaolinite, illite, and iron hydroxide, together with mixtures of kaolinite with iron hydroxide and illite with iron hydroxide, prepared initially at pH 9.5. The reasons for the significant differences for some of the mixtures will be discussed in a later section when we deal with the nature of the surface reactive forces of the soil fractions.

2.8 CONCLUDING REMARKS

For natural attenuation of contaminants in soils to occur, physical, chemical, and biological interactions between the contaminants and soil constituents are required. This means that the properties and characteristics of the contaminants and the soils must be known in order to determine whether contaminant assimilation by the soil will occur. Assessments and evaluations of natural attenuation of contaminants need, also, to account for abiotic and biotic reactions between contaminants and soils.

This chapter deals with the nature of soils and pays particular attention to those soil fractions that contribute directly to the assimilative capacity of soils. Only those properties of soils deemed pertinent in the various processes associated with transport of leachates in the soils are considered. A more thorough discussion of the properties and behavior of soils can be found in Yong et al. (2012).

The main items and issues considered in this chapter include:

- Nature of soil—origin, geologic and regional controls, soil fractions and compositional control on development of pertinent soil features, properties and characteristics. These are important factors in the interactions between the soil solids and contaminants.
- Chemically reactive groups, defined as surface functional groups, associated with the surfaces of the various soil fractions make these surfaces reactive. The presence of soil particles with reactive surfaces allows these particles to react chemically with solutes and other dissolved matter in the porewater. The soil fractions with reactive surfaces include layer silicates (clay minerals), soil organics, hydrous oxides, carbonates and sulfates.
- Mechanisms associated directly with contaminant assimilation and attenuation (i.e., soil uptake) can be optimized when all the available soil solids' surfaces interact with the solutes in porewater.
- To determine the reactive surfaces of soil particles, the techniques used lead to results that are operationally defined. Values of SSA determined are directly dependent on the procedure used and the analytical technique used to reduce the data obtained.

REFERENCES

Bolt, G.H., and Bruggenwert, M.G.M. (1978), Composition of soil, in G.H. Bolt and M.G.M. Bruggenwert (Eds.), *Soil Chemistry: Basic Elements*, Elsevier Scientific Publishers, Amsterdam, the Netherlands, 281p.

Brunauer, S., Emmett, P.H., and Teller, E. (1938), Adsorption of gases in multimolecular layers, *J. Am. Chem. Soc.*, 60:309–319.

Carter, D.L., Mortland, M.M., and Kemper, W.D. (1986), Specific surface, in A. Klute (Ed.), *Methods of Soil Analysis, Part. 1: Physical and Mineralogical Methods*, American Society of Agronomy, Madison, WI.

Flaig, W., Beutelspacher, H., and Reitz, E. (1975), Chemical composition and physical properties of humic substances, in J.E. Gieseking (Ed.), *Soil Components*, Springer-Verlag, Berlin, Germany, Vol. 1, pp. 1–219.

Gast, R.G. (1977), Surface and colloid chemistry, in J.B. Dixon and S.B. Weed (Eds.), *Minerals in Soil Environments*, Soil Science Society America Publication, Madison, WI, pp. 27–44.

Greenland, D.J., and Mott, C.J.B. (1985), Surfaces of soil particles, in D.J. Greenland and M.H.B. Hayes (Eds.), *The Chemistry of Soil Constituents*, John Wiley & Sons, New York, 409p.

Grim, R.E. (1962), *Applied Clay Mineralogy*, McGraw-Hill, New York, 422p.

Mortland, M.M., and Kemper, W.D. (1965), Specific surface, in C.A. Black (Ed.), *Methods of Soil Analysis*, America Society of Agronomy, Madison, WI, pp. 532–544.

Newman, A.C.D., and Brown, G. (1987), *The Chemical Constitution of Clays*, Longman Scientific and Technical, Harlow, UK, pp. 1–128.

Parks, G.A. (1965), The isoelectric points of solid oxides solid hydroxides and aqueous hydroxo complex systems, *Chem. Rev.*, 65:177–198.

Sposito, G. (1984), *The Surface Chemistry of Soils*, Oxford University Press, New York, 234p.

Yong, R.N. (2001), *Geoenvironmental Engineering: Contaminated Soils, Pollutant Fate and Mitigation*, CRC Press, Boca Raton, FL, 307p.

Yong, R.N., and Mourato, D. (1988), Extraction and characterization of organics from two champlain sea subsurface soils, *Can. Geotech. J.*, 25:599–607.

Yong, R.N., and Mourato, D. (1990), Influence of polysaccharides on kaolinite structure and properties in a kaolinite-water systems, *Can. Geotech. J.*, 27:774–788.

Yong, R.N., Nakano, M., and Pusch, R. (2012), *Environmental Soil Properties and Behavior*, CRC Press, Boca Raton, FL, 435p.

3 Soil-Water Relations

3.1 INTRODUCTION

Chapter 1 introduced the subject of the use of the natural attenuation (NA) of contaminated soils procedure as a passive remedial treatment process for contaminated sites. The preceding discussions have highlighted the importance of a good understanding of the two interacting partners—namely, contaminants and soil—when utilization of this natural remediation process is contemplated, and that careful monitoring of the progress of natural remediation should be conducted during the course of the remediation process. Because of the highly variable properties of the two interacting partners, instead of rigid specifications, regulatory agencies have issued guidelines and protocols to guide users on the use of NA as a remediation technique. The key element of such guidelines and protocols on the use of NA is the monitored natural attenuation (MNA) treatment procedure.

The discussions in Chapter 2 regarding the nature of the kinds of constituents in natural soils show how complex an ordinary soil mass can be—especially in regard to the types of soil particle surfaces and properties. These surface features are key elements in the interactions between soil particles and contaminants. These interactions occur when contaminants are brought into contact with the soil solids—meaning that the transport of contaminants is required, and for this to occur, an aqueous phase is required. Contaminants, once already fixed in place in soil, will move by diffusive and advective means when an aqueous medium is present. The discussion in this chapter focusses on the relations between soil particles in the presence of water, with particular attention to the properties and attributes resulting therefrom. Knowledge of the various mechanisms of interactions between the soil particles and the solutes in water will tell us how the reactive surfaces of the soil particles react with the chemical nature of the water in the soil pores. This will allow us to better understand the nature of interactions between contaminants and soil particles and determine the type of surface complexation model that best describes the retention and attenuation of contaminants in the soil-water system.

3.2 FUNCTIONAL GROUPS AND ELECTRIC CHARGES

Section 2.6 in the previous chapter began our discussions on surface functional groups as integral features of the surface properties of soil reactions. Interactions between contaminants and soil fractions resulting in physical adsorption of the contaminants (*physisorption*) by the soil fractions is by and large electrostatic in nature—attraction established between positive and negative charges. Other interactions involving ionic and covalent bonding that result in the chemical mass transfer of contaminants are considered as chemical adsorption (*chemisorption*). The basic structure of many of the soil fractions and their surface functional groups provide the basis for formation of complexes between the functional groups and the contaminants in the porewater. The nature and extent of surface complexation depend on both the reactive properties of the soil particles and the contaminants themselves.

The discussions in Sections 2.5 and 2.6 in Chapter 2 on the reactive surfaces of layer lattice clay mineral particles such as kaolinites, illites, chlorites, etc. and the information provided in Table 2.1 showed that isomorphous substitution of the ions in the tetrahedral and octahedral layers by lower valence ions results in the development of electric charges on the siloxane surfaces. The various functional groups at the basal and edge surfaces of inorganic soil fractions, together with isomorphous substitutions in the lattices of the layer-lattice clay minerals, result in the development of negative and positive charges distributed on the surfaces and edges of the soil particles. Electric charges can also be developed on the faces of the edges of clay particles due to broken bonds at the edges of the particles—resulting in production of hydrous oxide-types of edge surfaces. Broken octahedral sheets in a kaolinite mineral particle, for example, will provide for Lewis acid sites [Al(III) · H_2O] which can bind OH groups in single coordination.

The OH groups on the silica surface of siloxane upper bounding surfaces, where both silanol and siloxane functional groups can exist together, are the centers of adsorption of the water molecules. The surface silanol groups are weak acids and will bond with internal silanol groups which are capable of establishing hydrogen bonding with water. The surface will be hydrophilic when surface silanol groups dominate and will be hydrophobic when siloxane groups dominate. The Al^{3+} in exposed edges of octahedral sheets of particle edges containing silanol and aluminol groups will complex with both H^+ and OH^- in the coordinated OH groups—with Si^{4+} complexing only with OH^-.

3.2.1 Point of Zero Charge and Variable Charges

We noted in the previous chapter that the nature and magnitude of charges for hydrous oxides and oxides are dependent upon their basic structure and the pH of the immediate environment characterizing the system. The general term *oxides* is commonly used to include the hydrous oxides, oxyhydroxides, and oxides. A significant distinguishing feature of the various oxides such as iron, aluminum, manganese, titanium and silicon is the fact that their surfaces essentially consist of broken bonds—in contrast to the clay minerals described in the previous chapter. Interaction of the oxide surfaces with water will be between the unsatisfied bonds on the oxide surfaces and the hydroxyl groups of dissociated water molecules. Both negative and positive charges will exist on the oxide surfaces, the predominance of each being dependent on the pH of the system. The surfaces of the oxides of iron and aluminum, for example, show coordination to hydroxyl groups which will protonate or deprotonate depending upon the pH of the surrounding medium. When the sum of the negative charges equals the sum of the positive charges, the *point of zero charge* (*pzc*) is attained. This means that one could switch from a net positive charge to a net negative charge when one changes the pH of the system from pH ranges below the *pzc* to values above the *pzc*.

We see, in Figure 2.10 in Chapter 2, various types of surface functional groups associated with soil organic matter. These can protonate or deprotonate depending on the pH of the aqueous environment. They will develop positive or negative charges depending on the pH of the soil and their respective acidity or basicity constants (i.e., pK_a or pK_b). The acidic properties associated with the soil organics are due to the carboxyl group and the hydrogen in the oxygen-containing functional groups which can be dissociated. The carboxyl and phenolic OH groups are considered to be responsible for a significant portion of the source of negative charge. Hayes and Swift (1985) report a range of 2–4 meq/g for soil organics—in comparison with the charge range of 0.01–2 meq/g for clay minerals.

From the previous discussions, we will note that there are two kinds negative charges associated with the surfaces of soil particles; constant or permanent negative charges, and pH-dependent negative charges. The constant or permanent negative charges are attributed to isomorphous substitutions and/or site vacancies in the layer-lattice structures of the clay minerals and non-crystalline hydrous oxides. In the case of pH-dependent negative charges, while the functional groups of soil organic matter such as carboxyls and phenolic groups possess pH-dependent negative charges, both non-hydrolyzable and hydrolyzable types of inorganic fractions also show pH-negative charge dependency. The non-hydrolyzable pH-dependent negative charges are associated with the hydrous-oxide coatings on clay mineral particles. In the case of hydrolysable pH-dependent negative charges, this characteristic is obtained from the structural OH_2 groups that result from protonation of hydroxyls at defects from removal of structural Al, Fe, and Mg (Huang and Jackson, 1966). Strictly speaking, the surfaces of soils fractions associated with these charges should be referred to as *variable-charge surfaces*, and soils that show such characteristics should be identified as variable-charge soils.

The positive charges associated with the surfaces of soil fractions are most commonly found in the hydrous oxides and hydroxides of Al, Fe and Mn, and in the broken edges of clay minerals. *Charge reversal* refers to the situation where the net charges on a particle surface (i.e., charge density) change in sign from a positive to a negative or vice versa when one progresses from a system pH below the *pzc* to above the *pzc*. This is a characteristic of soil fractions such as hydrous oxides and kaolinites. We should distinguish between the point of zero charge (*pzc*), and the *isoelectric point* (*IEP*). We define the *pzc* to mean the instance (pH_{pzc}) when the positive and negative surface charges are equal in number. In essence, the *pzc* is the pH at which titration curves of the candidate soil solution intersect at the zero point of adsorption densities, i.e., when the sum of the adsorption densities of H^+, (Γ_H), and OH^-, (Γ_{OH}) is zero. We recall that in portraying the data from titrations that the curves begin on the positive portion of the $\{\Gamma_H - \Gamma_{OH}\}$ ordinate at low pH values, and progressively decrease to minus portion as one increases the pH as seen in Figure 3.1 for a kaolinite soil at constant KCl concentration and with Pb^{2+} added as $Pb(NO)_3$.

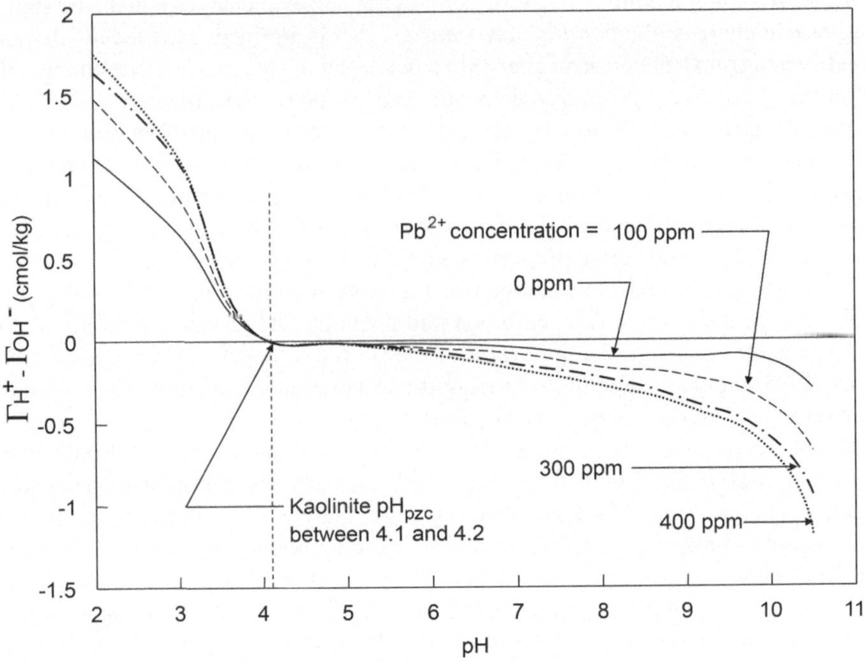

FIGURE 3.1 Titration curves for kaolinite at constant KCl concentration, with Pb^{2+} added in the form of $Pb(NO)_3$. (From Yong, R.N., *Geoenvironmental Engineering: Contaminated Soils, Pollutant Fate, and Mitigation*, CRC Press, Boca Raton, FL, 307, 2000.)

The isoelectric point (*IEP*) refers to the pH at which the electric potential developed at the solid-liquid interface, as a result of movement of colloidal particles in one direction and counterions in the opposite direction, becomes zero. At this point, one obtains a balance between positive and repulsive energies. While one can sometimes obtain accord between the pH*pzc* and pH*iep*, this should be considered to be unusual since these values are operationally defined.

3.2.2 Electrified Interface and Interactions

The net charges on the surfaces of the clay particles constitute an electrified interface in reaction with a solution containing anions and cations. Since these charges are more likely to be negative, reactions with cations in solution (porewater) will be both electrostatic and chemical. The nature of the distribution of the various ions (solutes) in the solution is conditioned by the various reactions between these ions and the charged surfaces—as shown in simplified form in Figure 3.2. The partly hydrated cations and anions in the inner Helmholtz plane are called potential determining ions (*pdis*) and are bonded to the reactive surfaces by ionic and covalent bonds. Potential determining ions contribute directly to the charge and potential on the surfaces of reactive particles. While there exists some minor conceptual differences concerning the exact nature and distribution of the various ions adjacent to the reactive surfaces

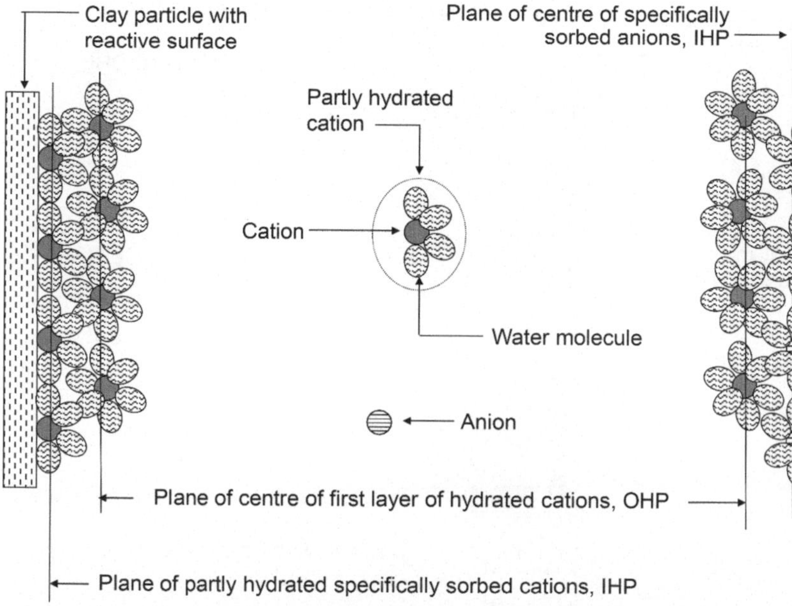

FIGURE 3.2 Specifically sorbed ions, inner Helmholtz plane (IHP) and outer Helmholtz plane (OHP). Left-hand drawing shows partly hydrated specifically sorbed cations and fully hydrated cations. Right-hand drawing shows anions as specifically sorbed anions and hydrated cations. The specifically sorbed ions are potential determining ions (*pdis*).

(e.g., Stern, 1924; Grahame, 1947; Kruyt, 1952; Sposito, 1984; Greenland and Mott, 1985; Ritchie and Sposito, 2002), there is general agreement on the existence of: (a) the altered or structured water layers adjacent to the reactive surfaces, and (b) the swarm of counterions forming a diffuse layer of ions.

The modification of the Stern layer model by Grahame (1947) to include the inner Helmholtz plane (IHP) and outer Helmholtz plane (OHP) as shown in Figure 3.2 is depicted in a more generic form in Figure 3.3 to illustrate the results of various interactions emanating outward from the surface of the reactive particle. The structured water adjacent to the surface of the particle is due to the specifically adsorbed ions at the interface. Ritchie and Sposito (2002) maintain that the complexes formed between the surface functional groups and the ions at the interface are inner sphere complexes. This assumes direct contact between them without interruption from any water molecule. When a layer of water molecules interrupts contact between these, the complexes are classified as outer sphere complexes. The IHP and OHP shown in Figure 3.3 identify the positions of the inner sphere and outer sphere complexes with corresponding distances of χ and β, respectively. The thickness of the Stern layer is obtained as $\delta = \chi + \beta$. Electrostatic bonding mechanisms for the counterions beyond the Stern layer required to satisfy the net negative charge of the reactive particles result in the formation of a diffuse ion layer.

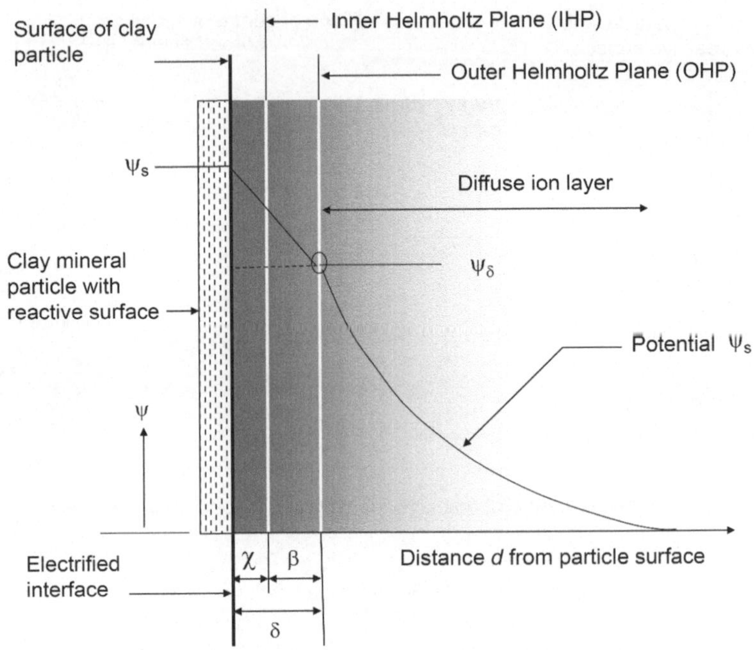

FIGURE 3.3 Generalized model of electrified interface with aqueous solution containing dissolved solutes.

The surface charge σ_s associated with the surface of the particle shown in Figure 3.3 is balanced by the sum of the Stern layer charge σ_δ at the OHP and the diffuse-layer charge σ_{ddl}. The surface potential ψ_s, also associated with the surface of the particle, varies with electrolyte concentration and the nature of the charge of the soil particle, i.e., constant charge surface or a pH-dependent charge surface. This surface potential drops from ψ_s at the surface to a potential ψ_δ at the OHP—as shown in Figure 3.4. The potential ψ (electric potential) beyond the OHP can be described by the Gouy-Chapman diffuse double layer model—as will be seen in the next section. From the viewpoint of electrokinetics, ψ_δ is considered to be equal (or almost equal) to the zeta potential ζ.

The Coulombic interaction established between the ions and forces associated with the negative charged sites on the reactive surfaces of the particles in the IHP can be calculated as E_{ihp} as (Yong, 2000):

$$E_{ihp} = \frac{z_i e^2}{\varepsilon R} \tag{3.1}$$

For the interaction energy E_{ohp} in the OHP, the four energy components (Yong, 2000) include: (a) Coulombic interaction energy E_c, (b) ion-dipole interaction E_{id}, (c) dipole-dipole interaction E_{dd}, and (d) dipole-site interaction E_{ds}.

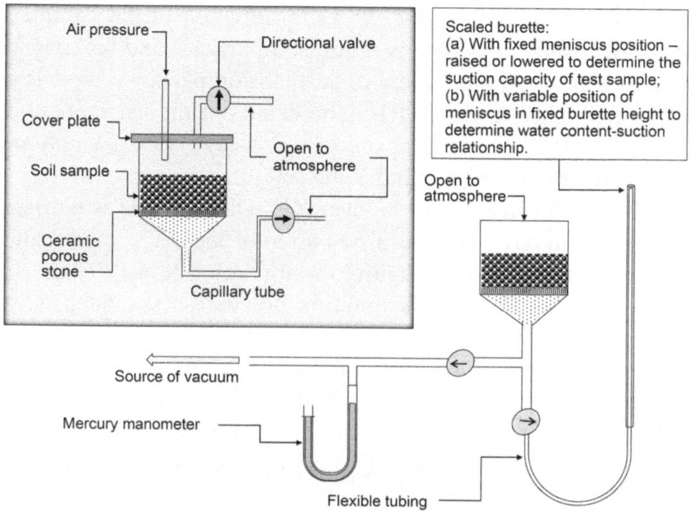

FIGURE 3.4 Buchner-type apparatus for determination of soil water-holding capacity using the pressure method (top left inset). Right-hand sketch shows the tensiometer method with two options for applying the negative pressure. (From Yong, R.N. et al., *Environmental Soil Properties and Behavior*, CRC Press, Boca Raton, FL, 434, 2012.)

$$E_{ohp} = \frac{z_i e^2}{\varepsilon R} + \frac{\mu e}{\varepsilon r^2}\left(1 - \frac{D_n \mu^2}{\varepsilon r^3}\right) + \frac{\mu e}{\varepsilon r_1^2} \tag{3.2}$$

where:

z_i = Valence of the *i*th species of ion in the bulk solution
e = Electronic charge
ε = Dielectric constant
R = Distance between the center of ion i and the corresponding negative charge site on the reactive particle
r = Sum of the radii of the ion and water molecule
r_1 = Distance between center of the dipole and corresponding negative charge site on the particle
μ = Dipole moment of a water molecule = 10^{-10} esu cm
D_n = Geometrical factor (0.334 for 3 water molecules and 1.188 for 6 water molecules).

3.2.3 Interactions and Surface Complexation Models

The degree of interaction between contaminants in the porewater and soil particles, and the degree of interaction between proximate particles depend on how one views the interactions immediately adjacent to the reactive soil particle surfaces. The energy status of a representative element of a soil-water system is defined by the level of interactions established. The total potential ψ shown in Figure 3.3 is a characteristic of the interaction energy. Determination of the complexes formed between the surface functional groups of the soil particles and the ions in the porewater can be performed using a variety of surface complexation models such as the single-, double-, and triple-layer models. The details of the more popular complexation models can be found in several texts, e.g., Kruyt (1952), Singh and Uehara, (1986), Bockriss and Reddy (1970), and Yong (2000).

The Gouy-Chapman diffuse double layer (GC-DDL) model is perhaps the most familiar model of the interaction and distribution of ions. The model allows one to compute the potential ψ shown in Figure 3.3 in relation to the distance d from the charged particle surface. Because of bonding processes associated with chemical bonding and complexation in the δ region shown in Figure 3.3, the simple electrostatic interaction calculations do not apply in the region within the vicinity of the charged particle surfaces, i.e., the GC-DDL model does not provide an accurate description of ψ immediately adjacent to the charged particle surface. The following assumptions and conditions for determination of ψ are generally used:

1. Interaction between the charged ions and solutes (cations and anions) in solution and the charged (soil) particle surfaces are Coulombic in nature.
2. The ions in solution are considered to be point-like in nature, i.e., zero-volume condition.
3. The density of the charges ρ due to the assumed point-like ions which contribute to the interactions (i.e., space charge density) can be described by the Boltzmann distribution.

The relationship for P can be obtained (e.g., Kruyt, 1952; van Olphen, 1977; Yong, 2000)—as shown in Equation (3.3):

$$\psi = -\frac{2\kappa T}{e} \ln\coth\left(\frac{d}{2}\sqrt{\frac{8\pi e^2 z_i^2 n_i}{\varepsilon\kappa T}}\right) \tag{3.3}$$

where:
κ = Boltzmann constant
n_i = Concentration of the ith species of ion in the bulk solution
T = Temperature.

The relationship between the surface charge density σ_s and surface potential ψ_s is:

$$\sigma_s = \left(\frac{2n_i\varepsilon\kappa T}{\pi}\right)^{\frac{1}{2}} \sinh\frac{z_i e}{2\kappa T}\psi_s \tag{3.4}$$

3.2.4 Applications and Chemical Speciation

For those whose interests lie primarily in "knowing" whether natural attenuation of contaminants can or will occur in candidate sites and soils, there are controlled laboratory attenuation tests that can provide direct experimental evidence on samples obtained from the candidate sites. How representative these tests are, and whether or not these tests and the results obtained can be replicated in the field are questions that cannot be easily answered without a better understanding of the interactions and processes involved. For those who are interested in determining (a) the degree (scale) of natural attenuation, (b) the time scale required to reach certain levels of attenuation, (c) the effectiveness of natural attenuation, and (d) the lateral extent or plume generation required to reach satisfactory attenuation remediation criteria, knowledge of the basic interactions between contaminants and soil particles will provide them with the tools to evaluate and predict transport performance and partitioning and chemical mass transfer. For abiotic processes, much of the information required can be obtained via determination of chemical speciation.

Prediction of chemical speciation in respect to contaminants in soils, using specially developed computer models is a subject that has received considerable attention in recent years. These models which include PHREEQE (Parkhurst et al., 1980; Parkhurst and Appelo, 1999), PHREEQM (Appelo and Postma, 1993), GEOCHEM (Sposito and Mattigod 1979), SOILCHEM (Sposito and Coves, 1988), EQ3 (Wolery, 1983) and MINTEQA2 (Allison et al., 1991) are well described by the respective model developers. The key elements of the models include calculation procedures based on assumption of thermodynamic equilibrium for speciation of soluble complexes, dissolution and precipitation of minerals, and sorption of ions onto soil particles' surfaces using a variety of surface complexation models. Knowledge of the outcome of competition between the functional groups on the reactive surfaces of soil particles and the various species of ions in solution for formation of complexes is fundamental to the success of determination of attenuation via soil assimilation.

3.3 SOIL-WATER ENERGY CHARACTERISTICS

Interactions between soil solids, water, and the various dissolved solutes (contaminants) in the porewater can be determined in relation to soil-water energy relationships—generally known as soil-water characteristics. These energy relationships, which are the outcome of the geochemical interactions of the porewater and the surface properties of the soil solids, describe the water- holding capability of a soil. They are key indicators of the water-holding capability (known as the *water-holding capacity*) of a soil. The energy relationship of soil for water describes the energy with which water is held in a soil mass in relation to its water content and is highly dependent on the chemistry of the porewater and the surface properties of the soil solids. The two commonly used terms to describe the energy with which water is held in soils are *soil-water tension* and *soil suction*—depending on whether one uses thermodynamics terminology or soil-water engineering applications. The use of the term *soil-water tension* indicates that the water in the soil is in equilibrium with a less than atmospheric pressure, and usage of the term *soil suction* is taken to mean that the soil exerts a suction force to take in (and hold) water.

A useful simple test procedure using a Buchner-type pressure membrane system is shown in Figure 3.4, and in the case of clays and fine silts, the pressure-membrane

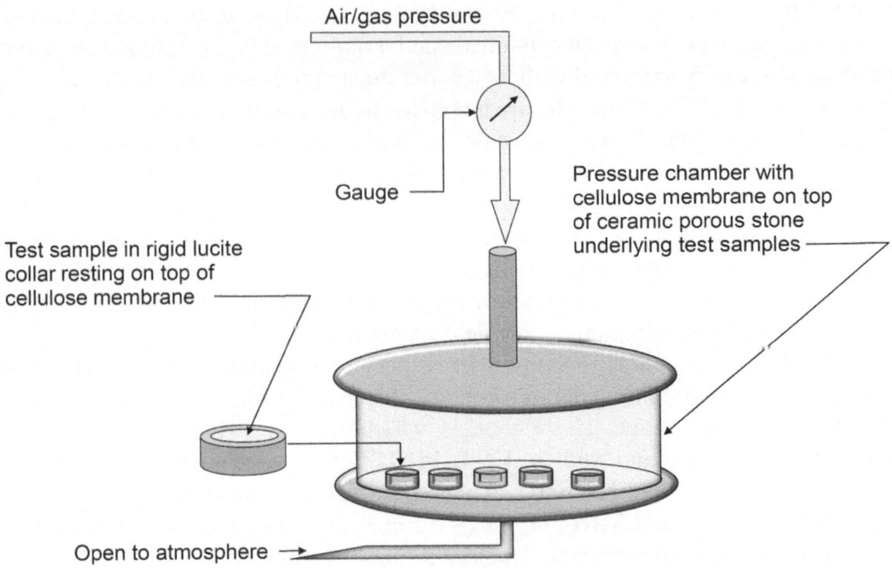

FIGURE 3.5 Pressure membrane apparatus used for determination of soil-water potential for silts and clays. (From Yong, R.N. et al., *Environmental Soil Properties and Behavior*, CRC Press, Boca Raton, FL, 434, 2012.)

type of apparatus shown in Figure 3.5 is used. In these test systems, different kinds of soils are permeated with water with and without dissolved solutes.

The water-holding capacity, water retention, energy relations between soil solids and water, and soil-water potential ψ, are all indicative of the geochemical and physical reactions involving the surface functional groups of soil solids and contaminants. As we have seen from the discussions in the previous chapter and previous sections, the surfaces of soil fractions with functional groups provide the stage for chemical reactions between the contaminants (ions, dissolved solutes, etc.) in the porewater of the soil.

In respect to the bottom water retention curve for sand in Figure 3.6, the absence of surface functional groups in sands means that the primary mechanism for water retention in sands is the availability of capillary forces. If we conduct a simple "height of capillary rise" experiment using a vertical column of clean sand, it will become clear that the height of capillary rise h in the column of sand will be determined by the grain-size distribution and density or packing of the sand grains (i.e., by the effective radius r of the pores in the sand). Assuming that the shape of the soil voids can be well approximated by capillary tubes of consistent bore, the relationship between h and r is: $h = \frac{2\sigma_w \cos\alpha}{r\gamma_w}$, where σ_w is the surface tension of water, γ_w is the density of water, and α represents the contact angle between the wetted soil particle surface and water. The capillary tube model shows that the smaller the value of r, the greater the height of capillary rise h. In the case of the other two water retention curves shown in Figure 3.6, i.e., the curves for the loamy soil and clay, the surface forces associated with the soil solids are responsible for the water retention and water

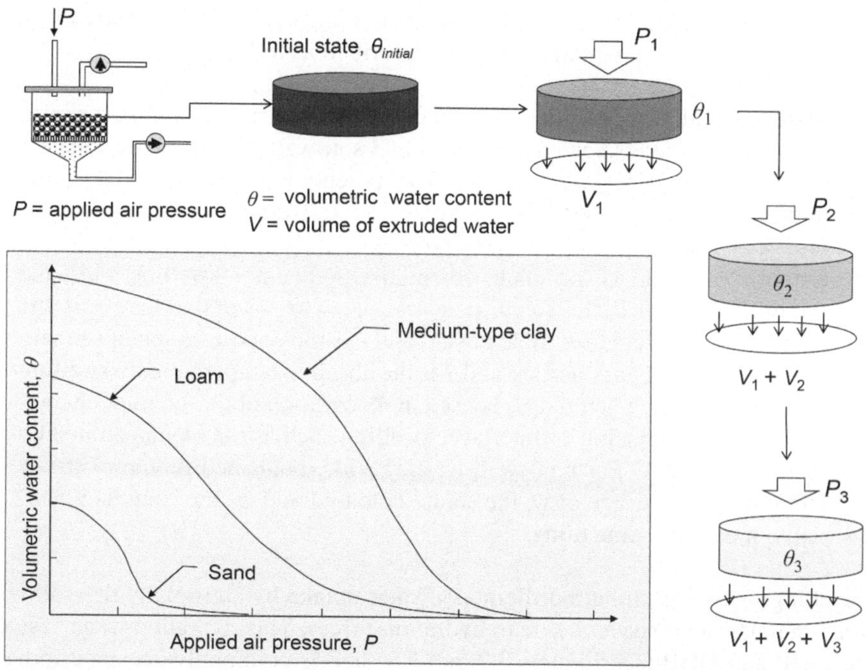

FIGURE 3.6 Schematic illustration of desorption-type experiment using the Buchner-type pressure apparatus. The right-hand sketches illustrate the steps in the test procedure, and the bottom left-hand graph shows typical desorption curves obtained using the apparatus. P = applied air pressure; θ = volumetric water content; V = volume of extruded water. (Adapted from Yong, R.N. et al., *Environmental Soil Properties and Behavior*, CRC Press, Boca Raton, FL, 434, 2012.)

holding capacity of the soils. The energy relationships between soil particles and water can be described or determined by the work required to move water into and out of a soil mass—using the porewater in soil as the frame of reference. In essence, the total work required to move the water into and out of the soil can be determined as the soil-water potential ψ.

3.3.1 Components of Soil-Water Potential

The soil-water potential ψ, which characterizes the energy state of soil water, varies throughout a soil mass because the forces acting on the soil water at any one point in the soil are never uniformly distributed. Using a pool of free water at the same elevation and temperature of a soil mass under one atmospheric pressure as a reference base, we define ψ as the total work required to move a unit quantity of water from the reference pool to the point under consideration in the soil. It is a negative number. The different components that make up a total soil-water potential ψ include two major components and three other components that are generally deemed to be of lesser importance. These two major components of the soil-water potential ψ, which have the greatest influence on the water-holding capacity of soils are the matric ψ_m and osmotic ψ_π potentials. They are described as follows:

- Matric potential ψ_m—this is a property of the soil matrix and pertains to sorption forces between soil fractions and soil-water. For granular materials, the matric potential is the capillary potential. However, for clay soils this is not the case because of microstructural units, surface functional forces, and reactive surfaces (see Figure 3.7).
- Osmotic potential ψ_π or solute potential ψ_s—For non-swelling soils, the osmotic potential is the solute potential, $\psi_\pi = \psi_s = nRTc$ where n is the number of molecules per mole of salt, and c is the concentration of the salt, R is the universal gas constant and T is the absolute temperature. In swelling soils, the osmotic potential ψ_π is used in recognition of the singular characteristic of interlamellar or interlayer swelling such as that demonstrated by montmorillonites. For non-swelling clay soils, published literature shows interchangeable usage of ψ_s the solute potential and ψ_π the osmotic potential to mean the same thing.

In swelling soils (e.g., montmorillonites), water uptake by an initially dry soil will be into the interlayer spaces—due to hydration forces. This hydration water resides in the IHP and OHP (see Figures 3.2 and 3.3) and is commonly accepted to be a

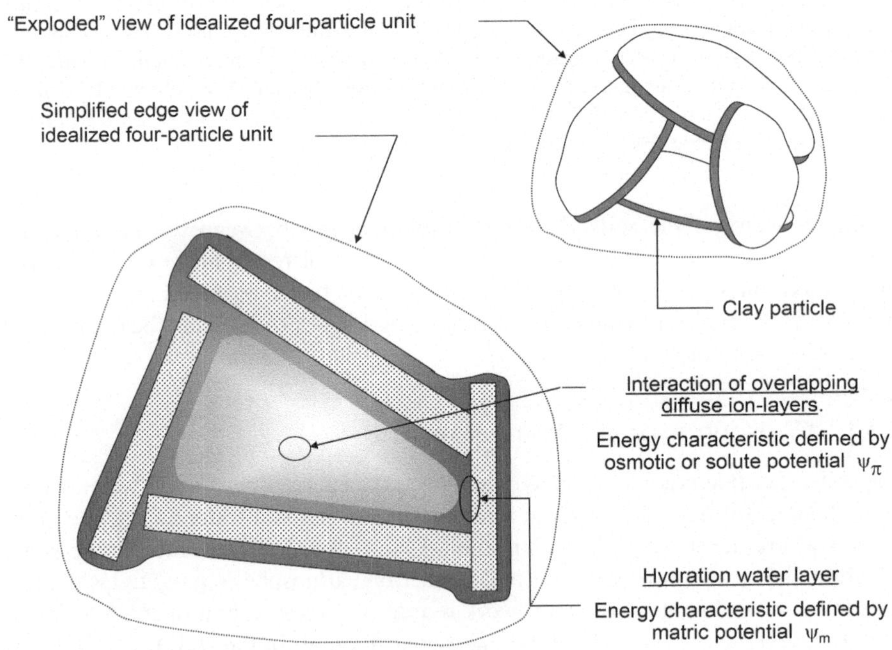

FIGURE 3.7 Schematic drawing showing exploded view of idealized four-particle unit illustrating hydration water layer and interaction of overlapping diffuse ion-layers. The "location" of the associated components of soil-water potential (matric and osmotic/solute) are shown.

different form of water structure. The common expression for this water is *structured water*. The expansion of soil volume associated with this hydration water is commonly defined as *crystalline swelling*, with water uptake attributed to the matric potential ψ_m. It should be noted that interlayer or interlamellar swelling beyond crystalline swelling stems from interactions described by the osmotic or solute potential ψ_π or ψ_s.

3.4 CHEMICAL REACTIONS IN POREWATER

The surface functional groups associated with the various soil fractions and the ions and other dissolved solutes such as naturally occurring salts in the porewater of a soil will react chemically when brought together to form a wet soil mass. As in the previous chapters, we will continue to refer to the various soil fractions as soil solids. The mechanisms leading to chemical mass transfer of contaminants (partitioning and attenuation) are intimately linked to the chemistry of the contaminants and the reactive surfaces of the soil solids. Interactions between contaminants (solutes) and soil solids involve many different sets of chemical reactions—including biologically mediated chemical reactions. The pH of the soil-water system and the various other dissolved solutes in the porewater influence the various interaction mechanisms such as acid-base reactions, speciation, complexation, precipitation and fixation.

3.4.1 Acid-Base Reactions, Hydrolysis

The porewater in a wet soil, without dissolved solutes and other contaminant ions, is by itself a solvent that can be either a *protophilic* or a *protogenic* solvent. It can function as an acid or as a base. Through self-ionization, it can produce a conjugate base OH^- and a conjugate acid H_3O^+. The definition of an *acid* as an aqueous substance which dissociates to produce H^+ ions, and a *base* as an aqueous substance which dissociates to produce OH^- ions, is better viewed in terms of proton donation or acceptance, thus allowing one to consider non-aqueous substances in terms acids and bases. The Brønsted-Lowry definition of an *acid* as a substance which has a tendency to lose a proton (H^+), and a *base* a substance which has a tendency to accept a proton allows us to do just that. Accordingly, the Brønsted-Lowry definition of an *acid* is a *proton donor* (*protogenic* substance) and a *base* is a *proton acceptor* (*protophilic* substance). Since water has the capability to both donate and accept protons (i.e., both protogenic and protophilic), it is called an *amphiprotic* substance.

Hydrolysis, which is an acid-base reaction, refers to the reaction of H^+ and OH^- ions of water with the solutes and ions present in the porewater and is basically a neutralization process. The presence of ionized cations and anions associated with soil solids in a soil-water system results in pH levels in the soil-water system that vary from below neutral to above neutral pH values dependent on the strength of ionization of the ions. Hydrolysis reactions in a soil-water system will continue so long as the reaction products are removed from the system—for example, through processes associated with precipitation, complexation, and sorption.

3.4.2 Oxidation-Reduction (Redox) Reactions

Abiotic and biotic oxidation-reduction (redox) reactions occur in the porewater in soils. Oxidation-reduction reactions involve the transfer of electrons between the reactants. The activity of the electron e^- in the chemical system typified by the reactive soil particles and contaminants in the porewater is of particular importance. A measure of this electron activity is the *redox potential Eh* and is given as:

$$Eh = pE\left(\frac{2.3RT}{F}\right)$$

where:

E = Electrode potential
R = Gas constant
T = Absolute temperature
F = Faraday constant.

The mathematical term pE is the negative logarithm of the electron activity e^-, and the relationship between Eh and pE at a standard temperature of 25°C is determined as:

$$Eh = 0.0591pE = E^0 + \left(\frac{RT}{nF}\right)\ln\frac{a_{i,ox}}{a_{i,red}} \tag{3.5}$$

where:

E^0 refers to the standard reference potential
n = number of electrons, and the subscripts for the activity a refer to the activity of the ith species in the oxidized (ox) or reduced (red) states.

The redox capacity is a measure of the number of electrons that can be added or removed from the soil-water system without a measurable change in the Eh or pE, and is comparable to the buffering capacity, which measures the amount of acid or base that can be added to a soil-water system without any measurable change in the system pH.

Because bacteria in the soil utilize oxidation-reduction reactions as a means to extract the energy required for growth, they essentially function as catalysts for reactions involving molecular oxygen and soil organic matter and/or organic chemicals. Electron transfer in a redox reaction is generally accompanied by proton transfer.

Redox reactions result in the decrease or increase in the oxidation state of an atom in the case of inorganic solutes. This has some significance particularly for those contaminants that have multiple oxidation states (e.g., chromium and arsenic) where increases or decreases in the oxidation state will alter the toxicity of the inorganic contaminant. In the case of organic chemicals, redox reactions will result in the gain or loss of electrons in the chemical.

3.5 INTERACTIONS, EXCHANGES, AND SORPTION

The interactions between dissolved solutes (ions and compounds) in the porewater and the soil solids, described in the previous sections, include such phenomena as transformations, precipitations, dissolution, exchanges, complexation, and sequestering. Referring to the dissolved solutes, ions, and compounds in the porewater as contaminants, the mechanisms involved in partitioning of the contaminants result in the chemical and physical mass transfer of the contaminants from the porewater to the surfaces of the soil solids—with the primary mechanism being that of sorption. The term *sorption* is used to include the various processes such as physical adsorption, chemical adsorption (chemisorption), and precipitation involved in the partitioning process.

We define physical adsorption of contaminants in the porewater by soil solids as the attraction of the contaminants to the surfaces of the soil solids due to the charge deficiencies of the soil solids. The previous discussions on the diffusive double layer (DDL) model (Section 3.2.2) indicated that because of the reactive surfaces of the soil solids, counterions are attracted to these reactive surfaces because of the requirement for electroneutrality. Cations and anions are specifically or non-specifically adsorbed by the soil solids depending on whether their interactions are in the diffuse ion-layer or in the Stern layer—as shown in Figure 3.2. Counterions in the diffuse ion-layer are non-specifically adsorbed. They are held primarily by electrostatic forces and will reduce the magnitude of the potential ψ, but will not influence the sign of ψ. They are sometimes referred to as *indifferent ions*. Adsorption of alkali and alkaline earth cations by clay minerals is a good example of non-specific adsorption. Adsorption of cations is related to their valence, and crystalline unhydrated and hydrated radii. Cations with smaller hydrated size or large crystalline size would be preferentially adsorbed—everything else being equal.

Specific adsorption of counterions occurs within the Stern layer (Figure 3.2), and the forces involved are those associated with the electric potential within the Stern layer. Specific adsorption of cations involves bonding by covalent bonds through the O and OH groups. Specifically adsorbed ions, which are referred to as *specific ions*, have the ability to influence the sign of ψ. Examples of specifically adsorbed cations include the adsorption of various heavy metals such as Pb, Cu, Cr and Cd by the oxides and hydrous oxides of Al, Fe, and Mn. Sposito (1984) refers to specific adsorption as inner-sphere surface complexation of the ions in solution by the surface functional groups of the soil fractions.

Physical or non-specific adsorption of anions are held by electrostatic or Coulombic forces in the diffuse double layer or in the Stern layer. By and large, the kind of sorption is considerably less than the adsorption capacity for cations since the soil fractions which have positive charge sites are primarily the oxides and edges of some clay minerals. Adsorption of anions is influenced by the pH of the soil-water system and the electrolyte level. Specific adsorption of anions is more a ligand exchange reaction than a sorption, and involves anion displacement of OH^- from the surface and incorporation as a ligand in the coordination of the structural cations (Bolt, 1978).

3.5.1 Bonding and Sorption Mechanisms

Bonds between charged particles of the various soil fractions and also between them and the charged contaminants are developed as a result of interactions between the two parties. This is one of the principal means for sorption of contaminants. The interactions between positively and negatively charged atoms and molecules result in development of interatomic bonds which include ionic, covalent and coordinate covalent bonds. In ionic bonding, electron transfer between atoms results in an electrostatic attraction between the resulting oppositely charged ions. Covalent bonding involves more or less equal sharing of electrons between the partners. In coordinate covalent bonding, the shared electrons originate only from one partner.

Other interactions such as ion-ion interactions between the soil solids and contaminants are Coulombic in nature. A significant portion of the bonding of contaminants or even between different types of soil solids occurs as a result of interactions between nuclei and electrons. These are basically electrostatic in nature. For example, electrical bonding occurs: (a) between the negative charges on clay mineral surfaces and positive charges on organic matter, and (b) also between negatively charged organic acids and positively charged clay mineral edges. The hydrogen bonds formed between soil-organic matter and clay particles are electrostatic or ionic bonds. Bonding between the oxygen from a water molecule to the oxygen on the clay particle surface is a strong bond in comparison with other bonds between neutral molecules. This type of bonding is not only responsible for holding layers of clay minerals together, but also in bonding organic molecules to clay surfaces and holding water at the clay particles' surfaces. The presence of polyvalent exchangeable cations permits adsorption of organic anions such as those in organic chemicals to clay mineral surfaces through the formation of polyvalent bridges.

Interactions such as those between instantaneous dipoles, and dipole-dipole interactions also produce forces of attractions which may be responsible for sorption of contaminants. For non-polar molecules (e.g., organic chemicals) these types of interaction are the most common type of sorption bonding mechanism established between non-polar molecules such as organic chemicals and the soil solids. The interaction forces developed, which are categorized as *van der Waals* forces, include: (a) Keesom—forces developed as a result of dipole orientation, (b) Debye—forces developed due to induction, and (c) London dispersion forces. Another type of bond is the steric bond, which develops as a result of ion hydration at the surface of the soil solids.

Chemical adsorption or *chemisorption* refers to high affinity specific adsorption which occurs in the inner Helmholtz layer (see Figures 3.2 and 3.8) through covalent bonding. The ions penetrate the coordination shell of the structural atom and are bonded by covalent bonds via O and OH groups to the structural cations. The valence forces bind atoms to form chemical compounds of definite shapes and energies. Chemisorbed ions can influence the sign of ψ, and are called *potential determining ions* (*pdis*). They are also sometimes referred to as *high affinity specifically sorbed ions*.

3.5.2 Cation Exchange

Ion exchange can occur between the ions in the diffuse ion-layers and the surfaces of the reactive soil particles because of charge imbalance. In a soil-water system, *cation exchange* occurs when positively charged ions in the porewater are attracted to the

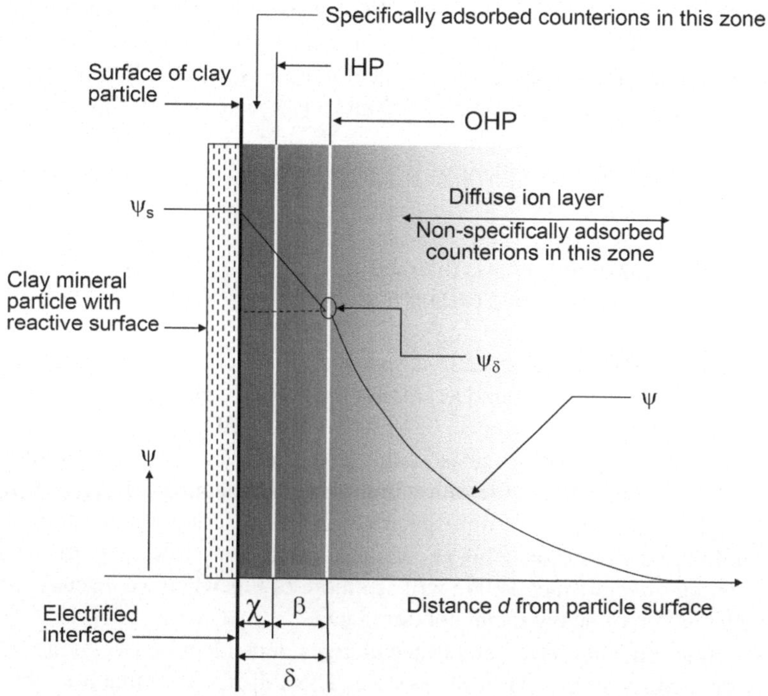

FIGURE 3.8 Generalized model of electrified interface with aqueous solution showing location of specifically and non-specifically adsorbed counterions. IHP = inner Helmholtz plane; OHP = outer Helmholtz plane.

surfaces of the clay fractions because of the need to satisfy electroneutrality. This process, which is stoichiometric, and which responds to the need for electroneutrality in the system, requires the replacing cations to satisfy the net negative charge imbalance of the charged reactive surfaces of the soil particles. Exchangeable cations are those cations associated with the charged sites on the surfaces of (primarily) clay minerals. By and large, the number of charged sites that are considered as exchange sites are determined by isomorphous substitution in the layer lattice structure of the clay minerals. The quantity of exchangeable cations held by the soil is called the *cation exchange capacity* (*CEC*) of the soil and is generally equal to the amount of negative charges. It is usually expressed as milliequivalents per 100 g of soil (meq/100 g soil). Although exchangeable cations are usually associated with clay minerals, we must note that other soil fractions also contribute to the exchange capacity of a soil. Amorphous materials and natural soil organics also contain surface functional groups that will contribute to cation exchange. Measured values for CEC can range from 15 to 24 meq/100 g soil for Fe-oxides, from 10 to 18 meq/100 g soil for Al-oxides, and from 20 to 30 meq/100 g for allophanes. The nature and distribution of the oxygen-containing functional groups of *soil organic matter* (*SOM*) are also influential in cation exchange. In particular, the carboxyl and phenolic functional groups appear to be the major contributors to the CEC of these soils. Appelo and Postma (1993) have linked the CEC of a soil to clay content, clay minerals, soil organic matter, and oxides/hydrous oxides. They provide CEC values of up to 100 meq/100 g soil for goethites and hematites, and from 150 to 400 meq/100 g soil for organic matter at a pH of 8. The empirical relationship cited by Appelo and Postma (1993) which relates the CEC to the percentage of clay less than 2 μm and the organic carbon is given as follows:

$$\text{CEC (meq/100 g soil)} = 0.7\ Clay\% + 3.5\ OC\%$$

where:
$Clay\%$ = Percentage of clay less than 2 μm
$OC\%$ = Percentage of organic carbon in the soil.

The predominant exchangeable cations in soils are calcium and magnesium, with potassium and sodium being found in smaller amounts. Exchangeable cations can be replaced by another of equal valence, or by two of one-half the valence of the original one. When the exchange cations possess the same positive charge and similar geometries as the replacing cations, the following relationship applies: $M_s/N_s = M_o/N_o = 1$, where M and N represent the cation species and the subscripts s and o represent the surface and the bulk solution. This exchange mechanism plays a significant role in the partitioning of heavy metals. We will see more of this when we discuss partitioning and attenuation of heavy metal contaminants.

For soil fractions that have net surface charges dependent on pH, values of CEC will vary depending on the pH of the system. The soil fractions that are included in this list are kaolinites, natural soil organics, and the various oxides or amorphous materials. In kaolinites, for example, the values of CEC can vary by a factor of 3 between the CEC at a pH of 4 (CEC = 2) to a pH of 9 (CEC = 6). Higher variations can be expected for oxides because the proportion of pH dependent charges is much higher for the oxides in comparison with the proportion of pH-dependent edge surface charges to planar surface charges in kaolinites.

The technique used for measurement of CEC in soils is such that operational differences will produce differing results—thus rendering the resultant CEC measurement as an operationally defined quantity. The condition that cation sorption should occur on all available sites requires one to ascertain that such indeed occurs without creation of artifacts. Saturation and subsequent exchange in the interlayers of layer-lattice clay minerals is a particular issue that highlights the problem. We also need to be aware that the reactions between the saturating cation solution and soil fractions can produce results that would be erroneous. Using ammonium acetate (NH_4OAc) as a saturation fluid, for example, for soils with significant amounts of carbonates, can cause dissolution of $CaCO_3$ and gypsum. This would result in extraction of excessive amounts of Ca^{2+} by NH_4^+, producing thereby, CEC measurements that would be unreasonably high.

3.6 CHEMICAL BUFFERING AND PARTITIONING

The *chemical buffering capacity* of soils is one of the mechanisms involved in the natural attenuation of contaminants. The chemical buffer capacity of soil, which is the reciprocal of the slope of the titration curve of the soil, is defined as the number of

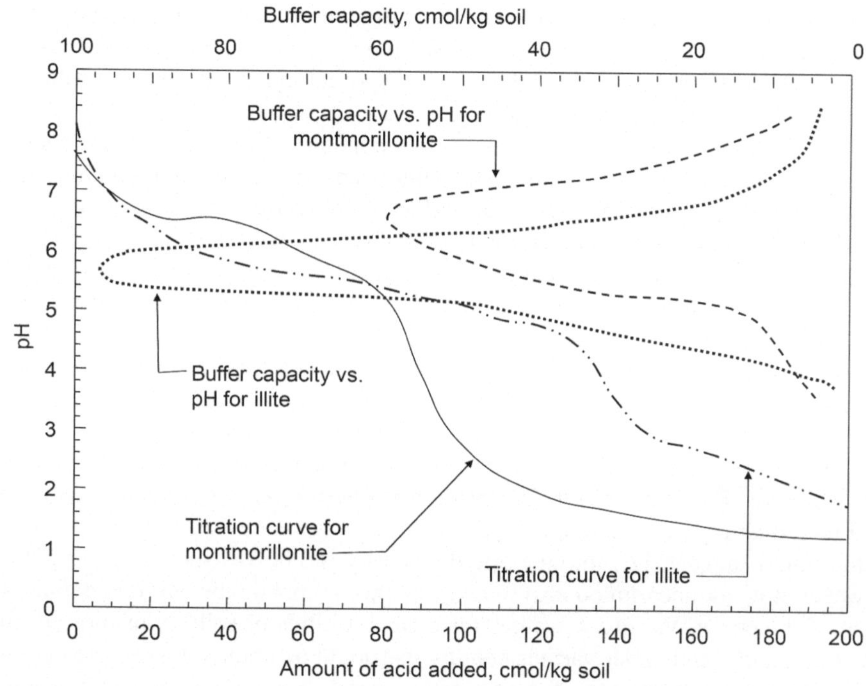

FIGURE 3.9 Titration and buffer capacity curves for illite and montmorillonite (clays). Note that the horizontal axis for buffer capacity curves begin with zero value on the right-hand side, and the relationships are given with respect to the pH ordinate axis.

moles of H^+ or OH^- that must be added to raise or lower the pH of 1 kg of the soil by 1 pH unit. Figure 3.9 shows the titration curves for two clay soils—a montmorillonite and an illite. The soil buffer capacity shown in the diagram is determined from the negative inverse slope of the respective titration curve and plotted in relation to pH. Expressing the buffering capability β of a soil in terms of changes in the amount of hydroxyl ions (OH^-) or hydrogen ions (H^+) added to the system, and in respect to the resulting pH changes we obtain: $\beta = \frac{dOH^-}{dpH} = \frac{dH^+}{dpH}$

We can obtain a clearer picture of how well a soil can perform as a chemical buffer by viewing its buffering capacity curve. Figure 3.9 shows that when the pH of illitic soil-water system is greater than 4, its capacity for chemical buffering is higher than the montmorillonite. This is very interesting and informative since the illitic clay (illite) has a smaller CEC than montmorillonite. What this tells us is that the high resistance in pH change in illite is due not only to adsorption of H^+ onto the exchange sites on the clay particles, but also by the neutralization of H^+ by the carbonates in the illitic clay.

3.6.1 Partitioning, Adsorption Isotherms, and Distribution Coefficients

One method for determination of partitioning of contaminants in a leachate stream or in the porewater by sorption mechanisms is to conduct batch equilibrium adsorption isotherm tests. Because the procedure is conducted with soil solutions, the results obtained are more than likely indicative of maximum sorption partitioning—since the soil particles in the soil solution are considered to be completely dispersed. All the soil particle surfaces in the soil solution are exposed and available for sorption of the contaminants in the aqueous phase.

Figure 3.10 shows the typical shapes of more popular adsorption isotherms characterized as *high affinity type, constant type* (linear adsorption curve), *Freundlich type* and *Langmuir type* isotherms. The relationships between the concentration of solutes adsorbed s*, and the equilibrium concentration of solutes in the aqueous solution *c*, are given adjacent to the respective curves. The constants k_n ($n = 1\ldots4$) and m are generally obtained from experiments and data fitting procedures. It is important to take note of the fact that both the constant type and Freundlich type isotherms will predict limitless adsorption and must be used with defined limits based on experimental information. The use of adsorption isotherms will be discussed in greater detail in the next few chapters dealing with attenuation of heavy metals.

The distribution coefficient k_d is generally obtained as the slope of the adsorption isotherm. For constant-type adsorption isotherms, $k_d = k_1$ (Figure 3.10). For the Freundlich type and Langmuir type adsorption isotherms, k_d is generally taken as the initial tangent, i.e., initial slope defined by the curves shown in the figure. However, it is not uncommon also to choose other points on the curve to define the slope of the curve. Various k_d values for different kinds of soils in relation to different pollutants and under various conditions should be obtained when predictions of transport and attenuation of contaminants need to be made. However, it must be noted that the use of batch equilibrium testing where soil solutions are used to

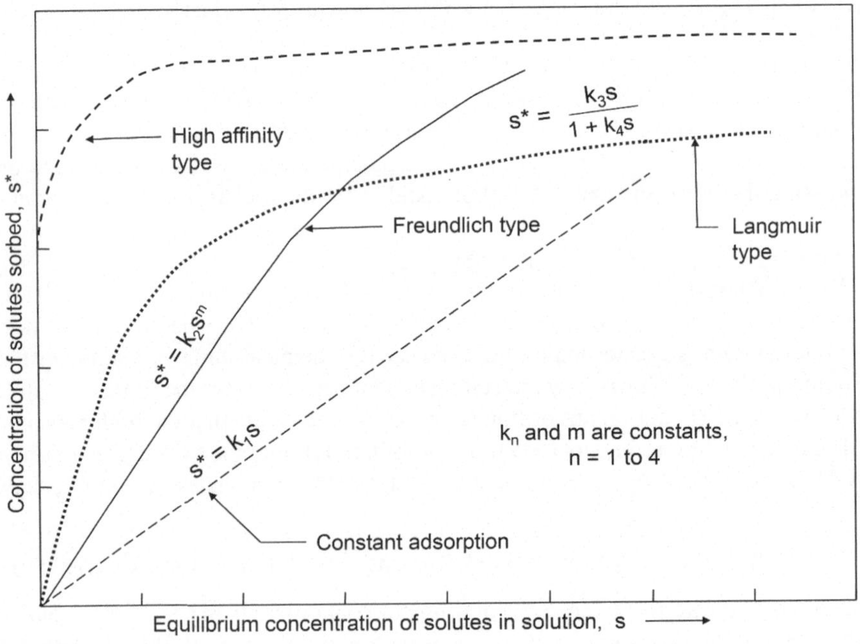

FIGURE 3.10 Different types of adsorption isotherms obtained from batch equilibrium tests.

determine the adsorption isotherms and k_d do not in any way represent compact soils found in the field. Soils in the field are generally in compact form, and do not present individual soil particles for interaction with contaminants in the porewater. What we need is to conduct tests that would expose the compact soils to contaminants in the porewater—i.e., leaching tests conducted on soil columns (discussed in detail in the next few chapters).

3.7 WATER UPTAKE AND MOVEMENT IN SOILS

Up to this point, we have considered the situation where interactions between soil solids and porewater are evaluated under conditions which involve the static state of soil water, i.e., no porewater movement is involved. Porewater in soils is rarely in a static state. Additions and subtractions of porewater (or groundwater) arise due to rainfall, snow melt, evaporation, transpiration, condensation, irrigation, and drainage—to name a few types of events that contribute to or subtract from the water content in soils. Flow of groundwater and porewater occurs through advective and diffusion forces. Distribution and migration of the water accompanying any of the previously mentioned events occurs in response to the many fluxes arising from the internal energy of the water itself and from internal and external mechanisms and driving forces due to thermal, ionic, osmotic, gravitational, hydraulic and other thermodynamic gradients. These gradients, forces, fluxes, together with the transmission properties of the soil will determine the rate and amount of water movement in the soil.

Except for liquid contaminants, we consider the movement of all non-liquid contaminants in soils to be directly associated with the presence and movement of porewater in soils. While the movement of water into soils can be broadly separated into three categories: (a) uptake of water by dry soil; (b) water movement in partly-saturated soil, sometimes referred to as unsaturated moisture movement or unsaturated flow; and (c) water movement in fully saturated soils, it is not always possible to "say" when uptake of water stops and when unsaturated moisture movement begins. In actual fact, it likely that a considerable overlap occurs between water uptake and the beginning of unsaturated moisture movement.

3.7.1 WATER UPTAKE AND UNSATURATED MOISTURE MOVEMENT

In this section, we will be concerned with water uptake into dry soils and water movement in partly saturated soils. We refer to the wetting of a dry soil as water uptake when it is clear that the water is being drawn into the soil by the internal forces of the soil. In sands for example, we would clearly associate these internal forces as the capillary forces. In clay soils however, we have internal forces associated with both the matric and solute potentials ψ_m and ψ_s.

The characteristics of water uptake (i.e., wetting) of a dry clay soil are different between non-swelling and swelling soils once we get beyond the hydration layer. The primary mechanism involved in initial wetting of dry soil solids' surfaces is the action arising from the adsorption energy of water. The thickness of the hydration water adsorbed onto the surfaces of the soil particles due to this mechanism will not exceed 1 nm for non-swelling soils. This water is generally called *crystalline water*—to reflect the fact that the structure of this water is unlike that of ordinary bulk water.

For swelling soils such as the 2:1 dioctahedral series of alumino-silicate clays (montmorillonites and nontronites for example—see Figure 2.7 in Chapter 2) water uptake in response to the matric potential ψ_m can seemingly exceed 1 nm. The further uptake of water (beyond the hydration layer) is not in response to ψ_m but is in fact due to the mechanisms represented in the diffuse double-layer (DDL) models. If we refer back to Figure 3.7, we can see this illustrated in the form of the respective energy characteristics. The interaction of the overlapping diffuse ion-layers and the associated energy characteristic will define the swelling phenomenon. When this occurs, the soil volume expansion that occurs is called *swelling*—thus giving the name of *swelling soils* to such soils. The interlayer or interlamellar expansion due to crystalline water uptake (hydration) is a function of the layer charge, interlayer cations, properties of adsorbed liquid, and particle size. The energy characteristic associated with this is shown in Figure 3.7. Water uptake beyond hydration which is due to double-layer forces results in increases in interlayer separation space in proportion to $\frac{1}{\sqrt{s}}$, where s is the electrolyte concentration in the liquid phase. Because of the popular use of bentonites for barrier and buffer-liner systems, it is important to pay attention to the interlayer spaces since these interactions play a dominant role in the assimilative processes. Furthermore, the characteristics of water uptake and uptake of porewater containing contaminants differ somewhat from those of non-swelling soils. In particular, we need to be aware of the energy requirements associated with the movement of water in the interlayers and between particles. Predictions of water and solute movement which rely on specification of the solute and matric potentials, i.e. ψ_s and ψ_m, must recognize how these measurements are made and the relevance of these in relation to control of water uptake and movement.

3.7.2 Unsaturated Flow

Because the range of water content in partly saturated soils can vary from relatively dry to relatively wet, the mechanism for water transfer for either situation will be different. The term water transfer is used in preference to water movement because the presence of a vapor phase and its movement will add or subtract water from any one location. Where high temperature gradients exist in soils, vapor transfer is greater than liquid transfer in the relatively dry soils. However, in relatively wet but partly saturated soils, liquid transfer will outweigh vapor transfer. Vapor movement is by convective flow of the air in the soil and/or by diffusion of water molecules in the direction of decreasing vapor pressure. Vapor–pressure gradients can develop not only because of temperature differences, but also because of salt concentration differences and differential suctions in the soil. As we can see, vapor transfer in partly saturated soils can also occur under isothermal conditions.

In natural soils, and especially in natural clayey soils, it is not unusual to find from 2% to 10% air content in so-called "saturated" soils. This is because of the presence of entrapped air in the soil voids filled with water. In the absence of external pressure gradients, most of water movement in the liquid phase is due to the gradients of the

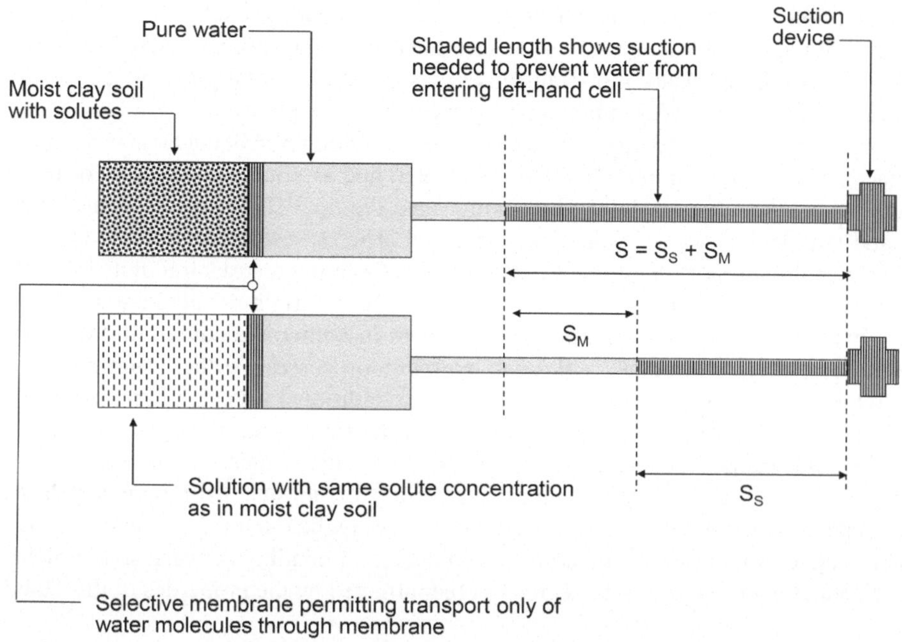

FIGURE 3.11 Osmometer-type cells showing development of suction required to counter flow of water into the left-hand side chambers because of the total potential ψ_T in the moist clay soil (top), and because of the solute potential ψ_s in the cell with the solution containing the same concentration of solutes (bottom). S, S_S and S_M refer to the total suction, solute suction and matric suction, respectively.

matric potential ψ_m and to physico-chemical forces associated with the interaction between the soil particles and water. Differences in concentrations of solutes between two points will create flow. Figure 3.11 shows an osmometer-type device where the right-hand cell contains pure water with access to: (a) a moist clay soil containing a specified concentration of solutes in the left-hand cell (top device), and (b) a solution containing the same solutes as in the moist clay soil and at the same concentration in the left-hand cell (bottom) replacing the moist clay soil. The water in the right-hand cell is connected to a suction measuring device. For simplicity in visualization, the length of the shaded horizontal column represents the equivalent suction needed to prevent transport of the water in the right-hand cell to the left-hand cells in response to the potential gradients established by the moist soil or the solutes in the solution.

The components of the total potential ψ described in Section 3.3 produce gradients that will provoke flow (liquid transfer). Because of the concentration of solutes in the solution in the left-hand cell (bottom), there is a tendency for the pure water in the right-hand cell to diffuse into the left-hand cell in response to the gradient set up by the solute potential ψ_s in each of the cells. The required suction to prevent the diffusion of water is shown in the diagram as S_s. When the left-hand cell contains a moist clay soil with the same kind and concentration of solutes, the required suction to prevent diffusion of water into the moist clay cell (top device) is much higher—due to the addition of the matric potential ψ_m in the moist clay soil. The schematic illustration shown in Figure 3.11 demonstrates the fact that even in the absence of external forces or pressures, liquid water moves in soils in response to internal gradients developed because of the differences in potentials between adjacent points.

Water movement in partly saturated soils occurs along film boundaries in soil pore spaces that are not completely filled with water, and as pore channel flow for those pore spaces that are completely filled with water. Figure 3.12 shows a sequence schematic that depicts water uptake into a dry soil. The representative volume element shown in the top portion of the diagram depicts three microstructural units interacting with each other. The schematic drawing in the top right of the diagram shows an "exploded" view of a 4-particle arrangement in a microstructural unit taking on water. Initially, water uptake is through hydration processes. Further water transport into the soil is from film boundary transport. As additional water is drawn in, the air within the soil must escape to the surface. At "full" saturation, if trapped air in some of the voids cannot escape, they will remain in the microstructural unit. Figure 3.13 shows the activation energy requirements for movement of ions in boundary layer associated with wetted particles. It is important to realize that the processes associated with assimilation of contaminants and degree of bonding between soil particles and contaminants are to some degree or other affected by the properties of this layer.

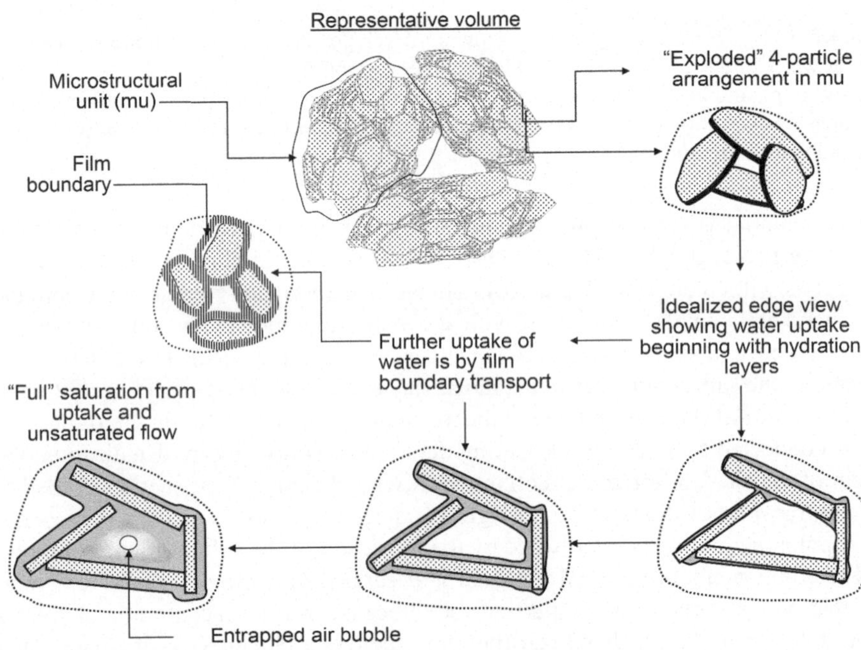

FIGURE 3.12 Water uptake into a dry soil. The schematic drawing shows an "exploded" view of a 4-particle arrangement in a microstructural unit taking on water through hydration processes and further transport into the soil from film boundary transport. At apparent full saturation, it is possible to entrap air bubbles in the structural units because of the inability of the air bubbles to escape.

We need to be very aware of the influence of the hydration (adsorbed) layer and surface charge distribution of clay particles on pollutant interaction and assimilative processes

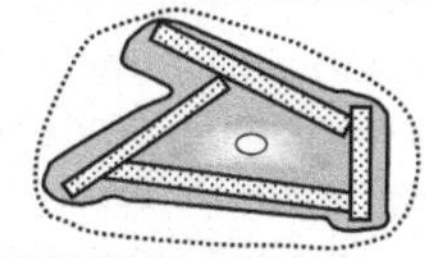

Higher activation energies for movement of exchangeable ions in the hydration or boundary layer due to bond breakage with neighbours and charge sites, and also in "shoving" aside of water molecules.

Activation energy

$$E_i = \frac{z\varepsilon^2}{D}\sum_{i=1}^{i=n}\frac{1}{r_i}$$

E_i = electrical energy of ion; r= distance between ion and charge site i; n = number of charge sites on particle;

z = valence; D = dielectric constant; ε = electronic charge

Boundary layer movement in wetting process responds to forces associated with reactive surfaces, and is more an IHP phenomenon – distinctly different from movement in the regions outside the OHP

Soil structure, size and distribution of units in macrostructure become important players in control of transport processes – e.g. proportion of film boundary to "regular" transport, and abundance of "menisci barriers" in units.

FIGURE 3.13 Activation energy requirements for movement of ions in boundary layer associated with wetted particles. Assimilation of contaminants and pollutants, and degree of bonding between soil particles and contaminants are all part of the equation for water uptake and boundary layer transport. IHP = inner Helmholtz plane; OHP = outer Helmholtz plane.

For a proper evaluation of the attenuation performance and capability of soils in actual field conditions, it is important to recognize that initial "degree of saturation" state of the field soil is a significant parameter. For soils located in the upper horizon where wetting and drying occurs, the field capacity of the soil plays a prominent role. It is at this point that most (if not all) of the water in the partly saturated soil will be distributed in the soil as film boundaries, i.e., there will be no pore spaces that are completely filled with water. The rate of water movement will be extremely slow. Removal or movement of the hydration water layer associated with the soil particles will require energy input higher than those described in Equations (3.1) and (3.2). With respect to movement of contaminants that require water as the transport agent, diffusion of the contaminants as solutes will occur along connecting film boundaries. So long as connected film boundaries exist, diffusive transport of contaminants along the film boundaries or boundary layers can occur.

For ease in measurements and computations of flow in partly saturated soils, it is often easier to measure differences in volumetric water content θ, between neighboring points instead of rate-of-flow between the same two points. Using the macroscopic flow velocity approach, and assuming a no-volume change condition, the equation of continuity states that the flow of water into or out of a unit volume of soil is equal to the rate of change of the volumetric water content, i.e. $\frac{\partial v}{\partial x} = -\frac{\partial \theta}{\partial x}$, where v is the macroscopic velocity, and x is the spatial coordinate. Although not totally appropriate for unsaturated flow at low water contents, the Darcy relationship $v = -k(\theta)\frac{\partial\psi}{\partial x}$ is often used in conjunction with the continuity condition—where ψ is the total soil-water potential referred to in Section 3.3. This permits one to obtain the unsaturated flow relationship in terms of a changing θ with distance as follows:

$$\begin{aligned}\frac{\partial\theta}{\partial x} &= \frac{\partial}{\partial x}\left(k(\theta)\frac{\partial\psi}{\partial x}\right)\\ &= \frac{\partial}{\partial x}\left(k(\theta)\frac{\partial\psi}{\partial\theta}\frac{\partial\theta}{\partial x}\right)\end{aligned} \tag{3.6}$$

Assuming that for any one soil, ψ is a single-valued function of θ, we can introduce the water diffusivity coefficient as $D(\theta) = k(\theta)\left(\frac{\partial\psi}{\partial\theta}\right)$ and rewrite Equation (3.6) as:

$$\frac{\partial\theta}{\partial t} = \frac{\partial}{\partial x}\left(D(\theta)\frac{\partial\theta}{\partial x}\right) \tag{3.7}$$

For vertical flow, using the coordinate z to designate the vertical spatial coordinate, we have:

$$\frac{\partial\theta}{\partial t} = \frac{\partial}{\partial z}\left(D(\theta)\frac{\partial\theta}{\partial z}\right) + \frac{\partial k(\theta)}{\partial z} \tag{3.8}$$

The wetting front profile obtained from a horizontal infiltration experiment conducted on a soil sample will show a typical form—as in Figure 3.14. Much can be learnt from the shape of the wetting front. Using the solutions provided by Yong and Wong (1973), specific cases governing θ and D can be studied, using Equation (3.7) to determine their influence on the shape of the wetting front profile. Figure 3.15 shows the various profiles for the case where D = constant to cases where D varies as θ varies. The shape of the wetting front shows how D varies as the volumetric water content in the soil continues to take in water.

3.7.3 Saturated Flow and Hydraulic Conductivity

Soil permeability to water (aqueous phase) is measured as hydraulic conductivity for saturated soils and diffusivity for unsaturated soils. For saturated soils the permeability is commonly expressed in terms of a Darcy permeability coefficient k, which is generally obtained via application of the Darcy model as a means of analysis of hydraulic conductivity data. From laboratory permeability measurements, the Darcy coefficient k is obtained from the relationship: $v = ki = k\frac{\Delta h}{\Delta L}$. The hydraulic gradient i is the ratio of the hydraulic head Δh and ΔL, the spatial distance. This form of expression of k does not consider the impact of the properties of the permeant on hydraulic conductivity, nor does it take into account the structure of the soil (soil micro and macrostructure). Since a unit soil mass is made up of an almost infinite number and arrangement of soil particles and peds (fabric units), it would not be surprising to note that soils with similar compositions can have different densities and correspondingly different hydraulic conductivities.

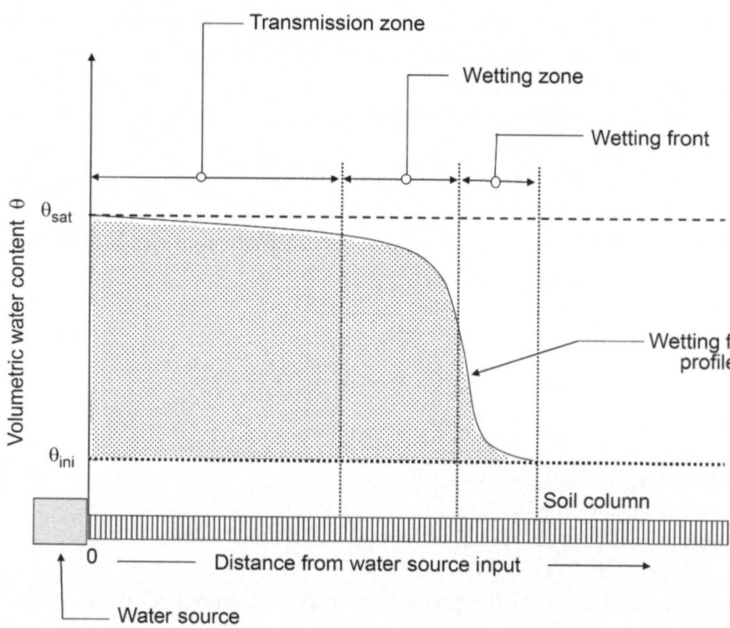

FIGURE 3.14 Characteristics of a wetting front profile. Shaded area represents the volumetric water content in the zone behind the wetting front. θ_{ini} and θ_{sat} are initial and saturated volumetric water contents, respectively.

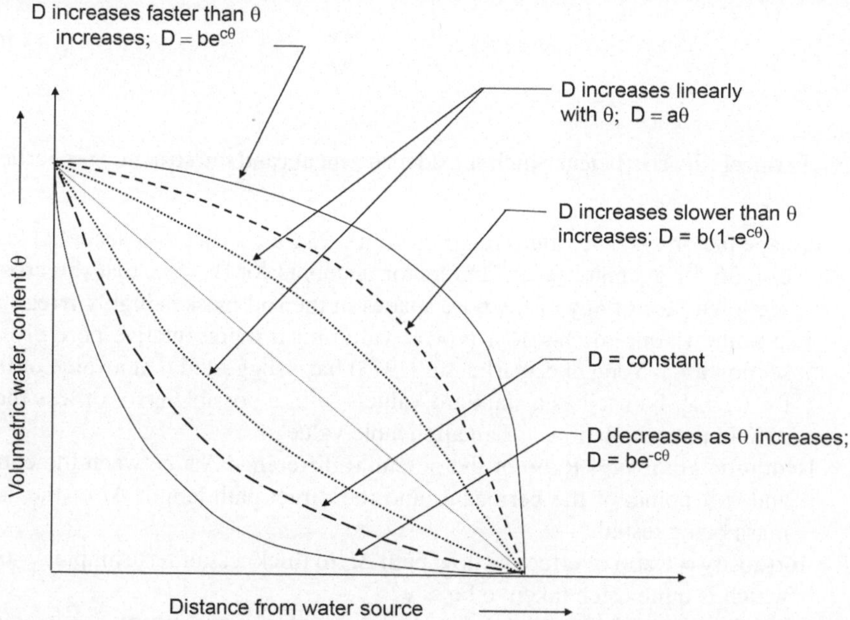

FIGURE 3.15 Analytically obtained wetting front profiles for various cases of D varying with θ. (a, b, and c are constants.)

For that matter, it should also not be surprising to learn that soils with similar densities and composition could also possess differing hydraulic conductivities—attributable to differing soil macro and microstructures.

To accommodate the influences of the permeant and soil structure such as permeant viscosity, soil-voids' features, tortuosity and shape of pore space cross section, a different type of permeability coefficient will be needed. To obtain this new coefficient, the Poiseuille relationship for flow through fine-bore tubes shown in Equation (3.9) is used in an adapted form for determination of the link between soil structure and soil permeability.

$$v^* = \frac{r^2\gamma}{8\eta}\frac{\Delta\psi}{\Delta l} \tag{3.9}$$

where:

v^* = Mean effective flow velocity through a narrow tube of radius r, γ and η are the density and viscosity of fluid or permeant respectively

$\Delta\psi$ = Potential difference between the ends of a tube of length Δl.

To account for the influence of the properties of pore channels defined by the structure of a soil, and the fact that the wetted soil particles' surface area is controlled by the structure of the soil, Yong and Mulligan (2003) have used a modification of the combined form of the Poiseuille and Kozeny-Carman model—as shown in Equation (2.2).

$$v = k^* i = \frac{C_s n^3 \gamma}{\eta T^2 S_w^2}\frac{\Delta\psi}{\Delta l} \tag{3.10}$$

where:

k^* = Permeability coefficient which considers permeant and soil structure properties, $= \frac{C_S n^3 \gamma}{\eta T^2 S_w^2}$

C_s = Shape factor, and has values ranging from 0.33 for a strip cross-sectional face to 0.56 for a square face. This factor accounts for the fact that the cross-sectional face of any of the pore spaces in the soil mass is highly irregular and allows one to choose a typical value for a representative pore cross-section area. Yong and Warkentin (1975) have suggested that a value of 0.4 for C_s may be used as a standard value—with a possible error of less than 25% in the calculations for an applicable value of k^*

i = Hydraulic gradient = Ratio of the potential difference $\Delta\psi$ between the entry and exit points of the permeant, and the direct path length Δl of the soil mass being tested

T = Tortuosity = Ratio of effective flow path Δl_e to thickness of test sample Δl and which is quite often taken to be $\approx \sqrt{2}$

γ and η = Density and viscosity of the permeating fluid respectively

n = Porosity of the unit soil mass

S_w = Wetted surface area per unit volume of soil particles.

Equation (3.10) contrasts with the standard Kozeny-Carman (K-C) relationship in that the wetted surface area consideration in the K-C model assumes that $S_w = S(1-n)$ and that the radius r of the Poiseuille tube is: $r = \frac{n}{S(1-n)}$.

This gives us the relationship for k from the K-C model as:

$$k = \frac{C_s \gamma n^3}{\eta T^2 S^2 (1-n)^2} \tag{3.11}$$

where S is the specific surface area of soil. With this measure of surface area, we assume that all particle surfaces are in contact with the permeating fluid. Experience with the use of the classical Kozeny-Carman relationship has been good when the pore sizes are relatively uniform, i.e., when the particle sizes are relatively uniform. However, experience has also shown that when pore sizes vary significantly, and especially when we have a high proportion of fine fractions in the soil, the relationship does not hold well. This should be intuitively obvious because the specific surface area of the soil will be proportionately higher because of the presence of the fines.

The adaptation introduced by Yong and Mulligan (2003) considers the surface area of the particles in terms of only the wetted surfaces, i.e., S_w. This allows one to have compositional differences and soil structure differences which can impact severely on the distribution of pore sizes and availability of soil particle surfaces for direct interaction with the permeant. We see from Equation (3.10) that C_s, T, and S are soil property parameters which are dependent on soil composition and soil structure. These can be expressed as a parameter $\beta = C_s/(TS)^2$. Along the same lines, the density and viscosity of the permeating fluid, γ and η respectively, are properties of the permeant and can be described by a parameter $\mu = \gamma/\eta$. Using these parameters, the relationship for k^* can be expressed as follows: $k^* = \mu\beta n^3$. Assuming that the physical properties of a leachate permeant are not too far distant from that of water at about 20°C, we can compute μ directly. Further assuming a tortuosity T value of $\sqrt{2}$, and $C_s = 0.4$, the relationship between soil permeability expressed as a coefficient k^* and the amount of surface area wetted in fluid flow through the soil can be shown in graphical form (Figure 3.16)—expressed in relation to the porosity of the soil. The wetted surface area shown in the diagram is the surface area of the pore channels in the soil through which fluid flow occurs.

The graphical relationship between soil permeability expressed as a coefficient k^* and the amount of surface area wetted in fluid flow through the soil determined in relation to the porosity of the soil is shown in Figure 3.17. The wetted surface area is the surface area of the pore channels in the soil through which fluid flow occurs. The relationships shown in the figure can be used to provide information on (a) wetted surface area, and (b) influence of soil structure in control of hydraulic conductivity. The wetted surface areas S_w are seen to be small fractions of the specific surface area of the various soils. This tells us that microstructure of the soils play a dominant role in the development of flow channels. Denoting the surface area of the soil solids comprising a unit volume of the soil being permeated as SSA_v, we can define the ratio S_w/SSA_v as the *wetted surface ratio* (*WSR*). The WSR provides an indication of the microstructure of the soil, and the extent of soil particle surfaces available for interaction with the fluid used for the permeability test, i.e., surface area wetted during hydraulic flow (Yong and Mulligan, 2003).

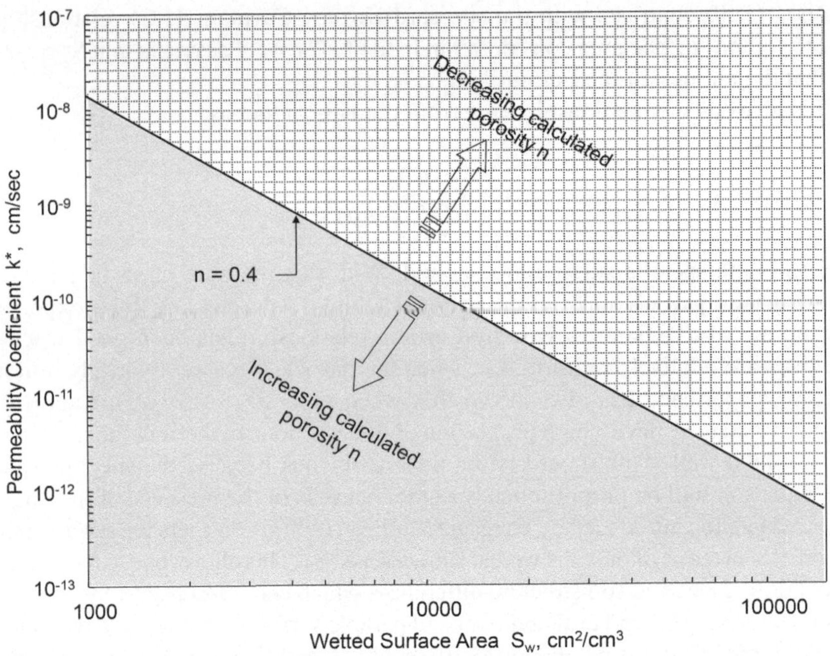

FIGURE 3.16 Variation of permeability coefficient k^* with wetted surface area S_w and calculated porosity n. (Adapted from Yong, R.N. and Mulligan, C.N., *Appl. Clay. Sci.*, 23, 179–186, 2003.)

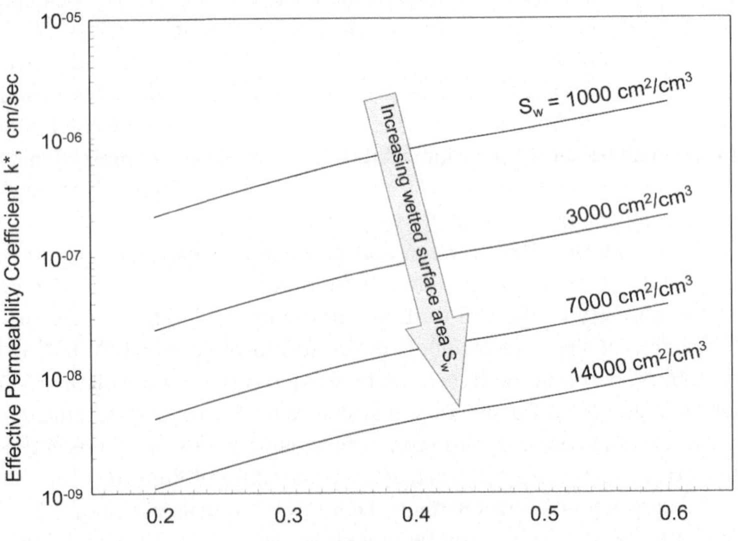

FIGURE 3.17 Variation of permeability coefficient k^* with calculated porosity—in relation to wetted surface area S_w. (Adapted from Yong, R.N. and Mulligan, C.N., *Appl. Clay. Sci.*, 23, 179–186, 2003.)

3.8 MOVEMENT OF SOLUTES

Movement of porewater solutes (i.e., solutes in the porewater) in soils occurs in saturated and partly saturated soils. The obvious necessary condition for movement of solutes is the presence of water within the pore spaces either as film boundaries or as water-filled pores. If continuity in film boundaries is established through contact between adjoining particle film boundaries, movement of the solutes will progress as diffusive movement. This is to say that in partly saturated soils, the predominant mechanism of transport or movement of solutes in the porewater is by diffusive means. Diffusive flow of solutes will also occur in the water-filled micropores of the microstructural units—as previously discussed—because of the forces of interaction between adjacent particles. In saturated soils, we will have both diffusive and advective flow, depending on whether the external hydraulic gradients are sufficiently large, i.e., diffusive flow of solutes will occur in the micropores and advective flow of solutes will more than likely occur in the macropores if the hydraulic gradients are sufficiently large (see next section).

3.8.1 Diffusion of Solutes and Diffusion Coefficient

Brownian activity of the solutes in the porewater results in a net diffusive flow of the solutes. When the porewater is in a more-or-less immobile state, e.g., hydration water layer and or water in the diffuse double layer, diffusion is the more likely mechanism of transport of the solutes. From a theoretical point of view, we can consider the diffusion coefficient of a target solute D_s to be equal to the effective molecular diffusion coefficient. A useful reference point is the infinite solution coefficient D_o which refers to a single ionic species in dilute solutions. From the studies of molecular diffusion of both Nernst (1888) and Einstein (1905) involving the movement of suspended particles controlled by the osmotic forces in the solution, we cite the following:

Nernst-Einstein

$$D_o = \frac{uRT}{N} = u\kappa T \tag{3.12}$$

Einstein-Stokes

$$D_o = \frac{RT}{6\pi N\eta r} = 7.166\times 10^{-21}\frac{T}{\eta r} \tag{3.13}$$

Nernst

$$D_o = \frac{RT\lambda}{F^2|z|} = 8.928\times 10^{-10}\frac{T\lambda}{|z|} \tag{3.14}$$

where:

u = Absolute mobility of solute
R = Universal gas constant
T = Absolute temperature
N = Avogadro's number
κ = Boltzmann's constant

λ = Conductivity of the target solute
r = Radius of hydrated solute
η = Absolute viscosity of the fluid
z = Valence of the ion
F = Faraday's constant.

Infinite solution diffusion D_o models incorporate such factors as: ionic radius, absolute mobility of the ion, temperature, viscosity of the fluid medium, valence of the ion, equivalent limiting conductivity of the ion, etc. Compiled values for D_o for various conditions can be found, for example, in Li and Gregory (1974) and Lerman (1979), amongst others, and experimental values for D_o for many major ions at various temperatures can be found in Robinson and Stokes (1959). Discussions of the effects of varied contaminant solutes on diffusion coefficients can be found in these same references.

To be sure that we use the appropriate transport in evaluation of transport of solutes in the porewater, we can use the Peclet number P_e as a screening tool. The Peclet number is defined as: $P_e = v_L d/D_o$, where D_o = diffusion coefficient in an infinite solution, d = average soil particle diameter, and v_L = longitudinal flow velocity (advective flow). Figure 3.18 shows the transport diagram using information reported by Perkins and Johnston (1963). This indicates that it is appropriate to consider transport of solutes as being diffusive when $P_e < 1$. In this range, diffusive transport of solutes in the porewater dominates any advective transport.

Partitioning of solutes during diffusive flow in the porewater can be considered in respect to: (a) association with the volumetric water content, and/or (b) the fluxes associated with the respective thermodynamic gradients. In the first approach using the volumetric water content association, we can specify:

$$\frac{\partial \theta s}{\partial t} = \frac{\partial}{\partial x}\left(k(\theta, s) \frac{\partial \psi_s}{\partial x} \right) \tag{3.15}$$

where:
s = Concentration of solute under consideration
$P_s = P(2,s)$ = Solute potential.

Assuming that the Darcy permeability coefficient k is relatively insensitive to the presence of solutes in the fluid phase, i.e., k is a function only of 2, and further assuming that P_s is also only a function of 2, we obtain from Equation (3.12) the following relationship:

$$\frac{\partial \theta s}{\partial t} = \frac{\partial}{\partial x}\left(D_s(\theta) \frac{\partial s}{\partial x} \right) - \frac{\partial (\rho^* s^*)}{\partial t} \tag{3.16}$$

where ρ^* and s^* represent the bulk density of soil divided by the density of water and the concentration of solutes sorbed by the soil particles.

We can identify the fluxes associated with the respective thermodynamic forces as follows:

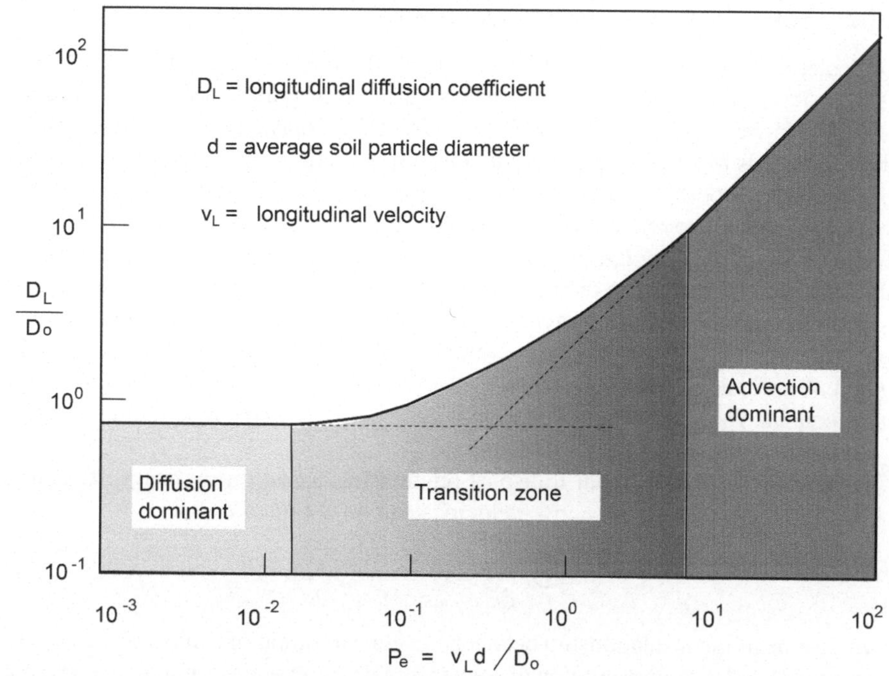

FIGURE 3.18 Diffusion and advection dominant flow regions for solutes in relation to Peclet number. (Adapted from Perkins, T.K. and Johnston, O.C., *J. Soc. Pet. Eng.*, 17, 70–84, 1963.)

$$J_\theta = L_{\theta\theta}\frac{\partial \psi_\theta}{\partial x} + L_{\theta s}\frac{\partial \psi_s}{\partial x}$$
$$J_s = L_{s\theta}\frac{\partial \psi_\theta}{\partial x} + L_{ss}\frac{\partial \psi_s}{\partial x} \tag{3.17}$$

where J_θ and J_s are the fluid and solute fluxes respectively, and L_{22}, L_{2s}, L_{s2}, and L_{ss} are the phenomenological coefficients. The coupled relationships can be expressed in conjunction with the various diffusivity coefficients as follows:

$$\frac{\partial \theta}{\partial t} = \frac{\partial}{\partial x}\left[D_{\theta\theta}\frac{\partial \theta}{\partial x} + D_{\theta s}\frac{\partial s}{\partial x}\right]$$
$$\frac{\partial s}{\partial t} = \frac{\partial}{\partial x}\left[D_{s\theta}\frac{\partial \theta}{\partial x} + D_{ss}\frac{\partial s}{\partial x}\right] - \frac{\rho}{\rho_w \theta}\frac{\partial s^*}{\partial t} \tag{3.18}$$

where $D_{\theta\theta} = L_{\theta\theta}\frac{\partial \psi_\theta}{\partial x}$, $D_{s\theta} = L_{s\theta}\frac{\partial \psi_\theta}{\partial \theta}$, $D_{ss} = L_{ss}\frac{RT}{s}$, $D_{\theta s} = L_{\theta s}\frac{RT}{s}$ are moisture, solute-moisture, solute, and moisture-solute diffusivity coefficients respectively (Elzahabi and Yong, 1997). The choice of the functionals for the phenomenological coefficients will be based on experimental information on the distribution of solutes along columns of test samples. Yong and Xu (1988) provide a useful identification technique for evaluation of these phenomenological coefficients.

3.8.2 Solute Movement in Saturated Soils

For Peclet numbers greater than 10^{-2}, i.e., $P_e > 10^{-2}$ (Figure 3.18), it is necessary to consider the effects of advective velocities on the transport of solutes in the porewater. When we reach the range of $P_e > 10$, advection plays a dominant role. When advection needs to be considered in the transport of solutes, the diffusion coefficient D_L is used. This coefficient is identified as the longitudinal diffusion-dispersion coefficient and is meant to reflect the advective velocity modification of the diffusive flow of solutes. The term *longitudinal* is used in conjunction with this coefficient to signify flow in the direction of the advective velocity. For convenience in terminology, D_L is often referred to as the *longitudinal diffusion coefficient*—even though advection is known to be involved in the movement of the solutes under consideration. The common expression used for D_L is: $D_L = D_m + \alpha v_L$. D_m represents molecular diffusion and is equal to $D_o \vartheta$, and where ϑ = tortuosity factor, α = dispersivity parameter, and v_L = advective velocity.

A general one-dimensional transport relationship based on Fickian diffusion of solutes and the effects of advective velocities can be obtained as:

$$\frac{\partial s}{\partial t} = D_L \frac{\partial^2 s}{\partial x^2} - v \frac{\partial s}{\partial x} - \frac{\rho^*}{n} \frac{\partial s^*}{\partial t} \tag{3.19}$$

If we assume a linear relationship between the concentration of solutes sorbed by the soil particles and the concentration of solutes remaining in equilibrium in the aqueous phase (i.e., constant adsorption isotherm as given in Figure 3.10), we can introduce the distribution coefficient k_d discussed in Section 3.6.1 into Equation (3.19) to obtain:

$$R \frac{\partial s}{\partial t} = D_L \frac{\partial^2 s}{\partial x^2} - v \frac{\partial s}{\partial x} \tag{3.20}$$

where:

$s^* = k_d s$

$R = \left[1 + \frac{\rho^*}{n} k_d\right]$ = retardation factor.

When a non-linear adsorption isotherm is used, Equation (3.19) should be used in conjunction with the full functional form for s^*.

Transport processes become somewhat more complicated when the soil is not a uniform homogeneous soil—as is the case for natural soils. We discussed the structure of soil in Chapter 2 in relation to some physical characteristics and properties of soils. Most, if not all, natural soils possess microstructural units (*mus*) which are the building blocks for the macrostructures that characterize such soils. The pore spaces in a natural soil are not uniform, not only because of the irregular shapes and sizes of the soil particles, but also because of the presence and distribution of the microstructural units. The microstructural units are varied in sizes. The pore spaces in these units which are classified as micropores will have different fluid and solute conducting characteristics. Many researchers suggest very strongly that we should pay more attention to the wide range of solute velocities within and between soil pores (Philip, 1968; Skopp and Warrick, 1974; Rao et al., 1980), and that we should recognize that some pore spaces could be non-conducting (stagnant). This is due to the fact that continuity between the pore spaces may not exist, and also because many micropores are too small and will not permit easy transmission of solutes because of the prohibitive energy requirements. However, because of the disparity in sizes between the macro- and micropores in the microstructural units, we can consider the microstructural units as sources/sinks for solutes (Figure 3.19). To account for this phenomenon, the D coefficient could be defined as follows (Paissioura, 1971; Rao et al., 1980, Wagenet, 1983):

$$D = (D_\pi + D_h + D_s) \tag{3.21}$$

where:

$D_\pi = D_o\tau$ = Effective molecular diffusion coefficient
D_o = Infinite solution diffusion coefficient
$D_h = \alpha v_L$
D_s = Dispersion coefficient accounting for dispersion effects caused by diffusion of solutes from stagnant to mobile regions, and is given as: $D_s = f\left[\phi, \frac{l^2}{D_{es}}, \frac{L}{v}, \ldots\right]$
τ = Physico-chemical tortuosity factor = $\theta\left(\frac{L}{L_e}\right)^2 \omega\chi$
ω = Coefficient relating to effect of charged soil particles on water viscosity
χ = Coefficient accounting for effects of anion exclusion
D_{es} = Effective diffusion coefficient in the stagnant region
φ = Porewater fraction in the conducting region.

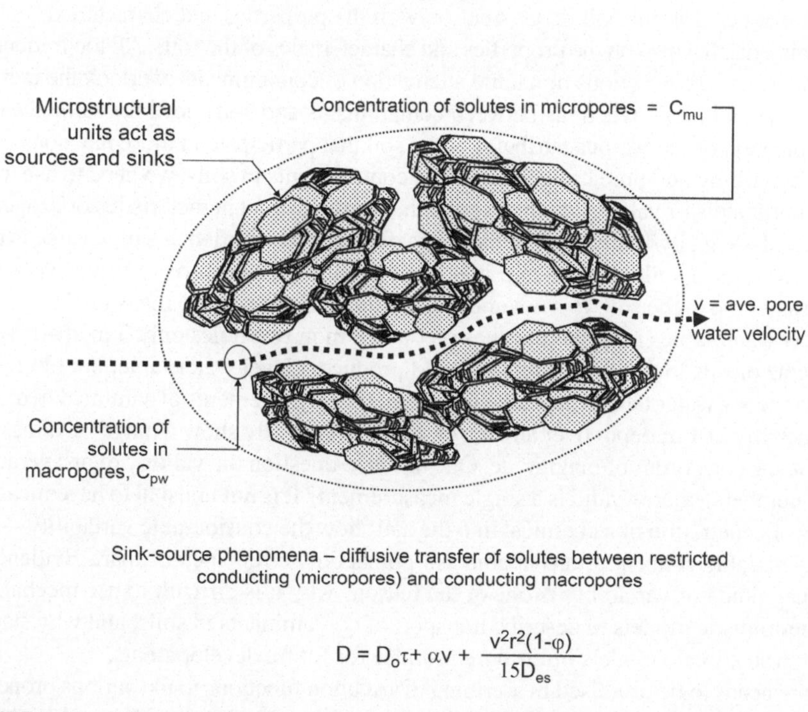

FIGURE 3.19 Schematic diagram showing sink-source phenomenon created by presence of microstructural units in diffusive flow of water through the soil-water system.

Using the expression for D_s given by Paissioura (1971) and Rao et al. (1980), the final relationship for the D coefficient given as Equation (3.18) has been obtained by Yong et al. (1992) as:

$$D = \left(D_o \tau + \alpha v + \frac{v^2 r^2 (1-\phi)}{15 D_{es}} \right) \tag{3.22}$$

where r is the average equivalent diameter of the soil particles.

The significance of the expression given in Equation (3.22) lies in the recognition of the influence of the wide differences in pore sizes in a natural soil. Leaching column tests and diffusion cell tests with natural samples will provide information which at best can be considered as representative of that particular sample in the test column or cell. Extrapolation or direct use of laboratory values for field predictions of transport of solutes will necessarily be dependent on the degree to which the laboratory samples have been able to replicate the structure of the field soils.

3.9 CONCLUDING REMARKS

For effective attenuation of contaminants in a candidate site, it is necessary for the soils in the site to possess attributes (characteristics and properties) that when interacting with the contaminant stream will achieve toxicity and/or contaminant concentration reduction. This requires one to develop a more detailed understanding of the interactions or processes that occur in the soil depend not only on the properties and characteristics of the contaminants, but also on the properties and characteristics of the soils. All too frequently, assessments and evaluations of natural attenuation of contaminants overlook the fact that the abiotic and biotic reactions between contaminants and soils need the same detail in determination of the various attributes of the soil that participate in those reactions.

In describing and predicting transport of contaminants in soils, we need to use transport coefficients or values which describe the transmission characteristics or properties of the soil—e.g., hydraulic conductivity or void ratio. Most often, a single value is used to characterize a soil, even though it is known that the characteristic values vary with space and time in the soil. It is not unusual to find that holes dug to measure the depth of shallow water tables may show values of 1 and 3 m in two holes only 3 m apart. While gradients of this kind of magnitude should produce a level water table, but obviously, instantaneous water table levels are very different. Measurements of saturated hydraulic conductivity at 1 m depth over an area of 1000 m^2 typically show a range of values that vary at least one order of magnitude. One needs to question the validity of an average of these numbers, or how valid is a single measurement? It is not unusual to have measurements of penetration of a chemical into the soil show the considerable variability—with highly variable concentrations at adjacent points only a few meters apart. Evidence of this magnitude of variability is one of the reasons why it is difficult to use mechanistic or deterministic models to describe transport of contaminants in soils, and why statistical or probabilistic models offer more promise for future developments.

Care needs to be exercised in ascribing distribution functions to the various properties and characteristics of soils. Experience has shown that the values of hydraulic conductivity and bulk densities for a typical site occupying say one hectare will have different distribution functions. By and large, bulk density values will generally be symmetrically distributed about the mean, in a Gaussian or normal distribution; this permits calculations of the mean and standard deviation. The same cannot be said for the hydraulic conductivity of the soil in the one-hectare site. Most often, one will obtain values of hydraulic conductivity that are not symmetrically distributed. What this says is that Gaussian statistics are generally not valid. Instead, distributions such as the Poisson may be more useful.

It is often argued that the modal value rather than the mean should be used for characteristics not fitted with Gaussian or normal distribution. The concept of modal characteristics of a soil for soil classification takes it cue from experience which suggests that this is the appropriate scheme for classification of most natural things with large variability. While a prediction of the disappearance of a chemical from a volume of soil may depend upon the modal value of hydraulic conductivity, contamination of a particular site will depend upon the larger value.

The previous chapter dealt with the nature of soils, paying particular attention to those soil fractions that contribute directly to the assimilative capacity of soils. This present chapter examines what happens when the soil fractions described in the previous chapter are exposed to water—as in a natural soil-water system. Only those properties of soils deemed pertinent in the various processes associated with transport of leachates in the soils are considered. While the many other physical properties of soils such as shear strength, compactibility, consolidation, etc. are to some extent important in considerations dealing with buffer-liner and barrier systems for landfills and impoundments, they are not addressed herein since they are secondary to the interactions considered in contaminant transport processes.

Knowledge of the basic interactions between soil solids and contaminants or dissolved solutes in the porewater is necessary if we are going to develop a better appreciation of the factors that contribute directly to the assimilative capacity of soils. The nature of the reactive surfaces in the soil-water system and how these surfaces are obtained will give us insight into how the soil conditions in the field will impact directly on the transport and fate of the contaminants under consideration.

The main items and issues considered include:

- The basic structure of soil fractions and their surface functional groups form complexes between the functional groups and the contaminants or solutes in the porewater.
- The presence of soil particles with reactive surfaces creates the situation where these particles will react chemically with solutes and other dissolved matter in the porewater.
- The soil fractions with reactive surfaces include layer silicates (clay minerals), soil organics, hydrous oxides, carbonates, and sulfates.
- It is important to remember that the nature of the bonding forces between the contaminants and the soil solids and the energy of interaction established depend on both the surface functional groups associated with the soil solids and the contaminants.
- The water-holding capacity of soil and the water retention characteristics of soils are direct functions of the energy of interaction between soil solids

and water. These tell us about the forces and interactions between water and the dissolved solutes in the water.

- The functional groups at the planar and edge surfaces of inorganic soil fractions, together with isomorphous substitutions in the lattices of the layer-lattice clay minerals result in the development of negative and positive charges distributed on the surfaces and edges of the soil particles. The nature and extent of surface complexation depend on the reactive properties of the soil particles and the contaminants themselves. Evaluation of the complexes formed can be performed using a variety of surface complexation models such as the single-, double-, and triple-layer models.
- Charge reversal occurs when the net charges on a particle surface (i.e., charge density) change in sign from a positive to negative or vice versa when one progresses from a system pH below the *pzc* to above the *pzc*.
- Interactions between soil solids, water and the various dissolved solutes in the porewater can be characterized in terms of energy relationships known as soil-water characteristics. These inform one about the water-holding capacity of a soil.
- The pH of the soil-water system and the various other dissolved solutes in the porewater contribute to interaction mechanisms such as acid-base reactions, speciation, complexation, precipitation and fixation.
- Bonds are developed as a result of interactions between charged particles of the various soil fractions and the charged contaminants. These bonds which constitute one of the principal means for sorption of contaminants include inter-atomic bonds such as ionic, covalent, and coordinate covalent bonds.
- Determination of partitioning of contaminants or solutes is generally obtained through batch equilibrium adsorption isotherm tests. The procedure is conducted with soil solutions and the results obtained are generally indicative of maximum sorption partitioning.
- Except for liquid contaminants, movement of all non-liquid contaminants in soils can be analyzed in terms of their association with the volumetric water content of the soil.
- When the porewater is in a more-or-less immobile state, e.g., hydration water layer and or water in the diffuse double layer, diffusion is the more likely mechanism of transport of the pollutant solutes.
- Water movement in partly saturated soils occurs along film boundaries in soil pore spaces that are not completely filled with water. Pore channel flow occurs for those pore spaces that are completely filled with water.
- Mechanisms associated directly with contaminant assimilation and attenuation need to be optimized in a soil-water system. This means that all the available soil solids' surfaces should be made available, resulting in maximization of surface area contact with the solutes in porewater and uptake—i.e., partitioning of the contaminants.
- To determine the reactive surfaces of soil particles, the techniques used lead to results that are operationally defined. Values of SSA determined are directly dependent on the procedure used and the analytical technique used to reduce the data obtained.

REFERENCES

Allison, J.D., Brown, D.S., and Novo-gradac, K.J. (1991), MINTEQA2/PRODEFA2: A geochemical assessment model for environmental systems, EPA/600/3-91/021.

Appelo, C.A.J., and Postma, D. (1993), *Geochemistry, Groundwater and Pollution*, Balkema, Rotterdam, the Netherlands.

Bockris, J.O'M., and Reddy, A.K.N. (1970), *Modern Electrochemistry*, Plenum Press, New York, Vols. 1 and 2.

Bolt, G.H. (1978), Adsorption of anions by soil, in G.D. Bolt and M.G.M. Bruggewert (Eds.), *Soil Chemistry: A Basic Elements*, Elsevier Scientific, New York, Chapter 5.

Einstein, A. (1905), Uber die von der Molekularkinetischen theorie der Warme Geforderte Bewegung von in Ruhenden Flussigkeiten Suspendierten Teilchen, *Annalen. Der. Physick.*, 4:549–660.

Elzahabi, M., and Yong, R.N. (1997), Vadose zone transport of heavy metals, in R.N. Yong, and H.R. Thomas (Eds.), *Geoenvironmental Engineering—Contaminated Ground: Fate of Pollutants and Remediation*, Thomas Telford, London, UK, pp. 173–180.

Grahame, D.C. (1947), The electrical double layer and the theory of electocapillarity, *Chem. Rev.*, 41:441–501.

Greenland, D.J., and Mott, C.J.B. (1985), Surfaces of soil particles, in D.J. Greenland and M.H.B. Hayes (Eds.), *The Chemistry of Soil Constituents*, John Wiley & Sons, New York, pp. 321–354.

Hayes, M.H.B., and Swift, R.S. (1985), The chemistry of soil organic colloids, in D.J. Greenland and M.H.B. Hayes (Eds.), *The Chemistry of Soil Constituents*, John Wiley & Sons. New York, pp. 179–320.

Huang, P.M., and Jackson, M.L. (1966), Fluoride interaction with clays in relation to third buffer range, *Nat. Commun.*, 211:779.

Kruyt, H.R. (1952), *Colloid Science*, Vol. I, Elsevier, Amsterdam, the Netherlands.

Lerman, A. (1979), *Geochemical Processes: Water and Sediment Environments*, John Wiley & Sons, New York.

Li, Y.H., and Gregory, S. (1974), Diffusion of ions in sea water and in deep-sea sediments, *Geochem. Cosmochim. Acta*, 38:603–714.

Nernst, W. (1888), Zur knetik der in losung befinlichen korper, *Z. Phys. Chem.*, 2:613–637.

Paissioura, J.B. (1971), Hydrodynamic dispersion in aggregated media: I Theory, *Soil Sci.*, 111:339–344.

Parkhurst, D.L., Thorstenson, D.C., and Plummer, L.N. (1980), PHREEQE—A computer programme for geochemical calculations, US Geological Survey, Water Resources Investigation Report, pp. 80–96.

Parkhurst, D.L., and Appelo, C.A.J. (1999), User's Guide to PHREEQC (version 2), a computer program for speciation, batch-reaction, one-dimensional transport and inverse geochemical calculations. US Geological Survey Water-Resources Investigations Report 99-4259.

Perkins, T.K., and Johnston, O.C. (1963), A review of diffusion and dispersion in porous media, *J. Soc. Pet. Eng.*, 17:70–84.

Philip, J.R. (1968), Diffusion dead-end pores, and linearized adsorption in aggregated media, *Aust. J. Soil. Res.*, 6:31–39.

Rao, P.S., Rolston, D.E., Jessup, R.E., and Davidson, J.M. (1980), Solute transport in aggregated porous media: Theoretical and experimental evaluation, *Soil Sci. Soc. Am. J.*, 44:1139–1146.

Ritchie, G.S.P., and Sposito, G. (2002), Speciation in soils, in A.M. Ure and C.M. Davidson (Eds.), *Chemical Speciation in the Environment*, Blackwell Science, Oxford, UK.

Robinson, R.A., and Stokes, R.H. (1959), *Electrolyte Solutions*, 2nd ed., Butterworths, London, UK.

Singh, U., and Uehara, G. (1986), Electrochemistry of the double layer principles and applications to soils, in D.L. Sparks (Ed.), *Soil Physical Chemistry*, CRC Press, Boca Raton, FL, pp. 1–38.

Skopp, J., and Warrick, A.W. (1974), A two-phase model for the miscible displacement of reactive solutes through soils, *Soil Sci. Soc. Am. Proc.*, 38:545–550.

Sposito, G. (1984), *The Surface Chemistry of Soils*, Oxford University Press, New York.

Sposito, G., and Mattigod, S.V. (1979), *GEOCHEM: A Computer Programme for Calculating Chemical Equilibria in Soil Solutions and Other Natural Water Systems*, University California, Davis, CA.

Sposito, G., and Coves, J. (1988), *SOILCHEM: A Computer Programme for the Calculation of Chemical Speciation in Soils*, The Kearny Foundation of Soil Science University, Davis, CA.

van Olphen, H. (1977), *An Introduction to Clay Colloid Chemistry*, 2nd ed., John Wiley & Sons, New York.

Wagenet, R.J. (1983), Principles of salt movement in soils, in D.W. Nelson et al. (Eds.), *Chemical Mobility and Reactivity in Soil Systems*, American Society of Agronomy, Madison, WI, Vol. 11, pp. 123–140.

Wolery, T.J. (1983), EQ3NR *A Computer Program for Geochemical Aqueous Speciation-Solubility Calculations*, University of California Lawrence Livermore Lab.

Yong, R.N., and Wong, H.Y. (1973), *Water Movement in Unsaturated Swelling Soils*, Soil Mechanics Laboratory, McGill University, Research Report 73-2.

Yong, R.N., and Warkentin, B.P. (1975), *Soil Properties and Behaviour*, Elsevier Publishing, Amsterdam, the Netherlands.

Yong, R.N., and Xu, D.M. (1988), An identification technique for evaluation of phenomenological coefficients in unsaturated flow in soils. *Int. J. Num. Anal. Methods Geochem.*, 12:283–299.

Yong, RN., Mohamed, A.M.O. and Warkentin, B.P. (1992) *Principles of Contaminant Transport in Soils*, Elsevier, Amsterdam, the Netherlands, 327p.

Yong, RN. and Xu, D.M. (1988) An identification technique for evaluation of phenomenological coefficients in unsaturated flow in soils. *Int. J. Num. Anal. Methods Geochem.* 12, 283–299.

Yong, R.N. (2000), *Geoenvironmental Engineering: Contaminated Soils, Pollutant Fate, and Mitigation*, CRC Press, Boca Raton, FL.

Yong, R.N., and Mulligan, C.N. (2003), The impact of clay microstructural features on the natural attenuation of contaminants, *Appl. Clay. Sci.*, 23:179–186.

Yong, R.N., Nakano, M., and Pusch, R. (2012), *Environmental Soil Properties and Behavior*, CRC Press, Boca Raton, FL.

4 Natural Attenuation of Heavy Metals

4.1 INTRODUCTION

The most common types of contaminants (pollutants) in contaminated sites considered as threats to human health and other biological receptors fall into two broad categories: (a) inorganic substances consisting of alkali and alkaline-earth metals and especially heavy metals such as lead (Pb), copper (Cu), cadmium (Cd), zinc (Zn), etc., and (b) organic chemicals such as polycyclic aromatic hydrocarbons (PAHs), petroleum hydrocarbons (PHCs), benzene, toluene, ethylbenzene and xylene (BTEX), etc. As discussed in the previous chapters, we will be using the term *contaminants* since this is understood to mean all types of contaminants that do or do not pose threats to biological receptors. The focus in this chapter will be on the various mechanisms involved in what is commonly described as the natural attenuation of inorganic contaminants and particularly heavy metals (HMs) in soils. Interactions between the inorganic contaminants will be between the surface reactive groups that characterize the surfaces of both the soil fractions and the contaminants, resulting in the assimilation of the contaminants by the soil in question.

We will be examining the interactions between the contaminants and soil solids, and the various reactions that contribute to the assimilative capacity of soils since these will impact directly on the movement and distribution of the contaminants in the soil. Particular attention will be paid to the heavy metals since these are by far the inorganic contaminants of concern—primarily because they constitute the greater proportion of inorganic contaminants found in contaminated sites. If natural attenuation is to be relied upon to provide a safe means for managing the presence and movement of contaminants in the ground, we need to be concerned with the processes associated with partitioning and environmental mobility of these heavy metals. The requirements for evaluation and prediction of their fate under site conditions considered for application of natural attenuation as a treatment or management procedure need to be explored.

4.2 INORGANIC CONTAMINANTS

The inorganic contaminants of note are alkali, alkaline-earth metals, transition metals, and heavy metals. The alkali and alkaline-earth metals are elements of Groups I and II in the Periodic Table. Alkali metals are strong reducing agents and are never found in the elemental state since they will react well with all nonmetals. The common alkali metals include Li, Na, and K, with Na and K being very abundant in nature. Other alkali metals in Group IA (Rb, Cs, and Fr) are less commonly found in nature. The metals in Group II include Be, Mg, Ca, Sr, Ba, and Ra. The more common ones, Mg and Ca are strong reducing agents, and they react well with many non-metals.

Transition metals occupy a position between Group II and Group III and are less reactive than the metals in Groups I and II. They are sometimes called *d*-block elements because of the addition of electrons to the sub-shells. Transition metals include V, Cr, Fe, Cu, Mo, Zn, Ag, and Hg. By and large, most (if not all) of the transition metals and heavy metals are listed as priority contaminants.

4.2.1 Heavy Metals

Those elements with atomic numbers higher than Sr (atomic number 38) are classified as *heavy metals (HMs)*. As Yong (2000) has indicated, common usage, however, includes elements with atomic numbers greater than 20 as heavy metals. Those elements having atomic numbers greater than 20 can be found in the lower right-hand portion of the periodic table, (*d*-block of the Periodic Table). These include 38 elements falling into three groups of atomic numbers as follows:

- From atomic numbers 22 to 34—Ti, V, Cr, Mn, Fe, Co, Ni, Cu, Zn, Ga, Ge, As, and Se;
- From 40 to 52—Zr, Nb, Mo, Tc, Ru, Rh, Pd, Ag, Cd, In, Sn, Sb, and Te,
- From 72 to 83—Hf, Ta, W, Re, Os, Ir, Pt, Gu, Hg, Tl, Pb, and Bi.

Except for Zn and those metals in Group III to Group V, the metals are *transition metals* because these are elements with at least one ion with a partially filled *d* subshell. Their properties are related to their electronic structures and the relative energy levels of the orbitals available for their electrons. As will be seen later when we examine the concepts of hard and soft acids and bases in connection with electron donors and acceptors, we will find that this is a significant factor.

The more common HMs associated with anthropogenic activities such as disposal of wastes in landfills, generation of chemical waste leachates and sludges, and metal mining industries, include: lead (Pb), cadmium (Cd), copper (Cu), chromium, (Cr), nickel (Ni), iron (Fe), mercury (Hg), and zinc (Zn). Cadmium, which is a member of the IIB group in the Periodic Table, is relatively rare as a natural material. It can be found in argillaceous and shale deposits as greenockite (CdS) or otavite ($CdCO_3$) and is usually associated with zinc, lead, or copper in sulfide form (Cameron, 1992). It is a bluish-white soft metal or grayish powder and is most likely to be found as Cd(II). The primary bonding mechanisms between Cd and soil solids are electrostatic. While it has properties similar to that of Zn, it is more mobile than zinc at low pH values, particularly at pH values between 4.5 and 5.5. Above pH 7.5, cadmium is not very mobile. Its divalent form is soluble, but it can also complex with cyanides, amines, halides, and other organics and oxides which can increase its mobility.

It has only been recently (30 years or more) that cadmium has become a concern due to the extensive extraction for use as a coating material in plastic industries, steel plating, pigment stabilization, and nickel-cadmium batteries (Fassett, 1980). Sources of cadmium include alloys, polyvinyl chloride (PVC) plastic manufacture, solders, fungicides, enamels, motor oil, textile manufacturing, electroplating and rubber, sewage sludges, and phosphate fertilizers (Matthews, 1984). It enters the environment through industrial effluents and landfill leaching, spills and leaks at hazardous waste sites, mining, and household wastes.

Copper, which is a member of the IB group in the Periodic Table, occurs naturally in sandstones and in minerals such as malachite and chalcopyrite. It is a reddish-brown metal that binds strongly to organic matter and clay minerals, thus decreasing its mobility. The organic matter, however, can be degraded through anaerobic or aerobic means releasing the copper in its monovalent or divalent states, respectively. The average content of copper in soils is 2–100 ppm. Plants can accumulate copper. Average contents are in the range of 5–30 ppm. Toxic levels in plants are in the range of 20–100 ppm. Increased levels of copper are due to uses in fertilizers, building materials, rayon manufacture, pesticide sprays, agricultural and municipal wastes and industrial emissions (Cameron, 1992).

Lead is found naturally in soils, and although it has three oxidation states (0, 2+, and 4+), the most common state is 2+. Compounds of Pb(II) have ionic bonds whereas the higher valence state, Pb(IV) compounds, have covalent bonds. Pb is most commonly found in the ore gelena (PbS) and in smaller quantities in cerussite ($PbCO_3$), anglesite ($PbSO_4$) and crocoite ($PbCrO_4$). It is a bluish-white, silvery, or grey metal with a high density of 11.4 g/cm^3. Lead can be found in soils at the surface and organic matter in higher quantities. In general, lead can be found deposited in soils as soluble halides, lead chlorobromide and PbCl. Br (de Haas and Zwerman, 1978). Sources of lead include lead-zinc smelters, ammunition, solder, glass, piping, insecticides, paints, and batteries (Jaworsky, 1978). The divalent form, which is the most common, is capable of replacing calcium, strontium, barium, and potassium in soils. In general, levels less than 10 ppm are found and mobility of lead in soils is low. Organic matter can adsorb substantial quantities of lead. Lead is released into the air from burning of wastes and fossil fuels and subsequently lands onto the soil. It also reaches the soil from landfills and paints.

Although not as toxic as cadmium, zinc is quite often associated with this metal. Zinc is a soft, white metal with bluish tinge. Soil texture, pH, the nature of the parent rocks and organic content all affect the natural content of zinc in the soil. Under acidic conditions, zinc is usually divalent and quite mobile. In the divalent state, the sorption mechanisms include electrostatic interactions and sequestering by organic matter. At high pH, zinc is bioavailable due to the solubility of its organic and mineral colloids. Zinc hydrolyzes at pH 7.0–7.5, forming $Zn(OH)_2$ at pH values higher than 8. Under anoxic conditions, ZnS can form upon precipitation, whereas the unprecipitated zinc can form $ZnOH^+$, $ZnCO_3$, and $ZnCl^+$. Natural levels of zinc in soils are 30–150 ppm. Levels of 10–150 ppm are normal while 400 ppm in plants are toxic. Sources of zinc include brass and bronze alloys, galvanized products, rubber, copying paper, cosmetics, pharmaceuticals, batteries, televisions, tires, metal coatings, glass, paints and zinc-based alloys (Cameron, 1992). It can enter the environment from galvanizing plant effluents, coal and waste burning, leachates from galvanized structures, natural ores and municipal waste treatment plant discharge.

Zinc is commonly found in wastes as zinc chloride, zinc oxide, zinc sulfate, and zinc sulfide (Agency for Toxic Substances and Disease Registry, 1995).

Nickel does not precipitate but does adsorb onto soils. Adsorption of nickel onto clays, oxides of manganese and iron, and organic material occurs. Mobility increases with the formation of complexes with organic and inorganic ligands.

Arsenic is found naturally in rocks and soil and originates in industrial sources (US EPA, 2000). Volcanoes, rock erosion, and forest fires transport arsenic into the environment. Average concentrations of arsenic in the soil are 5–10 μg/kg (Fergusson, 1990). Industrial sources include paints, dyes, wood preservation, herbicides, semi-conductors, pharmaceuticals, pulp and paper production, cement production, mining, and smelting operations. Fossil fuel and waste burning, and agricultural practices also release arsenic in to the environment. Arsenic is an inorganic semiconductor and metalloid. The most common valence states for arsenic are −3, 0, +3 (arsenite), and +5 (arsenate). Arsenic behaves in a manner similar to phosphate. The latter two are the most common in nature. Inorganic forms of arsenic are hydrides such as arsine, halides, oxides, sulfides, and acids (Kirk-Othmer, 1992). The valence form and speciation can change depending on pH, redox potential, microbial activity, and presence of other ions including sulfur, iron, and calcium (Vance 2001). Aluminum, calcium, and iron arsenates are highly insoluble and have shown specific or inner sphere adsorption on clay and oxide surfaces (Jacobs et al., 1970). Arsenate (As(V)) has a negative charge and arsenite (As(III)) has no charge in the pH range of 4–10. As(V) is found mainly as $HAsO_4^{2-}$ in alkaline soils and $H_2AsO_4^-$ in acidic soils. As(III) is more thermodynamically stable in a reduced environment, whereas As(V) is more stable in oxic soils. Organic forms of arsenic include mono- and dimethyl arsenic acid and other volatile methyl arsines. Mobility of arsenic in the environment is influenced by the presence of oxides of iron, aluminum, and manganese, carbonaceous materials and other soil components that enhance adsorption (Vance, 2001). Arsenite is more soluble and mobile than arsenate in soils and sediments when oxides are solubilized, arsenate is reduced to the more toxic and arsenite.

Sources of chromium are metal finishing, wood preserving, drilling, leather tanning, textile dyeing, ink manufacturing, and other industries (U.S. EPA, 1984). It is found frequently in soil in the Cr(V) and Cr(IV) oxidation states. The major form of Cr(VI) is CrO_4^{2-} at pH greater than 6.5 and $HCrO_4^-$ at pH less that 6.5. Both ions are very soluble. Cr(III) can form various stable, inert complexes. $Cr(H_2O)_6^{3+}$, $Cr\ (H_2O)_5(OH)^{2+}$, $Cr(H_2O)_3(OH)_3$, and $Cr(H_2O)(OH)_4^-$. The solubility of $Cr(OH)_3$ is very low at pH 7.0 and is in the order of 7×10^{-8} M (Ray et al., 1987). Cr(III) also complexes with Fe(III), forming $(Cr, Fe)(OH)_3)$ which has an even lower solubility (Rai et al., 1986). Thus, there is a significant decrease in the mobility upon the reduction of Cr(VI) to Cr(III).

Using M^{n+} as a notational scheme to identify metallic ions, e.g., Cu^{2+}, Zn^{2+}, etc., we note that they are generally coordinated (chemically bound) to water molecules, and in their hydrated form they exist as $M(H_2O)_x^{n+}$. The water molecules which form the coordinating complex are the ligands, and the metal ion M^{n+} occupies a position known as the central atom. *Ligands* are those anions which can form coordinating compounds with metal ions. Their characteristic feature is their free pairs of electrons. The number of ligands attached to a central metal ion is defined as the *coordination number*, and the coordination number of a metal ion is the same regardless of the type or nature of ligand.

Replacement of water as the ligand for M^{n+} can occur if the candidate ligand, generally an electron donor, can replace the water molecules bonded to the M^{n+}. Inorganic ligands that will form complexes with metals include:

$$CO_3^{2-},\ SO_4^{2-},\ Cl^-,\ NO_3^-,\ OH^-,\ SiO_3^{2-},\ CN^-,\ F^-,\ \text{and}\ PO_4^{3-}$$

Metal complexes can also be formed with molecules with lone pairs of electrons, e.g., NH_3 and PH_3. In the case of NH_3 for example, NH_3 is a Lewis base and a neutral ligand—e.g., $Co(NH_3)_6^{3+}$. Complexes formed between soil-organic compounds such as humic and fulvic acids and metal ions are generally chelated complexes. In essence, outer sphere complexes or ion pairs formed between metals and complexant ligands are generally seen to be weak electrostatic interactions. This is likely due to the fact that one or both of the charged species retains their hydration shell (Evans, 1989). When inner sphere complexes are formed, strong associations between the metals and complexant ligands are obtained via covalent bonding between the metal ions and the ligands.

Variability in oxidation states is a characteristic of transition elements (i.e., transition metals). Many of these elements have one oxidation state that is most stable, e.g., the most stable state for Fe is Fe(III) and Co(II) and Ni(II) for cobalt and nickel, respectively. Much of this is a function of the electronic configuration in the *d* orbitals. Unpaired electrons which comprise one half of the sets in *d* orbitals are very stable. This explains why Fe(II) can be easily oxidized to Fe(III) and why the oxidation of Co(II) to Co(III) and Ni(II) to Ni(III) cannot be as easily accomplished. The loss of an additional electron to either Co(II) or Ni(II) still does not provide for one half unpaired electron sets in the *d* orbitals. This does not mean to say that Co(III) does not readily exist. The complex ion $[Co(NH_3)_6]^{3+}$ has Co at an oxidation state of +3. Depending on the pH and redox potential of the porewater in soils, some heavy metals (HMs) can exist in more than one oxidation state,—e.g., selenium (Se) can occur as SeO_3^{2-} with a valence of +4, and as SeO_4^{2-} with a valence of +6, copper (Cu) can have valences of +1 and +2 for CuCl and CuS respectively, chromium (Cr) and iron (Fe) can exist as CrO_4^{2-} and $Cr_2O_7^{2-}$ for the valence state of +6, and Cr^{3+} and $Cr(OH)_3$ for the +3 valence state, and iron (Fe) can exist as Fe^{2+} and FeS for the +2 valence state and Fe^{3+} and $Fe(OH)_3$ for the +3 valence state.

In general, the pH of the micro-environment in a representative elementary volume which encompasses soil solids and porewater is a significant factor in the environmental mobility of heavy metal contaminants. To a very large extent, this is because of the influence of pH on the solubility of the heavy metal complexes. Nyffeler et al. (1984) shows that the pH at which maximum adsorption of metals occurs can be expected to vary according to the first hydrolysis constant of the metal (cationic) ions.

When the ionic activity of heavy metal solutes exceeds their respective solubility products, precipitation of these metals as hydroxides and carbonates is likely. Alkaline conditions are favorable conditions for precipitation of heavy metals, resulting generally in the formation of new substances in the porewater or as precipitates attached to soil particles. The process is a two-stage process: (1) nucleation and (2) particle growth. The main factors involved in precipitation of heavy metals include the pH of the soil-water system, concentration and species of heavy metals, nature of inorganic and organic ligands, and precipitation pH of the heavy metal contaminants. By and large, it is not easy to distinguish the various processes responsible for the removal, since the various sorption mechanisms and precipitation all result in the removal of heavy metals from the porewater.

An example of precipitation of heavy metals in aqueous solutions can be seen in the graphical results (Figure 4.1) of tests reported by MacDonald (1994). The curves shown in Figure 4.1 indicate that the pH value at which transition from soluble form to precipitate form is not the same for each of the metals tested. For instance, onset of precipitation for Pb starts at around a pH value of 3.5 for a single species metal, and appears to be complete at a pH value of about 7. This contrasts with Zn which does not appear to show much signs of precipitation until about a pH value of 6.4, whereupon precipitation appears to be near completion at a pH value of just over 7. The precipitation process for Zn appears to occupy one pH range in comparison to Pb which occurs over a pH of range of about 3.5.

The presence of other heavy metals affects the precipitation behavior of individual metals. The results shown in the bottom of Figure 4.1 represents precipitation

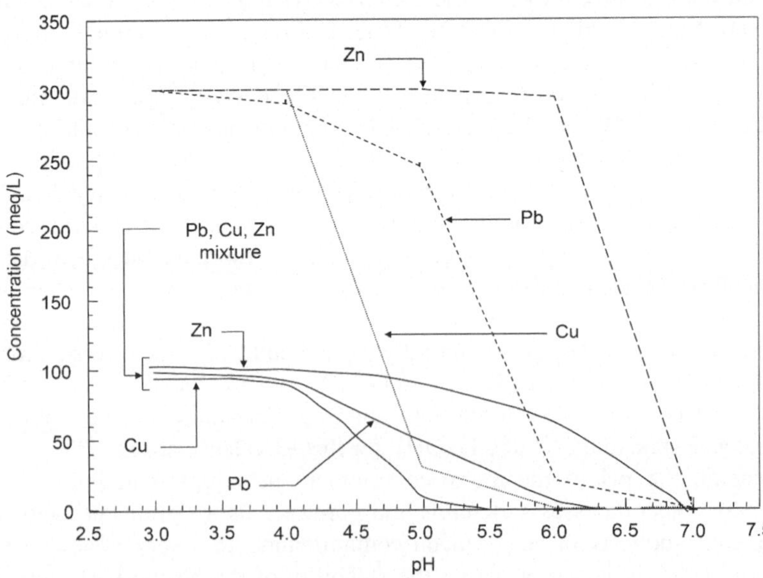

FIGURE 4.1 Precipitation of heavy metals Pb, Cu, and Zn in aqueous solution. Bottom curves are precipitation of individual metal from a mixture of Pb, Cu, and Zn in equal proportions of metal nitrate solution (100 meq each). Top curves are for single solutions of individual metals at 300 meq/L concentration. (Test results reported by MacDonald, E., Aspects of competitive adsorption and precipitation of heavy metals by a clay soil, ME Thesis, McGill University, 1994.)

behavior of the three heavy metals in a mixture with equal proportions of each of the metals. As we can see, onset of precipitation of Zn which appears to be at about pH 6.4 for the single component species has been lowered from pH 6.4 to about 4.4 when other metals are present. This represents a significant drop in the precipitation pH of Zn. However, the results in Figure 4.1 indicate that precipitation process for Pb and Cu does not appear to be significantly affected by the presence of other heavy metal contaminants.

4.3 MOBILITY, AVAILABILITY, AND SOURCE-PATH-RECEPTOR

The presence of heavy metals as contaminants in soils pose threats to biotic receptors if and when these metals are mobile and available to the receptors. The threat of environmental mobility and availability of heavy metals to biotic receptors is generally known as bioavailability. The term *bioavailability* is used by professionals in many different disciplines to mean the availability of contaminants that would be toxic when taken into the circulatory system of biotic receptors. The use of the *source-path-receptor* (*SPR*) approach or criterion as a valid and useful means for determining the need and requirements for remediation should not preclude one from overlooking the fact that unless we are able to determine the environmental mobility and availability (i.e., bioavailability) of contaminants in the contaminated ground; it would be difficult to establish the extent of remediation requirements. This means to say that not all detected contaminants in a contaminated site are mobile. Some of them, such as heavy metals, are fixed onto soil solids because of the various mechanisms shown in Table 4.1. As shown in the table, the various sorption mechanisms other than ion-exchange involve specific adsorption and

TABLE 4.1
Heavy Metal Retention by Some Clay Minerals

Clay Mineral	Chemisorption	Chemisorption at Edges	Complex Adsorption	Lattice Penetration[a]
Montmorillonite	Co, Cu, Zn		Co, Cu, Zn	Co, Zn
Kaolinite			Cu, Zn	Zn
Hectorite			Zn	Zn
Brucite			Zn	Zn
Vermiculite			Co, Zn	Zn
Illite	Zn	Zn, Cd, Cu, Pb		
Phlogopite			Co	
Nontronite			Co	

Source: Bolt, G.H., *Soil Chemistry, Part B: Physico-Chemical Models*, Elsevier Scientific, Amsterdam, the Netherlands, 479, 1979.

[a] Lattice penetration = lattice penetration and imbedding in hexagonal cavities.

other mechanisms such as chemisorption involving hydroxyl groups from broken bonds in the clay minerals, formation of metal-ion complexes, and precipitation as hydroxides or insoluble salts.

The bioavailability of heavy metals is of concern because of the possibility for uptake by plants, ingestion by humans and other biotic receptors. Availability of these metals comes generally as ionic transport in soil-water systems. In respect to heavy metal contaminants found in soils, there are at least five possible factors that can influence their environmental mobility and bioavailability, these are: (1) the hydrogeology of the site, (2) changes in acidity of the system, (3) changes in the system ionic strength, (4) changes in the oxidation-reduction potential of the system, and (5) formation of complexes. The principal mechanisms and processes involved in heavy metal sorption and retention include ion exchange, speciation, precipitation as a solid phase (oxide, hydroxides, carbonates), and complexation reactions (Harter, 1979; Farrah and Pickering, 1976a, 1976b, 1976c, 1977a, 1977b; 1979; Maguire et al., 1981; Yong et al., 1990; Yong, 2000). The use of the term *retention* means that sorbed contaminants (sorbate) remain fixed to the sorbent. We need to be aware that not all of the interactive processes result in retention. Strictly speaking, ion exchange adsorption should be considered as *retardation* because desorption of contaminants sorbed by ion exchange mechanisms can readily occur, thus releasing the originally sorbed contaminants.

Two kinds of surface charge reactions occur with most clay minerals: (a) reactions associated with the net negative charge developed from heterovalent cation substitution in the clay lattice structure, and (b) pH-dependent reactions due to hydration of broken bonds at the edges of the mineral particles. Taking kaolinite as an example, the two types of functional groups (silanol and aluminol) that populate the surfaces of the edges of the kaolinite particles are hydroxyl (OH^-) groups. The silanol group is singly coordinated to the Si^{4+} in the tetrahedral lattices whilst the aluminol group is singly coordinated to the Al^{3+} in the octahedral lattices. Both types of edges can be characterized as Lewis acid sites. These sites are capable of accepting at least one pair of electrons from a Lewis base. The variation of surface charge in relation to pH for a kaolinite with specific surface area of approximately 20 m^2/g at 25°C can be seen in Figure 4.2. Surface complexation modelling studies indicate that the Al sites are the principal proton acceptor sites, and that these sites are more acidic for kaolinite edges than for exposed Al-hydroxides.

Continuing our use of kaolinite as an example, if heavy metal concentrations in the contaminated porewater are less than the *cation exchange capacity* (*CEC*) of the kaolinite, desorption occurs easily because the mechanisms controlling initial sorption are mainly non-specific. However, if the heavy metal concentrations are greater than the CEC, desorption or release of the metals is more difficult because the total sorption processes will most likely include both non-specific and specific adsorption. The lyotropic series shows us that desorption of cations can also occur through exchange as a hierarchical protocol as follows:

$$Na^+ < Li^+ < K^+ < Rb^+ < Cs^+ < Mg^{2+} < Ca^{2+} < Ba^{2+} < Cu^{2+}\ Al^{3+} < Fe^{3+} < Th^{4+}$$

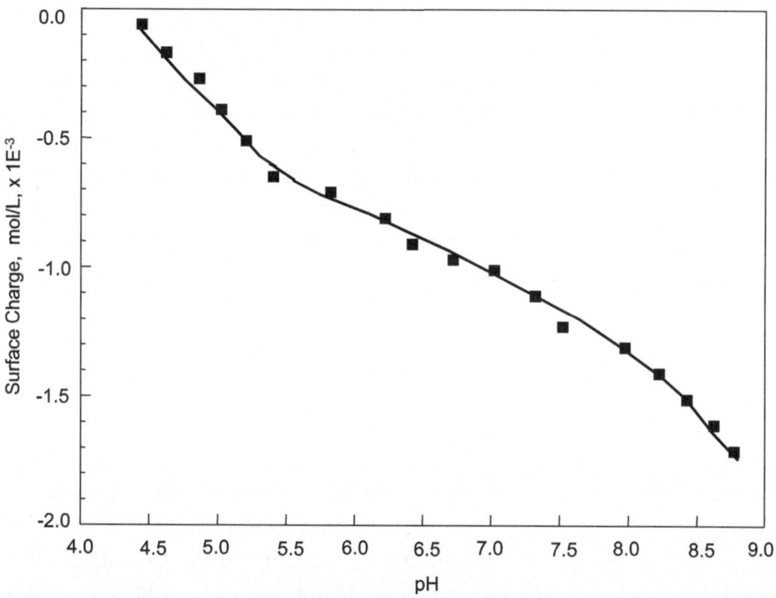

FIGURE 4.2 Surface charge for kaolinite in relation to pH, interpreted from some data reported by Brady et al. (From Brady, P.V. et al., Surface charge and metal sorption to kaolinite, Chapter 17, *Adsorption of Metals by Geomedia*, E.A. Jeanne (Ed.), Academic Press, San Diego, CA, 1998.)

4.4 INTERACTIONS AND ASSIMILATION OF HEAVY METALS

In the context of heavy metals contaminants, assimilation of heavy metals by soils means removal of these metals from the porewater. This phenomenon is also referred to as *partitioning* of contaminants. The processes contributing to assimilation are found in the reactions resulting from interactions in the soil-water system between soil particles and heavy metals—bearing in mind that these metals possess positive (cationic) charges and that the major sets of surface charges of soil solids are negative. By and large, these processes result in the complexation, precipitation, and sorption of heavy metals by the soil solids. The quantity of heavy metals removed from the porewater in a soil is dependent on both the assimilative or sorptive capacity of a soil and the types and concentrations of the metals in the porewater. As we have seen in the previous chapters, the surface properties of the soil particles and the composition of the soil are major factors in establishing its assimilative capacity.

4.4.1 Acid-Base Reactions

We briefly considered acid-base reactions in the previous chapter when we looked at the self-ionization capability of water. In this section, we will be considering these reactions with respect to heavy metals in the soil-water system. To do so, we follow the development previously given in Yong (2000). To broaden our concept of acids and bases, we first consider the Brønsted-Lowry concept. This concept considers a substance an acid if it has a tendency to lose a proton (H^+). Conversely, this concept considers a substance a base if it has a tendency to accept a proton. Accordingly, the Brønsted-Lowry acid-base concept considers an acid to be a *proton donor* (*protogenic* substance) and a base to be a *proton acceptor* (*protophilic* substance). Substances that have the capability to both donate and accept protons (i.e., both protogenic and protophilic), such as water and alcohols, are called *amphiprotic* substances.

We can consider reactions between HMs and soil solids and also ligands in terms of acid-base reactions. This can be done by considering the interactions and reactions in terms of electron transfer. We recall that acid-base reactions involve proton transfer between a proton donor (acid) and a proton acceptor (base) and that the transfer is a *protolytic* reaction. The process itself is called *protolysis*. The self-ionization of water for example is called *autoprotolysis*, and *neutralization* is the reverse of autoprotolysis. All bases have a lone pair of electrons to share with a proton. The donation of the electron pair in covalent bonding to an acid which accepts the electron pair will leave the electron donor (base) electron-deficient. The Lewis (1923) concept of acids and bases considers an acid as a substance that is capable of accepting a pair of electrons for bonding, and a base as a substance that is capable of donating a pair of electrons. In the context of *donor-acceptor* terminology, Lewis acids are electron acceptors, and Lewis bases are electron donors. As an example, all metal ions M^{nx} are Lewis acids that have vacant orbitals into which electrons can be introduced, and ligands behave as Lewis bases that have at least one pair of electrons not shared in a covalent bond (Evans, 1989). In the previous discussion on heavy metals and complexes formed with ligands, we see that the HMs are bonded with Lewis bases. This is explained by the fact that Lewis acids can accept and share electron pairs donated by Lewis bases. Whilst Lewis bases are also Brønsted bases, Lewis acids are not necessarily Brønsted acids since Lewis acids include substances that are not proton donors. However, the use of the Lewis acid-base concept permits us to treat metal-ligand bonding as acid-base reactions.

Pearson (1963, 1968) and Lester (1987) have classified Lewis acids and bases into categories according to their mutual behavior into categories of *hard* and *soft* acids and bases, based on demonstrated properties, e.g.:

A. *Hard acids*—generally small in size with high positive charge; high electronegativity; low polarizability; and no unshared pairs of electrons in their valence shells.
B. *Soft acids*—generally large in size with a low positive charge; low electronegativity; high polarizability; and with unshared pairs of electrons in their valence shells.
C. *Hard bases*—usually have high electronegativity; low polarizability, and difficult to oxidize.
D. *Soft bases*—usually have low electronegativity; high polarizability; and easy to oxidize.

Hydrolysis reactions of metal ions are influenced by: (1) pH of the active system; (2) type, concentration, and oxidation state of the metal cations; (3) redox environment; and (4) temperature. High temperatures favor hydrolysis reactions, as do low organic contents, low pH values, and low redox potentials.

The dissociation constant *k* expressed as a *pk* value (negative logarithm to base 10) of the dissociation constant, i.e., $pk = -\log(k)$ tells us much about the transport and adsorption of chemical species in the ground. The *pk* value is a good indicator of the strength of acids and bases. *pk* values can be determined using the Henderson-Hasselbalch relationship (Po and Senozan, 2001):

$$pk = \mathrm{pH} - \log_{10} \frac{\text{unprotonated form (base)}}{\text{protonated form (acid)}} \tag{4.1}$$

Strong acids are strong proton donors, and weak acids do not favor the formation of H^+ ions, i.e., they will show higher pH values than strong acids. Most highly charged cationic metals have low *pk* values and are strongly hydrolyzed in aqueous solution.

Hydrated metal cations can act as acids or proton donors, with separate *pk* values for each which decrease with dehydration of the soil. Water molecules are strongly polarized by exchangeable metal cations on the surfaces of clay particles. These are significant contributors to the proton donating process of clay particles. This results in a greater acidity of this water that would be predicted from considerations of the *pk* values of the hydrated metal cations in water (Mortland and Raman, 1968).

4.4.2 Redox and Eh-pH Relationships

Redox (oxidation-reduction) reactions involve electron transfer from one reactant to another, and as we have seen in Section 3.4.2, electron activity is characterized by the redox potential *Eh* given in Equation (3.5) as: $Eh = 0.0591pE$ at a standard temperature of 25°C. Free electron activity *pE* is a measure of how well a soil can be oxidized, and is generally expressed as a negative log: $pE = {}^{-}\log(e^{-})$. Hence, if we can establish the redox potential or redox capacity of a soil, we would be in a position to anticipate the electron transfer reactions involving the contaminants of interest in the soil-water system. The gain or loss of electrons by inorganic solutes such as heavy metals interacting in the soil-water system can change the valency of the elements. Electron gain by such a solute results in loss of a positive valence (reduction), and conversely, electron loss by a solute results in a loss of a negative valence (oxidation).

The stability or the state of inorganic solutes in the porewater is a function of several factors. Notable amongst these are the pH, *Eh*, or *pE* of the porewater, nature of ligands, temperature, and concentration of the inorganic solutes themselves. The influence of all of these can be calculated using the Nernst equation. If *A* and *B* represent the reactant and product respectively, we will have:

$$pE = 16.92Eh = E^0 + \left(\frac{RT}{nF}\right) ln \frac{[A]^a[\mathrm{H_2O}]^w}{[B]^b[H^+]^h} \tag{4.2}$$

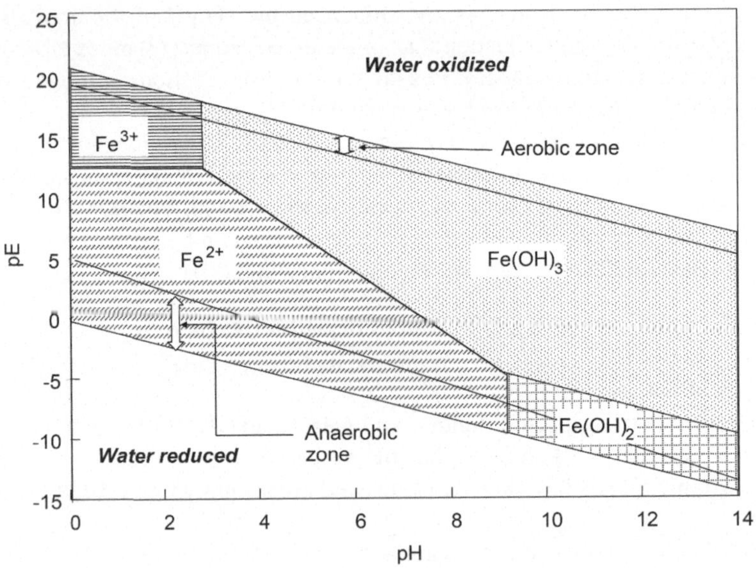

FIGURE 4.3 *pE*-pH chart for Fe and water with maximum soluble Fe concentrations of 10^{-5} M. The zone sandwiched between the aerobic and anaerobic zones is the transition zone. (From Yong, R.N., *Geoenvironmental Engineering: Contaminated Soils, Pollutant Fate and Mitigation*, CRC Press, Boca Raton, FL, 307, 2000.)

where E^0 is the standard reference potential, R, T, and F are the gas constant, absolute temperature, and Faraday constant respectively, and the superscripts *a, b, w, h*, refer to number of moles of reactant, product, water, and hydrogen ions, respectively. We see from Equation (4.2) that the stable product for a given set of reactants, or the valence state of the reactants is a function of the pH-*pE* status. Figure 4.3 (from Yong, 2000) shows a simplified *pE*-pH diagram for an iron (Fe)-water system for a maximum soluble iron concentration of $10^{-5}M$—using information from Manahan (1990), Sawyer et al. (1994), and Fetter (1993).

The uppermost sloping boundary defines the limit of water stability, above which the water is oxidized. Similarly, the lowest sloping boundary marks the limit of water stability below which the water is reduced. The *pE*-pH diagram shown in Figure 4.3 gives one a view of the various phases of Fe for the conditions noted. Thus, for example, we see that at a *pE* value of 4, Fe exists as Fe^{3+} at the lower pH values. Staying with a *pE* value of 4 and continuing with increases in pH, we note that as we approach a pH of about 6.4 and beyond, precipitation occurs resulting in the formation of Fe(III) hydroxides ($Fe(OH)_3$). A decrease in *pE* at the higher pH values will result in precipitates of Fe(II). More *pE*-pH diagrams can be found in specialized texts on redox reactions and in textbooks on aquatic chemistry, geochemistry and soil chemistry.

Clay minerals such as montmorillonite and illite can catalyze Cr(VI) reduction by organic compounds such as oxalate (Deng et al., 2003). The effect is particularly noticeable up to pH 4.7 for montmorillonite and pH 5.2 for illite. High organic content soils can thus be naturally attenuating materials for certain types of heavy metals—as for example the catalysis of chromium.

4.4.3 Molecular Interactions

Partitioning of contaminants refers to the mass transfer of the contaminants from the porewater of a soil-water system to the surfaces of the soil solids, i.e. the soil fractions. The processes involved in mass transfer include physical, chemical, biologically mediated processes, and precipitation. The term *sorption* is used to refer to the adsorption processes responsible for the partitioning of the dissolved solutes such as ions, molecules and compounds in the porewater to the surfaces of the soil fractions. Because of the many processes that contribute to partitioning of the dissolved solutes, and because it is difficult to determine the specific contribution of each process to the overall mass transfer, the term sorption is used as a general term to mean the overall process resulting in transfer of dissolved solutes to the soil solids. The mechanisms contributing to partitioning of contaminants as a result of sorption mechanisms include: (a) Coulombic interactions which are between nuclei and electrons and are essentially electrostatic in nature, and (b) interatomic bonds such as ionic, covalent, hydrogen, and van der Waals.

4.5 PARTITIONING OF HEAVY METALS

Whilst we have previously considered the basic sets of forces, mechanisms, and processes involved in interactions between contaminants and soil solids, a very pertinent part of the problem of determination of attenuation of contaminants in soils is the determination of actual partitioning of contaminants. In this particular case, we are interested in the partitioning of heavy metals, and wish to consider the means whereby partitioning is determined and/or assessed.

Of the various means for assessing partitioning, two types of tests appear to be the most popular: (a) batch equilibrium adsorption isotherm tests, and (b) column or cell leaching tests. Batch equilibrium adsorption isotherm tests work with soil solutions where all the soil particles are dispersed, and, therefore, have the greatest potential for all surfaces to interact with the contaminants. This has been discussed in general in Chapter 3. On the other hand, column or cell leaching tests work with compact or intact soil samples, and therefore rely on flow through the compact soil to establish interaction between the contaminants and soil solids. The proportion of soil solids' surfaces exposed to interaction with the contaminants in the porewater can vary anywhere from 15% to 70%, depending on the type of soil, density, and soil structure. Figure 4.4 shows a simplified sketch of the two types of sorption assessment tests.

The choice of test type to use depends on the questions to be addressed. Adsorption isotherms are generally obtained as characterization information, i.e., information concerning the adsorption potential of a candidate soil for a target contaminant species. The use of the candidate soil in the form of a soil solution essentially produces a dispersed soil-water system where all the soil particles or soil solids are available for interaction with the target contaminant. This gives one a standard base for

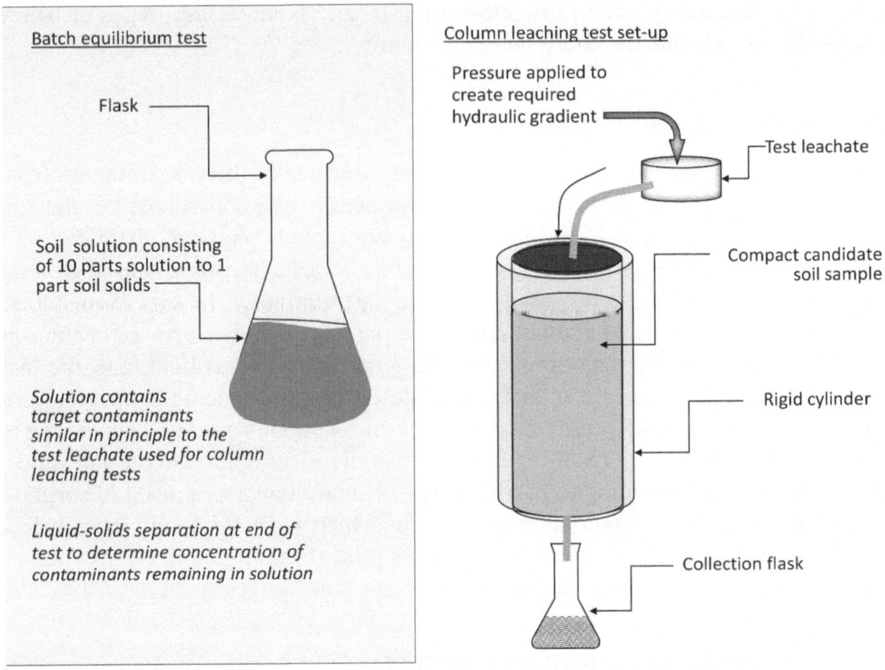

FIGURE 4.4 Batch equilibrium test for determination of adsorption isotherms and column leaching test set-up—showing the differences in test protocols and test sample. A choice of different soil solution ratios, ranging from 10:1 to 50:1 for batch equilibrium testing is used, depending on the composition of the soil and type of contaminants. (From Yong et al., 2012.)

comparison between different soils and/or different target contaminants. It is important to bear in mind that when multi-component contaminants are used as the target species, the proportions of each of the components comprising the multi-component mixture becomes very important. Selective sorption and speciation are some of the factors that can influence the sorption of each species. On the other hand, because leaching column or cell tests work with compact (i.e., intact) samples, information obtained using test leachates permit one to obtain an appreciation of "what might happen" in the field. The rationale for this type of test is founded on the assumption that the soil sample in the test column is representative of "what is in the field," and that leaching of a test leachate through the column will essentially replicate the field condition. One of the basic flaws with this type of reasoning is that it is frequently not possible to impose the field hydraulic gradients onto laboratory leaching columns. In most cases, the hydraulic gradients used for leaching column tests or conductivity cells will far exceed field hydraulic gradients—leaving one to ask the question as to how excessive gradients can possibly provide one with simulated field situations? A further complication in extrapolation of laboratory test results to field behavior is the variability of soil structure and properties—as discussed in Chapter 3. That being said, it is nevertheless useful for one to conduct leaching column tests, if for no other reason than to provide one with some insight into the probable performance of the soil material in the face of influent leachates. Testing of various kinds of soils under a variety of conditions will provide one with a base for comparison, and thus allow one to compare results on a qualitative basis.

4.5.1 Adsorption Isotherms

Adsorption isotherms play a major role in providing input for assessment of partitioning of heavy metals. Sorption of a chemical species (adsorbate) from the soil solution by the solid soil solids (adsorbent) is dictated not only by the composition of the soil, but also by the various kinds and distribution of chemical species (contaminants) in the soil solution. The sorption potential of the soil (adsorbent) for a particular adsorbate is determined by studying the characteristics of adsorption of the adsorbent for that particular adsorbate under specified conditions—i.e., ideal or under conditions simulating certain chemical mix species, etc. When multi-component species are used as the adsorbate, competition for adsorption sites will influence the amounts and proportions of the target species being studied.

In brief, *adsorption isotherms* refer to characteristic adsorption curves relating to the adsorption of individual target contaminants by candidate soils in relation to the available concentration of the target contaminant. These curves are obtained by conducting batch equilibrium tests, with test protocols shown schematically in Figure 4.5. The batch experiments usually utilize the soil as a soil suspension—with increasing concentrations of the chemical species to be sorbed in the different batches of soil suspension. The graphical representation of the relationship between the equilibrium concentration of adsorbate in solution and the equilibrium concentration of adsorbate sorbed by the soil particles is generally defined as the adsorption isotherm of the soil. A typical set of isotherms has already been shown in Figure 3.10 in Chapter 3. The relationship for the constant adsorption curve given as: $s^* = k_1 s$ in the figure relates the sorbed concentration of solutes s^* to the equilibrium concentration of solutes in solution s via a constant k_1. The obvious limitation to the use of the constant adsorption relationship is the fact that this relationship predicts limitless adsorption of the adsorbate. This is unlikely since there is a limited number of sorption sites available associated with the soil solids. The Freundlich and Langmuir adsorption isotherm models on the other hand are more realistic isotherms in comparison to the constant adsorption model. Whereas, these are improvements over the constant adsorption model, the models could still allow for unlimited adsorption.

Figure 4.6 shows the adsorption isotherms for an illite soil in relation to Pb as a contaminant species. A couple of interesting points should be noted: (a) the pH of the solution decreases as the amount of Pb (in nitrate form) is added to the solution, and (b) the initial high pH at low Pb input concentrations is due to the fact that the pH of the illite soil is high. It is important to take note that if the abscissa is plotted in terms of the concentration of solutes in the leachate solution used, as opposed to the equilibrium concentration (i.e., concentration remaining in solution), we will obtain two different adsorption isotherm curves. What makes this important is that the slopes of the curves will be different, and since the slope of the curve is used

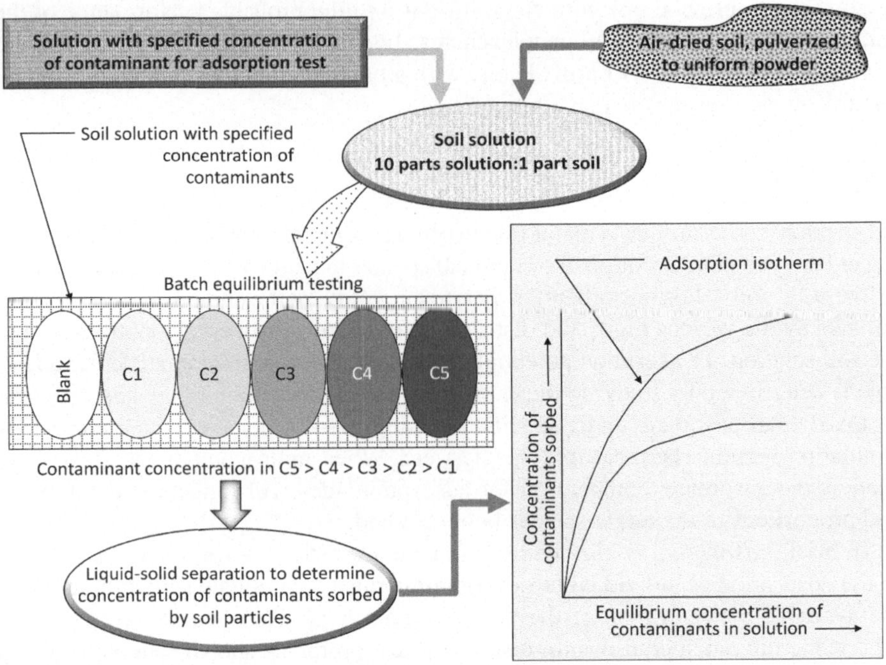

FIGURE 4.5 Batch equilibrium procedure for determination of adsorption isotherms for inorganic contaminants. The five batches soil solution shown on the left, starting with batch C1, have progressively increasing concentrations of contaminants in the solution. (From Yong, R.N. et al., *Environmental Soil Properties and Behaviour*, CRC Press, Boca Raton, FL, pp. 435, 2012.)

to provide information on the distribution coefficient k_d, one needs to be aware of the method of data portrayal. Yong (2000) has noted several pertinent points for consideration when adsorption isotherms are to be used for assessment of soil sorption capability.

- The units used for sorbed concentration (ordinate) and the equilibrium concentration (abscissa) must be consistent with each other. This is important in the subsequent use of the isotherm for determination of the distribution coefficient k_d—as will be seen later.
- We can control the pH of each batch of soil solution by adding buffering agents. However, when we choose to do so, we must be aware of the fact that the adsorption characteristics can also change. The amount and the nature of change will depend on both the type of soil and the contaminant.
- The adsorption isotherms will likely be different when multiple contaminants are used. In addition, the types of conjugate ions used for the contaminant(s) will also influence the character of the adsorption isotherm.
- Because of their higher specific surfaces and because of the types of functional groups, the oxides of iron, aluminum, and other amorphous materials have considerable influence on the development of adsorption isotherms.

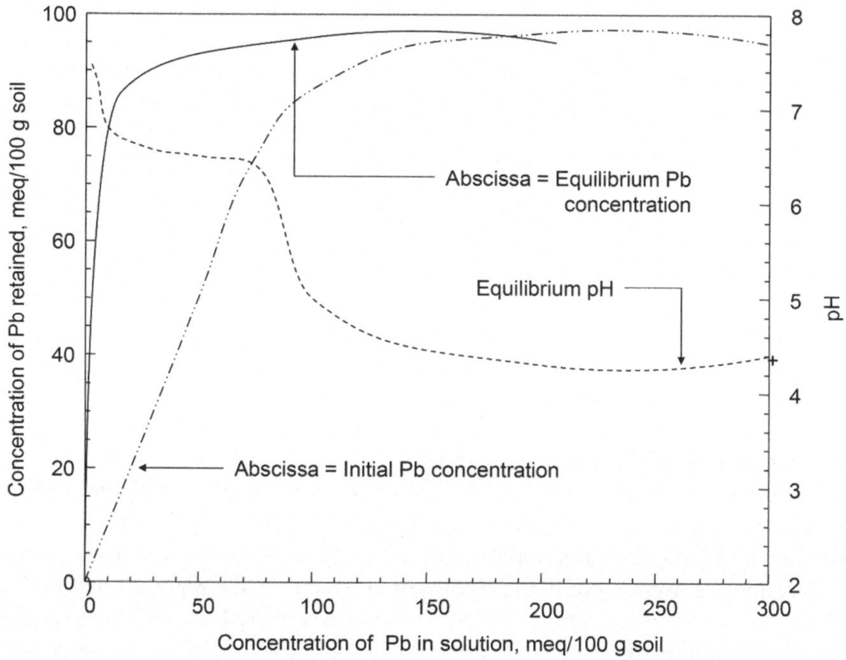

FIGURE 4.6 Adsorption isotherms for Pb by an illite soil. (Data from MacDonald, E., Aspects of competitive adsorption and precipitation of heavy metals by a clay soil, ME Thesis, McGill University, 1994.)

As has been used and noted before in Section 3.6.1 in Chapter 3, the distribution coefficient k_d is obtained as the slope of the linear adsorption isotherm, or as the tangent to prescribed points on non-linear adsorption isotherms. This describes contaminant partitioning between liquid and solids. We can develop various k_d values for different kinds of soils in relation to different contaminants and under various conditions. The primary requirement is the availability of batch equilibrium test data. The question as whether one uses batch equilibrium tests or leaching column tests to obtain partitioning and distribution coefficients should be noted at this point. The basis for the question lies in the nature of the test sample used to determine partitioning of the contaminant under investigation—i.e., the difference between soil solutions used for batch equilibrium tests and intact soil samples used for leaching column tests.

Irrespective of the nature of the soil samples used to determine partitioning of contaminants, it is important to bear in mind that most standard procedures assume that when interaction of the contaminants and soil particles occur, chemical equilibrium conditions are achieved within a 24-h period. For soils that do not contain *soil organic matter* (*SOM*), the 24-h period used is generally acceptable. Results quoted in the literature indicate that for clay soils without SOM, equilibrium is generally reached within a 3–4 h period. However, when SOM is present, equilibrium may not be reached for a few days, depending on the nature and proportion of SOM present in the soil. Under these circumstances, it is generally wise to obtain proper characterization of the soil to be used for the batch equilibrium tests so that the use of k_d obtained from an analysis of the test results can be properly assessed.

Figure 4.7 shows the results of batch equilibrium test with initial Pb concentration of 200 ppm and in relation to Pb removed from the aqueous phase and to pH. The Pb precipitation boundary separates the precipitation region from the "solution" and sorbed region. The diagram shows that when the pH of the system reaches about 4, precipitation of Pb begins to become an important component in the removal of Pb from the aqueous phase. This lends support to the information presented in Figure 4.1 with respect to the precipitation pH of Pb. The amount of Pb actually sorbed by the solids in the illite soil is shown in the bottom portion of the diagram. The diagram tells us about the state of Pb in relation to both the varying nature of the pH environment and the interactions with the soil solids. In the example shown, a heavy metal contaminant (Pb) is introduced into a soil solution as a $PbCl_2$ salt. A significant portion of the soluble ions will be sorbed by the soil solids as shown in the bottom part of the diagram, and the ions remaining in solution would either be hydrated or would form complexes, giving one Pb^{2+}, $PbOH^+$, and $PbCl^+$.

As shown in Figure 4.7, the importance of the pH of the porewater or soil water in the soil cannot be overstated. In the studies of arsenic sorption reported by MacLean and Bledsoe (1993), maximum arsenate adsorption occurred at a pH 5 on kaolinite and montmorillonite. Maximum adsorption on aluminum and iron oxides occurred at

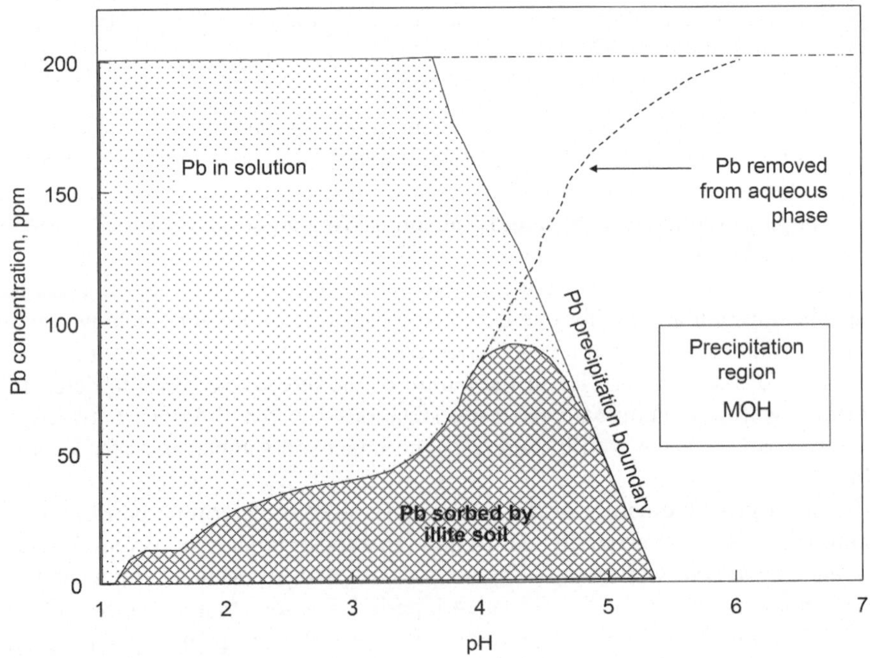

FIGURE 4.7 Solubility-precipitation diagram showing Pb removed from the aqueous phase, Pb in solution, and Pb sorbed by illite soil.

pH levels of 3–4. Sorption decreased as the pH increased, and it was determined that adsorption was specific—similar to the adsorption of phosphate on oxides. Sorption of arsenite, As(III), onto kaolinite and montmorillonite was shown to increase as one progressed from pH 3–9. In the case of sorption by iron oxide, it was shown that maximum adsorption of As(III) occurred at about pH 7.

4.5.2 Column Leaching Tests

The general test procedure for conducting leaching tests using columns is shown in Figure 4.8. Using replicate samples, we can introduce various amounts of the test leachate to individual soil samples thus permitting us to examine the distribution and retention of heavy metals in relation to amount of transport of influent leachate. By sectioning the leaching column sample at the end of the test, and by analyzing the porewater in each section for soluble ions and exchangeable and extractable ions in the soil solids, one can obtain various kinds of concentration profiles. Figure 4.9 shows a set of sorption characteristic curves similar in nature to the batch equilibrium adsorption isotherms. The values noted on the abscissa

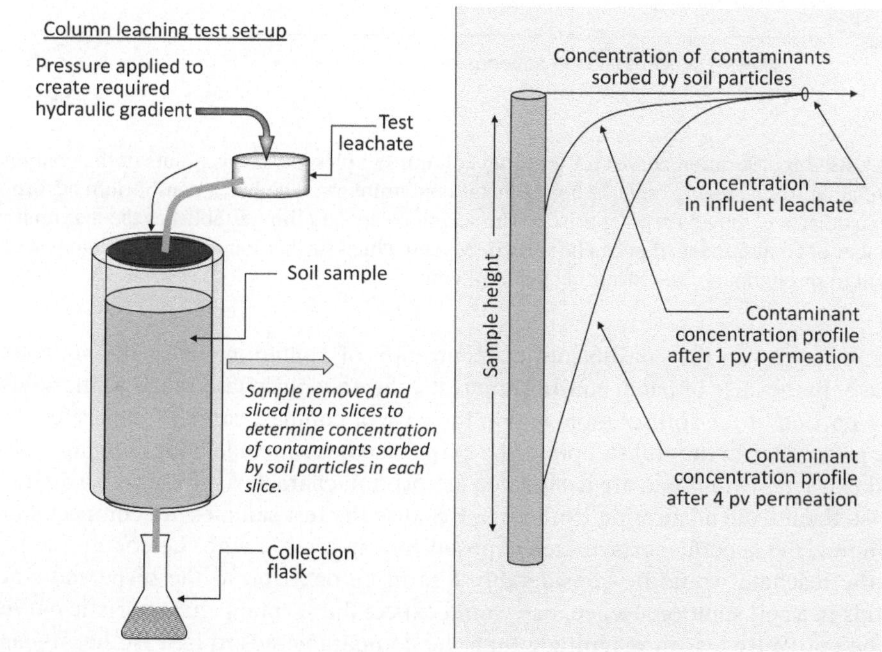

FIGURE 4.8 General procedure for column leaching tests to determine contaminant sorption profiles. Replicate samples are permeated with leachate and samples tested after permeation of 1 to 4 to 6 pore volumes (pv) of test leachate. Samples are removed after the pre-determined permeation and sectioned to determine the concentration of contaminants sorbed by the particles in each section. The sorption profiles shown in the right-hand side of the figure show typical profiles after 1 and 4 pore volume permeations of test leachate. One pv refers to the whole pore volume included in a test sample. (From Yong, R.N. et al., *Environmental Soil Properties and Behaviour*, CRC Press, Boca Raton, FL, pp. 435, 2012.)

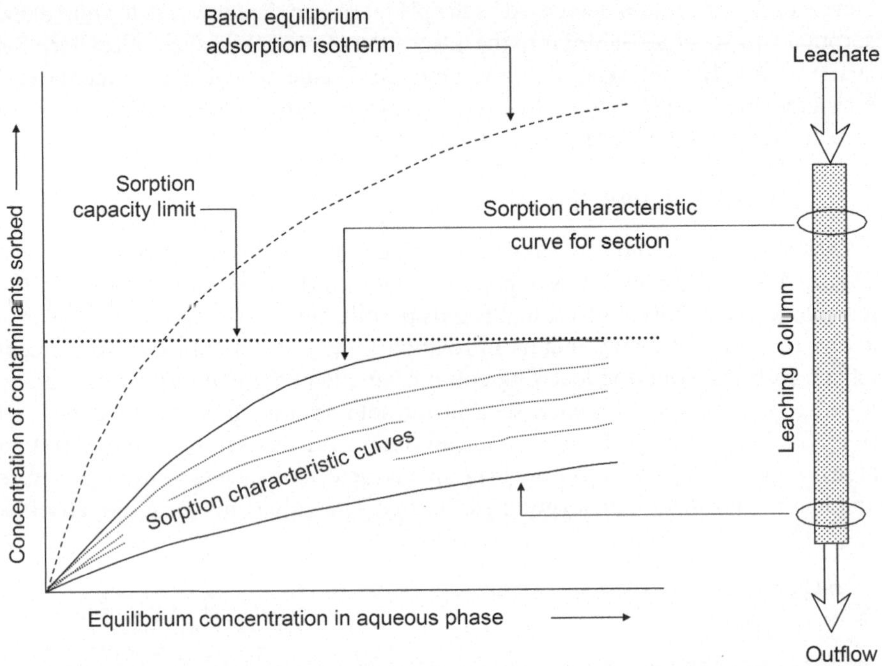

FIGURE 4.9 Sorption curves for leaching column samples at various points in the column in relation to increasing input of heavy metal contaminants. The batch equilibrium adsorption isotherm is shown for comparison. The sorption capacity limit establishes the maximum number of contaminants that can be sorbed, beyond which further input of contaminants will result in throughput of the incoming contaminants.

are in respect to the equilibrium concentration of contaminants in the aqueous phase. In the case of batch equilibrium tests, the aqueous phase refers to the solution portion of the soil solution, and in the case of compact samples, this refers to the pore water in the soil samples. The sorption curves include both exchangeable and extractable ions and are designated as sorption characteristic curves to distinguish them from adsorption isotherms. Because the test samples are compact soil samples, the specific surface area exposed for interaction with the contaminants in the leachate would be considerably less in comparison to the dispersed soil solids in a soil solution. Hence, one would expect the sorption characteristic curve to be markedly less in magnitude than the comparable adsorption isotherm—as illustrated in Figure 4.8.

The *sorption capacity limit* shown in Figure 4.9 essentially tells us that there is a limit to the number of contaminants (in this case heavy metal contaminants) that can be sorbed. The sorption characteristic curve for the top part of the sample shows that this limit is reached after a certain amount of input of contaminants. However, as one progresses downward in the sample, more input is required before the sorption capacity limit is reached. This is better shown in Figure 4.10 in terms of contaminant profiles along the length of the test column. As more influent

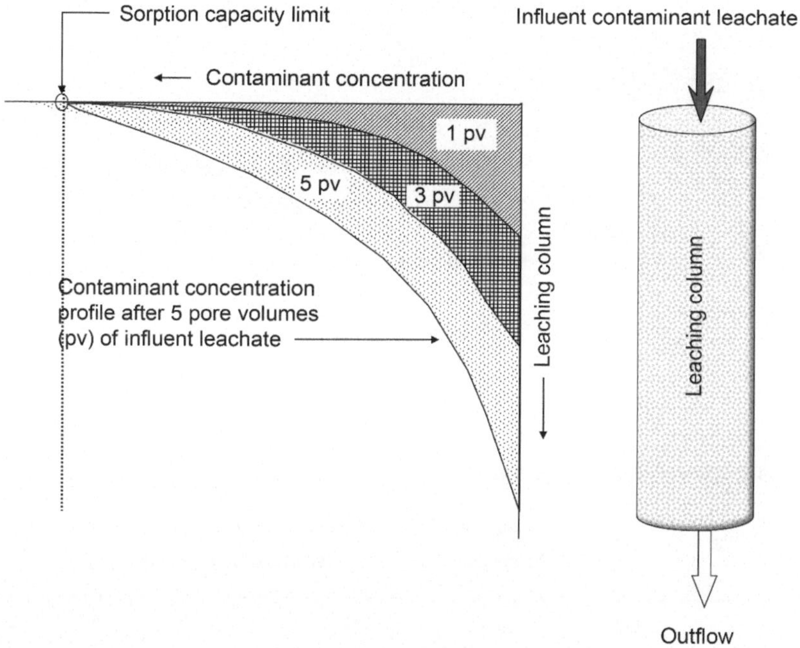

FIGURE 4.10 Contaminant concentration profiles along the length of column. Note increasing area under the profile curves as the number of influent contaminants increases. Given sufficient time and influent input, the sorption capacity limit shown at the left of the profiles will be reached.

leachate (containing contaminants) passes through the sample, the contaminant profile deepens—as shown by the 1 pv to 5 pv profiles. With time and more influent leachate, the point will be reached where the soil can no longer sorb the contaminants. At this point, the sorption capacity limit is reached, and the contaminant profile will be a vertical line originating from the capacity limit shown at the left of the drawing.

Figure 4.10 is a schematic illustration showing the sorption of contaminants along the length of a leaching column after undergoing 1, 3, and 5 pore volumes (pv) of incoming contaminant leachate. Actual experimental information of the contaminant Pb along the length of a kaolinite sample determined after 1, 3, and 5 pv leachate transport is shown in Figure 4.11. The portion sorbed or partitioned onto the soil solids is shown as "sorbed" Pb while the Pb remaining in the soil-water system is shown residing in the porewater. The "Pb in porewater" shown in the diagram represents the incremental amount. Addition of this Pb to the amount shown as adsorbed by the soil solids will give us the total Pb "sorbed" by the soil sample. These results are obtained using the procedure shown in Figure 4.8 for: (a) determination of the soluble ions in the portion dealing with pore fluid analysis, and (b) determination of exchangeable and extractable ions from the sections of soil making up the leaching column.

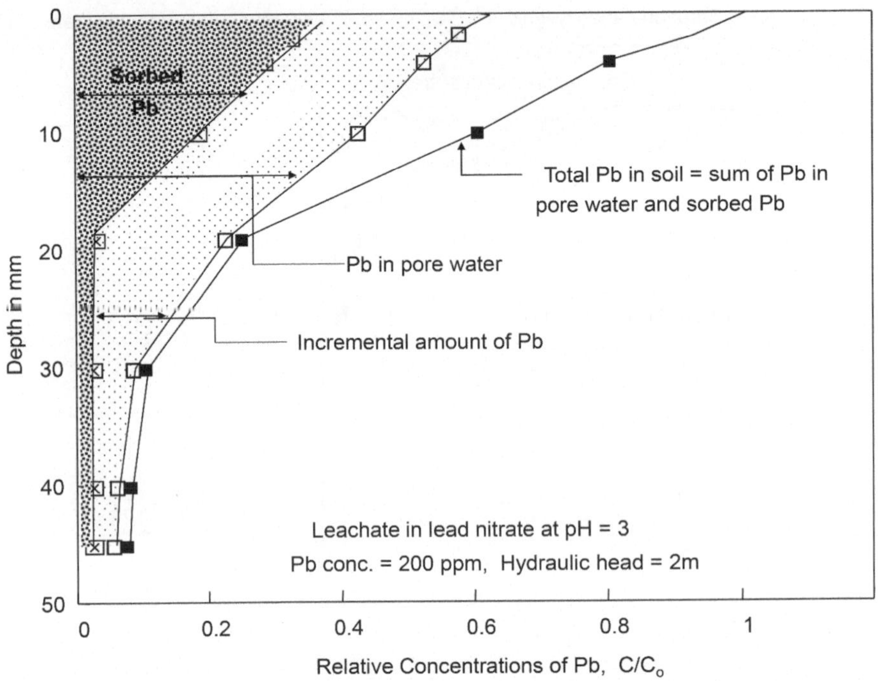

FIGURE 4.11 Pb concentration profile obtained from leaching column tests using a kaolinite sample after 5 pv of influent Pb leachate. Sorbed Pb and Pb in porewater contribute to total Pb in soil. (Data from Darban, A.K., Multi-component transport of heavy metals in clay barriers, Ph.D. Thesis, McGill University, Montreal, Canada.)

Evidence from other tests on heavy metals provides similar patterns of sorption performance. The retention profiles for total Pb, Cu, and Zn shown in Figure 4.12 have been obtained from acid digestion of estuarine alluvium samples from two different locations in Wales, UK—Neath (NEA) and Newport (PEA). Although both soils are classified as estuarine alluviums, there are some significant differences in the proportions of some of the soil fractions. For example, the PEA soil has 16.5% carbonates in comparison to the 3% shown in the NEA soil. The proportions of amorphous oxides/hydroxides and organics are also significantly different between the two. The PEA soil had 13.4% and 3.78%, respectively, for the oxides and organics as compared to 9.9% and 5.1%, respectively, for the NEA soil. Thus, even though the SSA for both soils are almost similar (75 m^2/g for PEA as compared to 73.3 m^2/g for the NEA soil), the CEC for the PEA soil was more than twice that of the NEA soil (39.4 compared to 14.8 meq/100 g, respectively). From the test results, the differences in the retention profiles of the metals are visibly different. What does this tell us? Obviously, even though the two soils are classed as estuarine alluvium soils, compositional differences matter a great deal in characterizing retention of the metals concerned. The nature of the functional groups of the different soil fractions combined with their influence on the development of soil structure contributed to the differences noted.

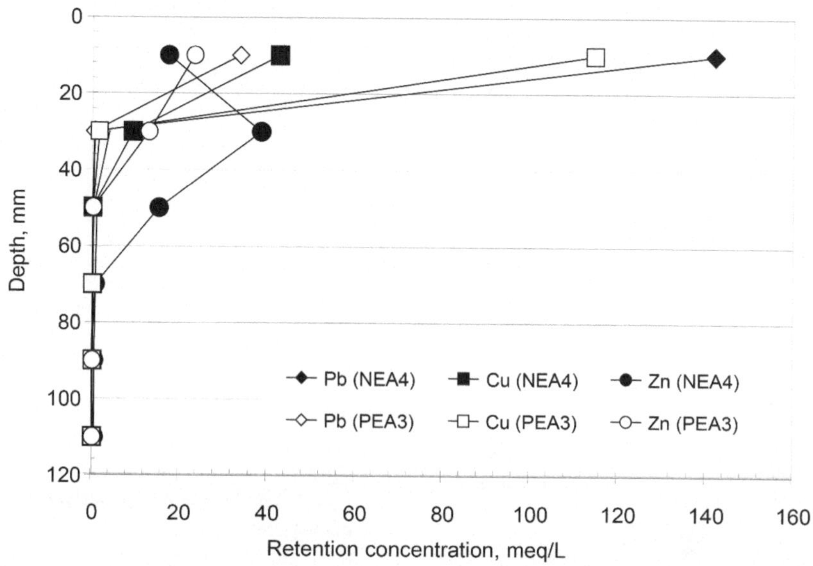

FIGURE 4.12 Retention profiles for Pb, Cu, and Zn for estuarine alluvium samples from Neath (NEA) and Newport (PEA), Wales, UK subject to leaching with a municipal solid waste leachate spiked with Pb, Cu, and Zn. Retention values are obtained from acid digestion of samples subject to 5 pv leaching column tests. (Data from Yong, R.N. et al., *Eng. Geol.*, 60, 307–322, 2001.)

4.5.3 Distribution Coefficient K_D

Section 3.6.1 in Chapter 3 touched on certain aspects of the distribution coefficient k_d. It is recalled that k_d was used in Equation (3.20) as part of the retardation factor R, in the form that stated: $R = \left[1 + \frac{\rho^*}{n} k_d\right]$, with the requirement that k_d was obtained from a linear adsorption isotherm. For adsorption isotherms that are nonlinear, piece-wise linear portions of the isotherms are chosen.

A comparison of the adsorption isotherm for kaolinite soil and its equilibrium sorption characteristic curve obtained through leaching column tests is shown in Figure 4.13. The separation distance denoted by "A" in the figure indicates the sorption differences between a fully dispersed soil particle system as obtained from soil suspension studies, and compact soils as used in leaching column tests. We have noted previously that these differences are in large measure due to the availability of particle surfaces to interact with the influent leachate. What this tells us is that specific surface area is a parameter of considerable importance in sorption and partitioning of contaminants—everything being equal. The reduction in available surface area in the compact samples is due not only to the packing of particles, but to the formation of tactoids and microstructural units that function primarily as sources and sinks for the transfer of contaminants. This has been discussed in Section 3.8.2 in Chapter 3 in relation to diffusion of solutes in soils. We should note that the greater or larger the value of "A" in Figure 4.13, the greater would be the presence of microstructural units.

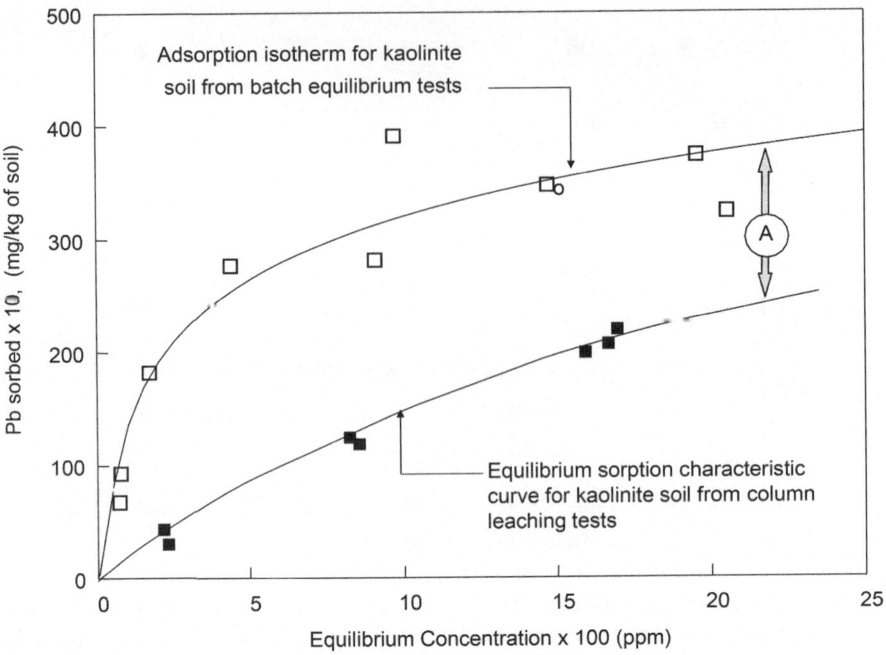

FIGURE 4.13 Comparison of Pb sorption curves obtained from batch equilibrium and column leaching tests for kaolinite soil. (From Yong, R.N., *Geoenvironmental Engineering: Contaminated Soils, Pollutant Fate and Mitigation*, CRC Press, Boca Raton, FL, 307, 2000.)

Viewing the kinds of curves portrayed in Figure 4.13, we can conclude that the distribution coefficient k_d obtained using the results of leaching column tests will be different from that obtained using the adsorption isotherm. We repeat the caution discussed earlier concerning the choice of curves to use for the determination of k_d values, i.e., calibration procedures using information from adsorption isotherms as opposed to actual field performance assessment and prediction using compact samples in laboratory leaching column tests. That being said, it is not unusual to see predictions of field transport performance using distribution coefficients obtained from adsorption isotherm information. This practice tends to predict greater contaminant partitioning capability than may actually exist.

The k_d values for divalent metals such as the heavy metals have been reported to be dependent on the total dissolved metal concentrations—namely the sum of the free metals (Me^{2+} and the inorganic ion pairs such as $MeOH^+$, $Me(OH)_2^o$, $Me(OH)_3^+$, $MeHCO_3^-$, $MeCO_3^o$, $MeNO_3^+$, $MeCl^-$, etc. (Sauvé et al., 2000). The presence of dissolved organic matter in soils is particularly important because of the formation of metal-organic complexes. These have considerable influence on the k_d values obtained for the heavy metals. We can conclude that the factors that influence the development of k_d values for metals include pH, concentrations of inorganic ligands, dissolved organic matter, and competition with other ions for sorption sites. Sauvé et al. (2000) suggest that k_d values for divalent metal ions such as heavy metals would be better predicted using empirical regressions that combine pH and either dissolved organic matter or total metal. They further suggest that the more direct approach which uses the semi-mechanistic competitive adsorption model would provide better model predictions of dissolved metal concentrations in solution. In this approach, the model accounts for pH, total metal content, soil organic matter (SOM), and competition between Me^{x+} and H^+ for sorption sites on the soil particles. The relationship obtained is given as follows:

$$\log_{10}(\text{dissolved metal}) = a + b \cdot pH + c \cdot \log_{10}(\text{total metal}) + d \cdot \log_{10}(\text{SOM}) \quad (4.3)$$

where a, b, c, and d are coefficients obtained from experimental information.

4.5.4 Breakthrough Curves and Retardation Factors

To better appreciate contaminant attenuation in soils, one can study the retardation of the contaminant or the sorption-attenuation characteristics of the target contaminant. A common form for presentation of information concerning attenuation of contaminants as the leachate passes through the leaching column is to plot the concentration sorbed by the soil column in relation to the amount of leachate transported through the column. Since many variables such as length of column, concentration and amount of leachate used, and time enter into the picture, a useful method for portrayal of results is to use a normalization procedure for data representation. Figure 4.14 shows a set of idealized breakthrough curves using relative concentration

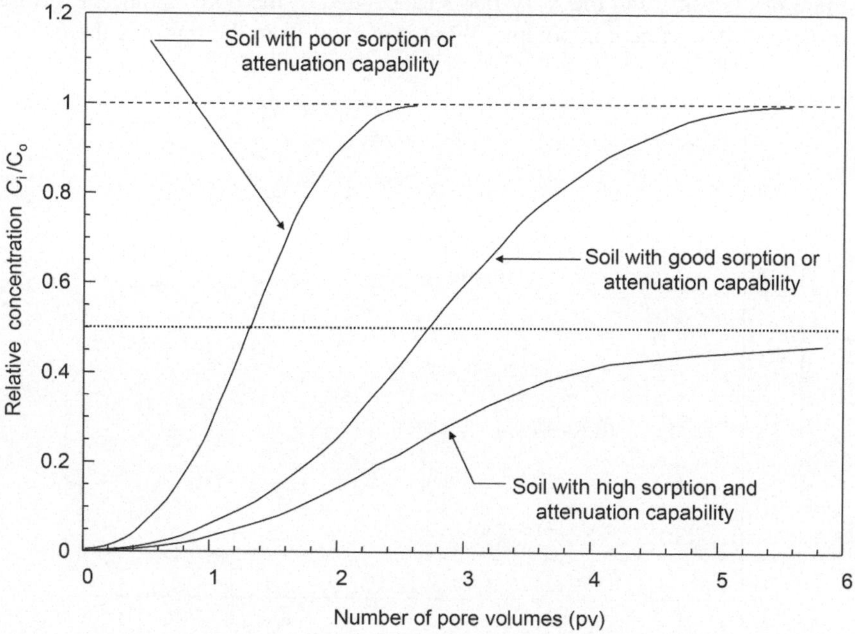

FIGURE 4.14 Idealized breakthrough curves showing soils with poor, good, and high sorption or contaminant attenuation capabilities.

C_i/C_o as the ordinate and pore volumes pv as the abscissa—where C_i represents the concentration of contaminants at any time, and C_o is the original concentration. With this normalization procedure for data portrayal, it is easy to compare the attenuation capabilities of different soils for various kinds of contaminants. Actual real-time data can be obtained by converting the pv values to actual time periods.

The breakthrough curves for three different kinds of are shown in Figure 4.15. The leachate used for the tests was a $Pb(NO_3)_2$ solution at pH values below the precipitation pH of the metal. With the absence of soil organics and amorphous materials in all three soils, the dominant attenuation (assimilation) process is most likely to be the same exchange mechanisms operative in characterizing the CEC of the soils. If one uses the CEC of a soil as a guide to sorption of metal ions, the low CEC for kaolinite (15 meq/100 g soil) would allow breakthrough to occur rapidly (at about 1.4 pv). The CEC values for the Lachenaie and the illite clay soils are relatively close to each other (60 and 50 meq/100 g soil, respectively). Breakthrough for the Lachenaie clay occurs at about 8.5 pore volumes whereas breakthrough for the illite occurs at about 10.7 pore volumes—in reverse order to their CEC values.

Figure 4.16 shows the influence of both the CEC and the wetted surface area of the pore channels in characterization of the breakthrough performance of three types of soils with distinctly different wetted particle surface areas (noted in Chapter 3, Table 3.1 and Figure 3.17). The wetted surface area (Chapter 3, Section 3.7.3) essentially provides the amount of particle surface area for interaction with the heavy metal contaminants and is one of the key factors in attenuating the contaminants. The abscissa shows the ratio of the CEC expressed in terms of the unit volume of soil particles of the soil at optimum dry density, and the S_w wetted surface area of the pore channels expressed as cm^2/cm^3 for the same unit volume. What is evident from the figure is the fact that

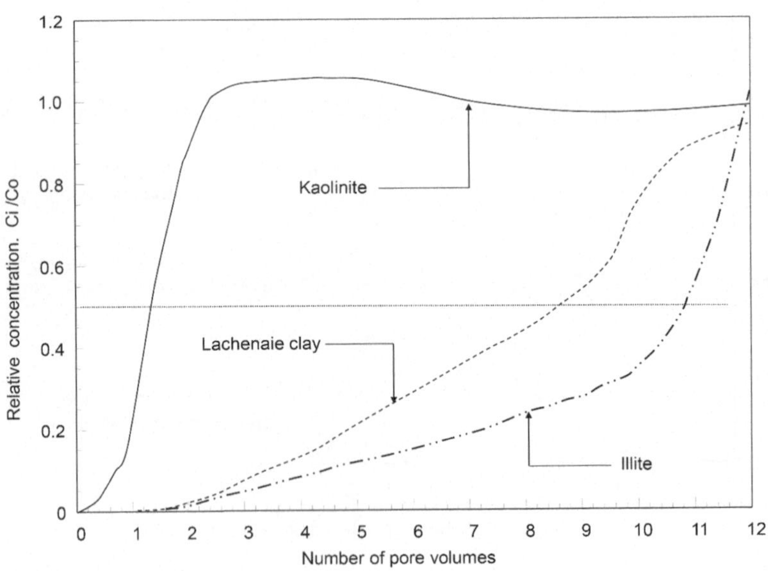

FIGURE 4.15 Pb breakthrough curves for three kinds of soil.

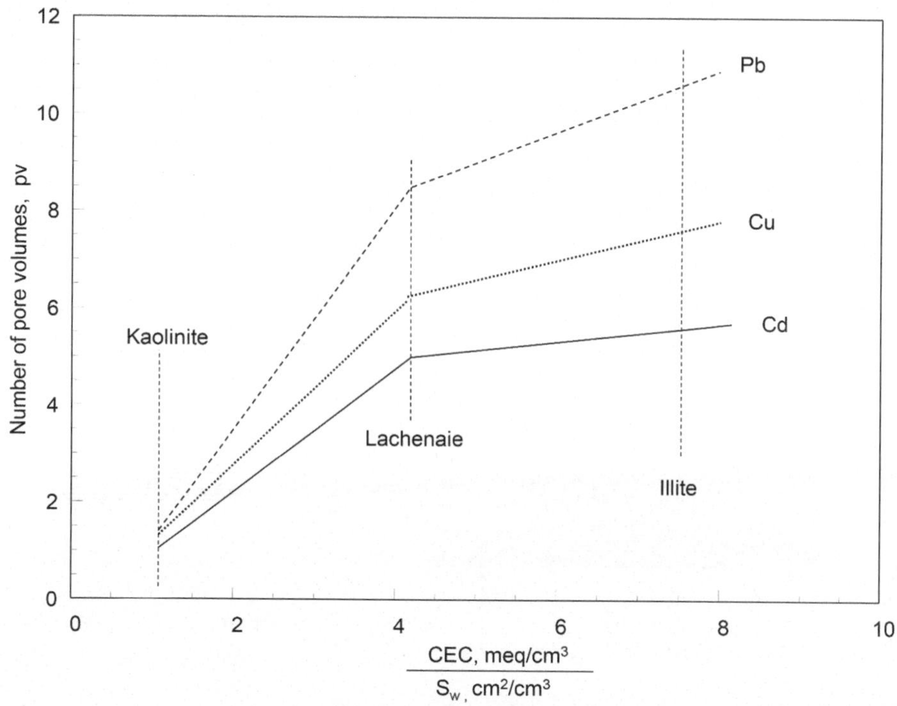

FIGURE 4.16 Breakthrough pore volumes (pv) for three soil types and three heavy metal species in relation to CEC (cation exchange capacity) and wetted surface area.

both CEC and S_w are important factors in the attenuation process. The positions of the curves, i.e., Pb in relation to Cu and Cd, are most likely influenced by the selectivity in sorption preference—as will be discussed in the next section.

The retardation factor is a parameter which describes the relationship between the speed of solute movement and the aqueous front. This parameter essentially describes the relative movement of the solute and its carrier—the leachate or porewater. Figure 4.17 shows a schematic of the relative movement of solutes and their aqueous carrier (leachate) after a specific time *t*. There are three "fronts" shown in the diagram. The uniform solute concentration front shown after time *t* is obtained when both aqueous carrier and solutes move at the same rate. This is the aqueous front after time *t*. This front is obtained if the solutes show no diffusive capability, and when there is no physico-chemical interaction between the solutes and the reactive surfaces of the soil particles. In the case of the solute diffusion front, the solutes diffuse along the porewater boundary layers and because of their diffusivity, the front can lead the aqueous front in the same time frame. The solute retardation front lags behind the uniform solute concentration front because of the effects of partitioning and interactions between solutes and soil particles.

Valocchi (1984) has used this retardation phenomenon as a parameter in the form of a relationship between the speed of solute movement and the aqueous front speed.

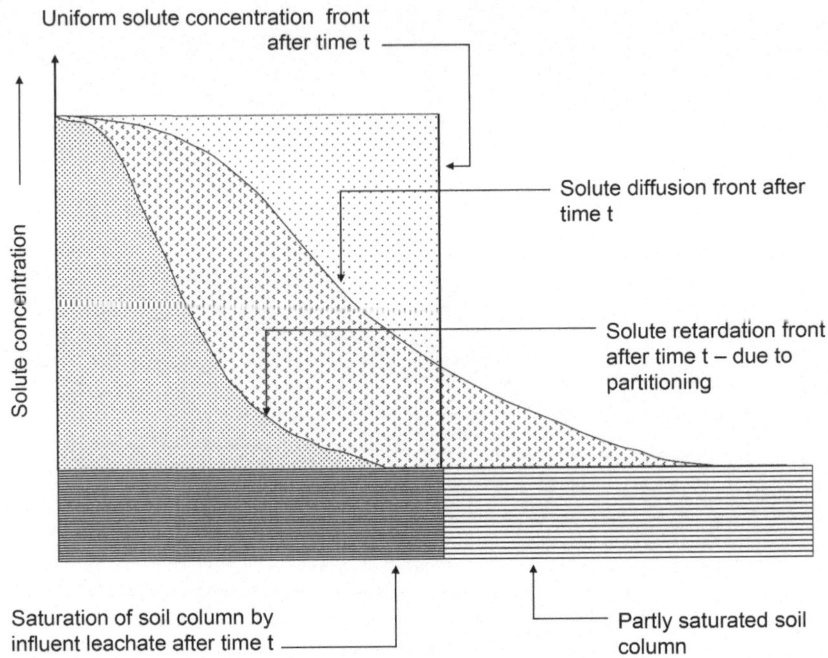

FIGURE 4.17 Movement of solutes and aqueous carrier (leachate) at a particular time *t*. The uniform solute concentration front is obtained when both aqueous carrier and solutes move at the same rate. In the case of the solute diffusion front, the solutes diffuse along the porewater boundary layers. The solute retardation front lags behind the uniform solute concentration front because of the effects of partitioning and interactions between solutes and soil particles.

The rationale underlying the use of the retardation parameter is founded on the reasoning that this is an indicator of interactions between the solution and the solid soil particles. de Matos et al. (2001) have studied the mobility of heavy metals in relation to soil chemical and mineralogical characteristics using this approach. Studying three Brazilian soils and horizons, they established that Zn showed the lowest values of retardation factors (parameters) in relation to Cd, Cu, and Pb. On a rating scale, Zn was lowest, with Cd rated slightly above it and Cu and Pb considerably higher. Pb showed the highest retardation factor. When we discuss selectivity and preferential adsorption, we will see a similar rating scale. What this means is that the mobility of Zn is greatest and that Pb is less mobile because of the greater partitioning of Pb onto the soil solids.

4.6 PREFERENTIAL SORPTION AND SELECTIVITY

4.6.1 Selectivity

We saw in Table 4.1 that clay minerals have various mechanisms for retention of heavy metal (HM) contaminants. In this section, we will examine the interactions between the heavy metal contaminants and clay mineral fractions with a view to determining the influence of different types of minerals and HMs on retention or sorption of the HMs.

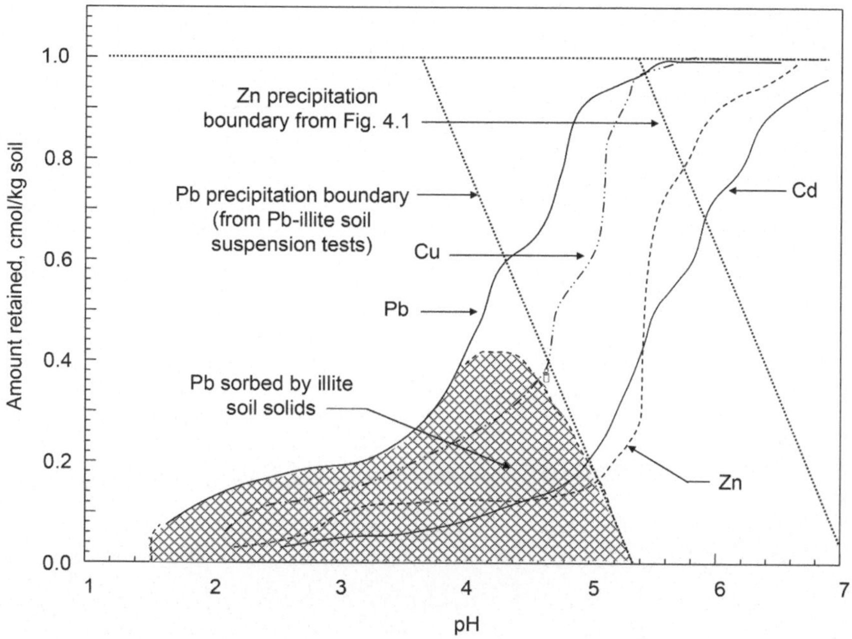

FIGURE 4.18 Retention of heavy metals by illite soil in soil suspension tests using single species HMs. Note actual amount of HM sorbed and amount removed from aqueous solution through precipitation mechanisms.

Figure 4.18 shows the retention data obtained by Phadungchewit (1990) in respect to an illitic soil and four different heavy metals: Pb, Cd, Cu, and Zn. The concentration of each HM contaminant used for the tests conducted was maintained at 1 cmol/kg soil. The precipitation information for Zn and Pb from Figure 4.1 and the Pb sorbed by the illite soil solids in Figure 4.6 are included in Figure 4.18. The influence of pH on retention of Pb, Cu, Zn, and Cd by the soil is well demonstrated, and as one might deduce from the precipitation boundaries shown, a major portion of the apparent HM retained can be attributed to precipitation of the HMs. Accordingly, one could argue that it would be more accurate to portray the "retained" portion as HM removed from the aqueous solution. Corresponding tests conducted with a mixture of the same heavy metals showed similar trends as can be seen in Figure 4.18. The slight differences in shapes of the "retention" curves can be attributed to competition for sorption sites. For comparison, we should look at the retention characteristics for Pb shown by different soils (Figure 4.19). The soils shown in the results of soil suspension tests for retention of Pb in the figure have reactive surfaces which are dissimilar in properties and characteristics—due in large measure to the differences in CEC and specific surface area (SSA) between the soils. These account for the differences in retention characteristics for Pb. We need to stress once again, that the "amount retained" indicated in the graphs refer to the amount of HM removed from the aqueous phase of the soil suspensions, and that it is not possible to distinguish between the various mechanisms participating in removal of the HMs.

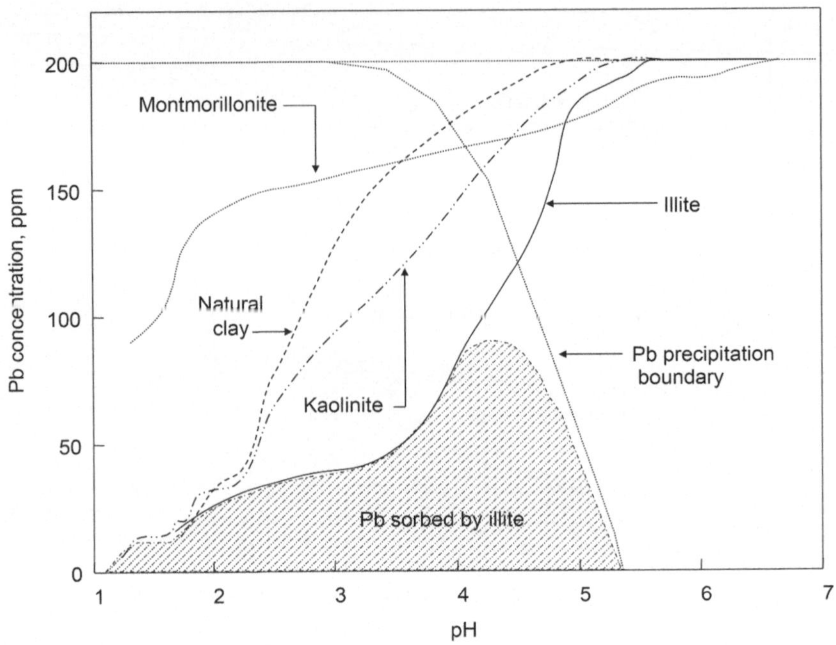

FIGURE 4.19 Influence of type of soil on retention of Pb in soil suspension tests.

The results shown in Figure 4.18 indicate higher retention of Pb at all the pH values—until full retention occurs. If we draw a vertical line from any pH value (abscissa), except for a small range around pH 5, we see what appears to be a preferential retention characteristic (*selectivity*) of the order of Pb > Cu > Zn ≈ Cd. Note that this order is in reverse order to that established by de Matos et al. (2002) for their retardation factors. This is as expected since selectivity refers to sorption preference. The selectivity order shows preference for Pb over Cu, Zn and Cd, meaning that Pb will be less mobile because it is preferentially sorbed. We also note that Zn will be more mobile because it is less preferentially sorbed, and thus is more available for movement in the porewater. Various studies by researchers have shown that sorption selectivity for HMs does exist and that this is in part determined by the type of soil fractions in the soil. Table 4.2 (from Yong, 2000) gives a brief representative account of the sorption selectivity for a representative group of soils.

The various test results reported by researchers in respect to retention of HMs in soil solution studies show that:

- Total retention (i.e., 100% retention) of HM at the higher pH values appear to be related to the precipitation pH of the HM. For example, Coles et al. (2000) show precipitation of Pb and Cd, forming $Pb(OH)_2$ and $Cd(OH)_2$ respectively, and that precipitation of Pb occurs at about 2 pH units lower than that of Cd.
- The presence of other HMs does not appear to change the total amount retained or the retention characteristics of most of the soils.

TABLE 4.2
Sorption Selectivity of Heavy Metals in Different Soils

Material	Selectivity Order	References
Kaolinite clay (pH 3.5–6)	Pb>Ca>Cu>Mg>Zn>Cd	Farrah and Pickering (1977a)
Kaolinite clay (pH 5.5–7.5)	Cd>Zn>Ni	Puls and Bohn (1988)
Illite clay (pH 3.5–6)	Pb>Cu>Zn>Ca>Cd>Mg	Farrah and Pickering (1977b)
Illite clay (pH 4–6)	Pb>Cu>Zn>Cd	Yong and Phadungchewit (1993)
Montmorillonite clay (pH 3.5–6)	Ca>Pb>Cu>Mg>Cd>Zn	Farrah and Pickering (1977b)
Montmorillonite clay (pH 5.5–7.5)	Cd=Zn>Ni	Puls and Bohn (1988)
Montmorillonite clay (pH~4)	Pb>Cu>Zn>Cd	Yong and Phadungchewit (1993)
(pH~5)	Pb>Cu>Cd~ Zn	
(pH~6)	Pb=Cu>Zn>Cd	
Al oxides (amorphous)	Cu>Pb>Zn>Cd	Kinniburgh et al. (1976)
Mn oxides	Cu>Zn	Murray (1975)
Fe oxides (amorphous)	Pb>Cu>Zn>Cd	Benjamin and Leckie (1982)
Goethite	Cu>Pb>Zn>Cd	Forbes et al. (1974)
Fulvic acid (pH 5.0)	Cu>Pb>Zn	Schnitzer and Skinner (1967)
Humic acid (pH 4–6)	Cu>Pb>Cd>Zn	Stevenson (1977)
Japanese dominated by volcanic parent material	Pb>Cu>Zn>Cd>Ni	Biddappa et al. (1981)
Mineral soils (pH 5.0), (with no organics)	Pb>Cu>Zn>Cd	Elliott et al. (1986)
Mineral soils (containing 20–40 g/kg organics)	Pb>Cu>Cd>Zn	Elliott et al. (1986)

Whilst there is considerable danger in making simple generalizations on HM retention based on soil suspension tests, we can nevertheless conclude that we need to consider a number of factors in assessment of metal-soil interaction such as (a) mechanisms contributing to sorption of the HMs, (b) types of soil fractions involved in interaction with the HMs, (c) types and concentrations of the HMs, and (d) pH and redox environments.

Preference or selectivity in metal species sorption is generally not the same for any two soils, since this is very closely related to the nature and distribution of the reactive surfaces available in the soil. The amounts of HM retained, and the pH influence on retention for any HM appears to be markedly affected by the presence of other metallic ions in the aqueous phase. The availability of reactive surfaces is a significant consideration in evaluation of sorption capacity and selectivity. Tests with *single species* and *composite species* (i.e., *multi-component species*), indicate that the selectivity order for the illite soil, such as that shown in Figures 4.18 and 4.19, would be: Pb > Cu > Zn ≈ Cd. The selectivity order for the montmorillonite soil appears to be sufficiently well defined for relatively higher pH values. For pH values below at about 4, the selectivity order appears to be: Pb > Cu > Zn > Cd. Results obtained from reactions at pH values below 3 are not quantitatively reliable because of dissolution processes, and should only be used for qualitative comparison purposes – i.e., dissolution processes can interfere with the HM sorption reactions. In general, selectivity is influenced by ionic size/activity, soil type and pH of the system.

Table 4.2 shows the selectivity order reported in some representative studies in the literature confirming that selectivity order depends on the soil type, pH environment of soil-contaminant interaction, and polarizing power of the metal cations. Evans (1989) maintains that the order of selectivity is primarily due to the increase in hydrated radii of the cations, i.e., metal cations with small ionic radii have larger hydrated radii because of greater attraction of the O–H dipoles in water to the charged ions. This reflects the polarizing capability of the cations. Elliott et al. (1986) reports that selectivity for divalent heavy metal ions applied with equal concentrations to a soil are sorbed according a selectivity rule governed by ionic size. Bohn (1979) states that the ease of exchange or the strength with which cations of equal charge are held is generally inversely proportional to the hydrated radii, or proportional to the unhydrated radii. Using this rule to the heavy metals shown in Figures 4.18 and 4.19, the predicted order of selectivity based on unhydrated radii should be:

$$Pb^{2+}(0.120\text{ nm}) > Cd^{2+}\ (0.097\text{ nm}) > Zn^{2+}\ (0.0.074°\text{m}) > Cu^{2+}(0.072\text{ nm})$$

However, Yong and Phadungchewit (1993) show a general selectivity order to be Pb > Cu > Zn > Cd. According to Forbes et al. (1974) sorption selectivity of heavy metals may related to the *pk* of the first hydrolysis product of the metals—where *k* is the equilibrium constant for the reaction in the above equation when $n = 1$. Using this *pk* selectivity ranking procedure, the heavy metals shown in the previous figures show a selectivity order as follows:

$$Pb(6.2) > Cu(8.0) > Zn(9.0) > Cd(10.1)$$

where the numbers in the brackets refer to the *pk* values.

4.6.2 Influence of Ligands

The kinds of inorganic and organic ligands present in the porewater are also important factors in the control of HM retention. Figure 4.20 shows the retention of Cd by kaolinitic clay using two control chloride solutions as a function of initial Cd concentration. What the results indicate is that speciation or formation of complexes has changed the amount of Cd sorbed by the kaolinite soil. Since mononuclear complexes formed between a central metal ion and a number of anions, or ligands (Butler, 1964) may be positive, negative, or neutral, we expect Cd to combine with chloride ions to form the following complexes: $CdCl^+$, $CdCl_2°$, $CdCl_3^-$, and $CdCl_4^{2-}$. The Cd results shown in Figure 4.20 (from Yong and Sheremata, 1991) refer to sorption of Cd from a solution made up as a $Cd(NO_3)_2$ salt with varying concentrations of Cd, together with an actual landfill leachate containing not only Cd, but also Mg, K, Na, SO_4, Zn, Cu, Pb, Cl, amongst other constituents in the leachate.

The *predicted sorption* curve shown in the figure refers to the theoretical amount of Cd sorbed if the mononuclear complexes $CdCl^+$, $CdCl_2°$, $CdCl_3^-$, together with

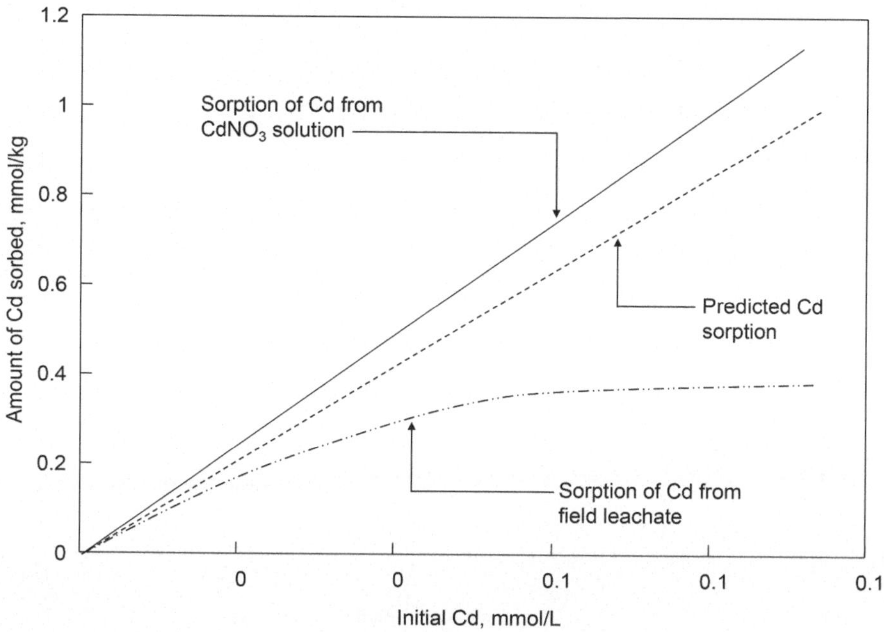

FIGURE 4.20 Sorption of Cd from $CdNO_3$ solution and from a landfill leachate by a kaolinite clay.

$Cd(OH)^+$ and $Cd(OH)_2$ (the hydrolysis of Cd) are formed. The difference between the predicted sorption curve and the actual Cd-leachate curve shown occurs because the other constituents in the landfill leachate also compete for sorption sites.

Speciation refers to the formation of complexes between heavy metals and ligands in the aqueous phase. When this happens, we obtain competition between the ligands and the soil solids for sorption of the heavy metals. In real waste leachates, we can expect to have all kinds of inorganic and organic ligands. These will most likely interfere with heavy metal sorption by the soil solids' surfaces because of the formation of complexes that are soluble, and yet unable to be sorbed onto the soil solids' surfaces. Thus, for example, Cl^- ions, sulfates, and organics may form complexes with heavy metals and interfere with their adsorption by the soil particles (Doner, 1978; Benjamin and Leckie, 1982; Garcia-Miragaya and Page, 1976). Doner (1978) shows that the transport of Ni, Cu, and Cd in leaching cell experiments can be as much as four times greater in the presence of Cl^- than in the presence of ClO_4^-, since ClO_4^- does not form complexes with these metals (Klanberg et al., 1963; Hester and Plane, 1964). Studies on Cd adsorption by soil particles indicate that the Cd which was not sorbed by the soil was in the form of $CdCl_2^\circ$, $CdCl_3^-$, and $CdCl_4^{2-}$.

Figures 4.21 and 4.22 show sorption of Cd by the same kaolinite soil from two control chloride solutions in relation to initial concentration and also in relation to pH.

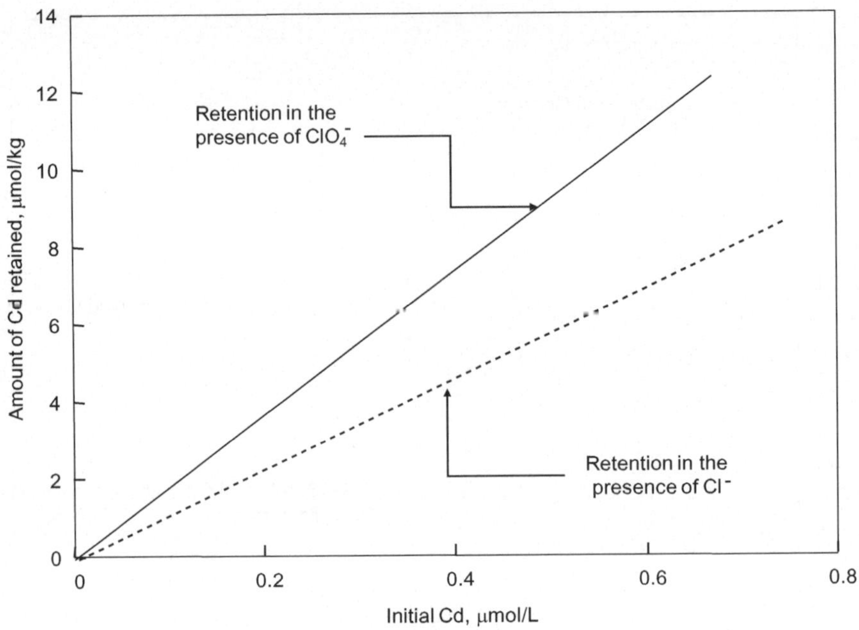

FIGURE 4.21 Retention of Cd by kaolinitic clay using two control chloride solutions—in relation to initial Cd concentration.

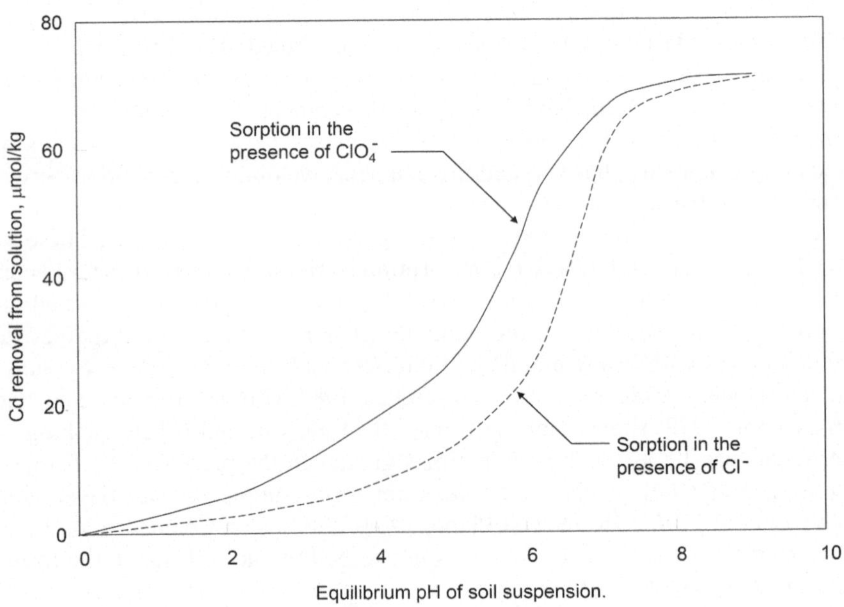

FIGURE 4.22 Sorption of Cd by kaolinite soil from two control chloride solutions in relation to pH.

The solutions of Cd made from $CdCl_2$ salt, used NaCl salt in the one instance and $NaClO_4^-$ in the other. The results demonstrate the effect of the mononuclear complexes formed. Research evidence shows that Cd^{2+} is sorbed more readily than $CdCl^+$ (Garcia-Miragaya and Page, 1976; Egozy, 1980; Benjamin and Leckie, 1982). There is evidence to indicate that ion pairs such as $CdCl^+$ are sorbed onto clay mineral surfaces (Sposito and Mattigold, 1979), and that neutral and negative ion pairs, such as $CdCl_2^\circ$ and $CdCl_3^-$ are not measurably sorbed by kaolinite. In regard to the speciation for the kaolinite clay, the amount of Cd not sorbed by the kaolinite soil particles in the presence of Cl^- is due to: (a) the decrease in activity due to NaCl, (b) competition from Na^+ for adsorption sites, and (c) complexation of Cd^{2+} as negative and neutral chloride complexes.

As the pH increases, there is a tendency for Cd to be removed from solution as hydroxides, or to precipitate onto the clay surface. The greater tendency for Cd to remain in solution in the presence of Cl^- than in the presence of ClO_4^- at pH values higher than the isoelectric point (IEP) arises because of the competition of Cl^- with OH^- for formation of complexes with Cd^{2+} which are unable to be sorbed by the kaolinite.

4.7 SOIL FRACTIONS AND SORPTION

The distribution of heavy metals (HMs) retained in a soil mass is a reflection of the sorption performance of each kind of soil fraction that comprises the total soil, i.e., the distributed HMs amongst the soil fractions is a direct function of the different sorption capabilities of the various soil fractions with exposed particle surfaces available for sorption. Knowledge of the distribution of partitioned HMs is useful since it allows us to assess or determine: (a) the sorption potential of candidate soils, (b) the fate and mobility of the sorbed HMs, and (c) the potential for removal of the sorbed HMs. It is clear that if we want to increase our capability for determination or assessment of the natural attenuation capability of soils, it is useful to obtain a better understanding and appreciation of how soil fractions participate in the many soil assimilative processes that result in the partitioning of heavy metals.

Common laboratory procedures available for determining the distribution of sorbed HMs include: (a) selective sequential extraction (SSE) techniques which are designed to selectively remove the HMs from target soil fractions, and (b) techniques relying on systematic inclusion or removal of soil fractions in heavy metal-soil interaction studies—systematic inclusion technique SSFA (selective sequential fraction addition), and systematic removal of soil fraction technique SSFR.

The SSE test is generally conducted using procedures similar to those for batch equilibrium studies. The SSFA test is a laboratory test which systematically "constructs" a soil sample to study the retention characteristics of the constructed or artificially prepared soil. Unlike the SSE samples, SSFA samples can be either compact soils or suspended solids. With this flexibility, assessment of the retention characteristics of the individual soil fractions can be made using leaching columns for the compact soils, and in a fashion similar to the SSE procedure for the suspended solids. This will be discussed in detail in the next section. The SSFR test samples can also be either compact soils or suspended solids. It is important to point out that removal of individual soil fractions from natural compact soils is fraught with considerable difficulty, and it is doubtful that complete removal of any one fraction can be realistically achieved.

4.7.1 Selective Sequential Extraction

The account given in Yong (2000) regarding SSE procedures provides the basis for the discussion in this Section. The underlying thought behind SSE procedures can be stated as follows: (a) If we have a way remove the HMs sorbed onto each kind of soil fraction, and if we could quantify the amounts of HM physically associated (bonded, retained, sorbed, etc.) with each type of soil fraction, then (b) we would have a better appreciation of how compositional features of a soil would influence the sorption or partitioning capability of a soil. Yong (2000) states that since the basic idea in application of the SSE procedure centers around the removal of sorbed HMs from individual soil fractions, the choice of chemical reagents is the key to selective extraction. Chemical reagents which are chosen are designed to selectively destroy the bonds established between heavy metal contaminants and specific individual soil fractions. There is no assurance that in destroying the contaminant-soil bonds, dissolution of individual soil fractions can be avoided.

In the general technique for sequential extraction, the extractant is mixed with the soil sample and the soil solution formed will be buffered to the desired pH as shown in the schematic illustration of the total procedure in Figure 4.23. Following

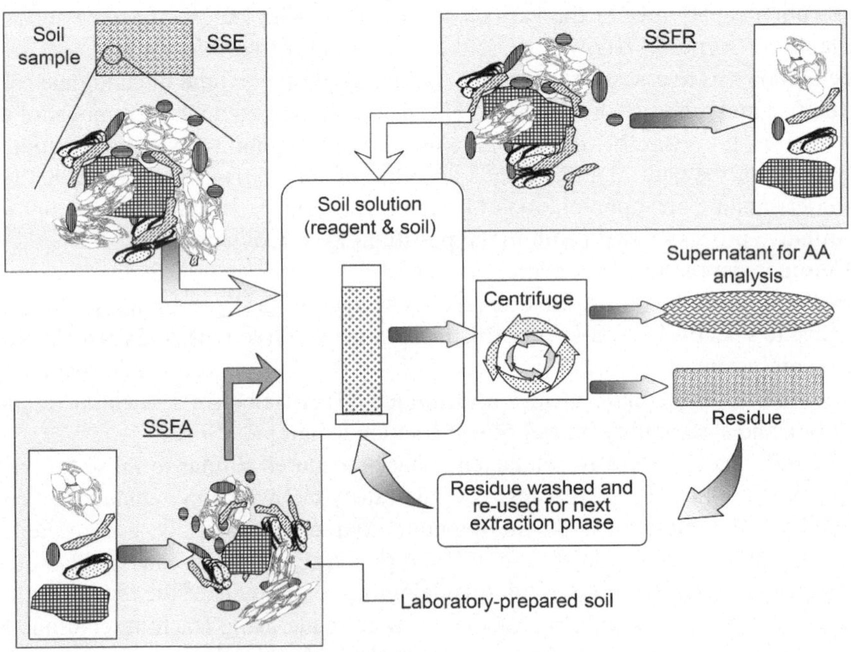

FIGURE 4.23 General laboratory procedure for implementation of SSE (selective sequential extraction), SSFA (selective sequential fraction addition), and SSFR (sequential fraction removal) procedures in determination of partitioning of heavy metals onto specific soil fractions.

equilibrium, the solution is centrifuged and the supernatant analyzed for heavy metals while the residue is washed with distilled water to ensure a "clean sample" is obtained for the next extractant treatment. The cycle is repeated until all the soil fractions are examined for their individual retention characteristics.

The published literature shows a variety of reagents used by different researchers in application of the SSE technique. Some of these are shown in Table 4.3. Also shown in the table is the technique proposed as a standard technique by the Bureau of Community Reference (BCR, now named the Measurements and Testing Programme). Ho and Evans (2000) have stressed the need for quality assurance in application of the SSE technique, and state that the BCR protocols nominally differentiates between: (a) acid-extractable (0.11 M acetic acid), (b) reducible (0.1 M hydroxylamine hydrochloride acidified to pH2 with nitric acid), and (c) oxidizable (1.0 M ammonium acetate extraction after oxidation by 8.8 M hydrogen peroxide). The numerals preceding each chemical treatment shown in each column of the table refer to the sequence of treatment. Figure 4.24 shows the treatment steps and the resultant product for Pb retention in an illitic soil. It is clear that the results obtained with respect to the distribution of partitioned HMs depend on the aggressiveness of the chemical reagents and the procedures used. As will be obvious, the results obtained from the SSE tests are operationally defined and should be considered only in a qualitative sense. These quantitative results are dependent on experimental techniques including sample preparation and conservation, solid to liquid ratio, extraction time and type and order of reactants. They can be used as a basis for comparison between different soils and contaminants, and particularly for developing an appreciation of the participation of the various soil fractions in the various processes leading to partitioning of the heavy metals.

There is some concern that redistribution of the extracted metals will occur during the sequential extraction stage. By this we mean that metals being extracted, for example, by the first reagent will be sorbed (re-sorbed) by the other soil fractions, thus giving lower values for extracted metals by the first reagent. It is further likely that a portion of the re-sorbed metals can and will be released upon application of the second reagent. This will mean higher amounts of released metals will be obtained in association with the second reagent—a result or an answer which is not really correct. The re-sorption study of sequentially extracted metals by Ho and Evans (2000) that included Pb, Cu, and Zn, showed that 20%–30% of the Cu and Zn released from the acid-soluble faction of the soil was re-sorbed by the reducible mineral oxide fraction. All of this reinforces the previous statement that the "measurement" of distribution of the partitioned HMs is operationally defined. It is nevertheless worthwhile repeating that while these measurements should be considered to be more qualitative than quantitative, they can nevertheless provide a good insight into the distribution of the partitioned heavy metals, and thus the mobility and transport of contaminants.

The chemical reagents used in the SSE technique are selected on the basis of their capability to release the heavy metal contaminants from specific soil solids, by destroying the bonds binding the heavy metals to the target soil fractions. These reagents are essentially extractants, and are classified as: concentrated inert electrolytes, weak acids, reducing agents, complexing agents, oxidizing agents, and strong acids (Tessier et al., 1979). It must be reiterated that the results obtained using the

TABLE 4.3
Summary of Some Extraction Procedures

Authors	Exchangeable	Bound to Carbonates	Bound to Fe-Mn Oxides	Bound to Organic Material	Residual
Tessier et al. (1979)	1-$MgCl_2$	2-NaOH/HOAc	3-$NH_2OH.HCl$ in 25% HOAc	4-H4-H_2O/HNO_3 + NH_4OAc	5-HF + $HClO_4$
Chester and Hugues (1967)	1-$NH_3OHCl + CH_3OOH$	2-$NH_3OHCl + CH_3COOH$	3-$NH_3OHCl + CH_3COOH$		
Chang et al. (1984)	1-KNO_3	4-Na_2EDTA		3-NaOH	5-HNO_3 (70°C–80°C)
Emmerich et al. (1982)	1-KNO_3	4-Na_2EDTA		3-NaOH	5-HNO_3
Gibson and Farmer (1986)	1-CH_3COONH_4 pH 7	2-CH_3COONa pH 5	3- 4-Hydroxylammonium + HNO_3/Acetic Acid	5-H_2O_2 + HNO_3 85°C	6-Aqua regia + HF + Boric acid
Yanful et al. (1988)	1-$MgCl_2$ + Ag thiourea	2-CH_3COONa + CH_3COOH	3-$NH_2OH.HCl$	4- + sulfides $H_2O_2 + HNO_3$	5-$HNO_3 + HClO_4$ + HF
Clevenger (1990)	1-$MgCl_2$	2-NaOAc/HOAc		3-HNO_3/H_2O_2	4-HNO_3 (boiled)
Belzile et al. (1989)	1-$MgCl_2$	2-CH_3COONa/ $NH_2OH.HCl/HNO_3$ Room temp.	3- Mn Oxide $NH_2OH.HCl/HNO_3$, NH_4OAC/HNO_3	4- + Sulf H_2O_2/HNO_3, NH_4OAc/HNO_3	
Guy et al. (1978)	1- (exch.+ adsor. + organic) $CaCl_2 + CH_3COOH$ + K-pyrophosphate	4- (carb. + adsor. + Fe-Mn nodules) $NH_3OHCl + CH_3COOH$	2- (metal oxides + org.) H_2O_2 + Diothinite + Bromoethanol		
Engler et al. (1977)	1- (exch.+ adsorb.) NH_4OAc	2-$NH_2OH.HCl$		3-H_2O_2/HNO_3	4-$Na_2S_2O_4$/HF/ HNO_3
Yong et al. (1993)	1-KNO_3	2-NaOAc pH 5	3-$NH_2OH.HCl$	4-H_2O_2 (3 steps)	5-HF/$HClO_4$ + HCl
BCR technique Ure et al. (1993) and Quevauviller et al. (1994)		1- HAc	2- NH_2OH-HCl pH 2	3-H_2O_2 pH2 -3, NH_4Ac pH 2	

Source: Yong, R.N., *Geoenvironmental Engineering: Contaminated Soils, Pollutant Fate and Mitigation*, CRC Press, Boca Raton, FL, 307, 2000.
1, 2, 3, 4, 5 indicates the sequence of the extraction.

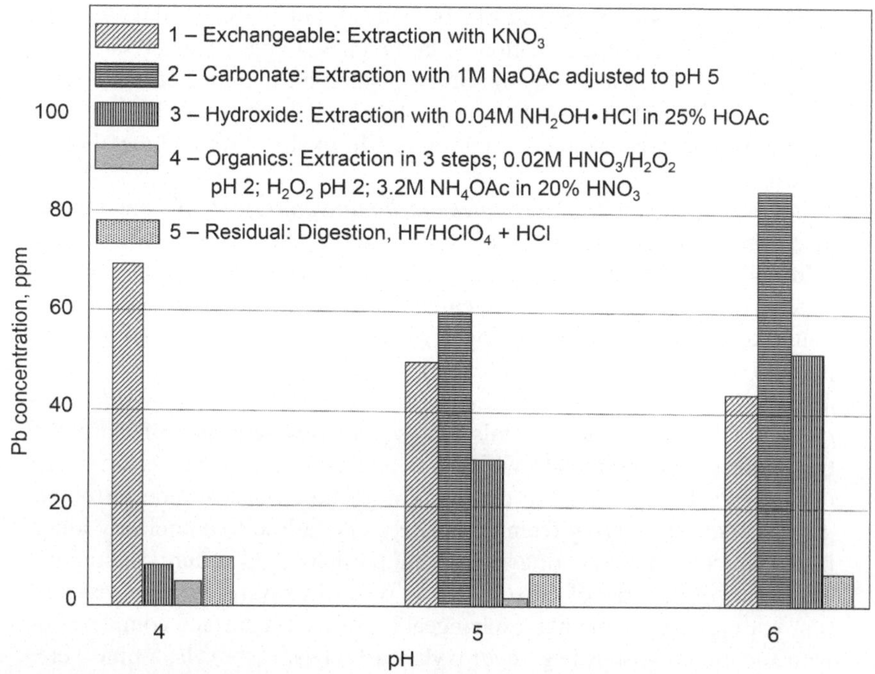

FIGURE 4.24 Distribution of retained Pb concentration for illite soil using SSE (selective sequential extraction). The various sequential extraction steps, from 1 to 5 are shown at the top part of the diagram. Note different amounts of retained Pb as a function of pH.

SSE technique are operationally defined. The general procedure in implementation of the SSE procedure requires one to begin extraction with the least aggressive extractant. The five different HM-bonding groups obtained as the operationally defined groups or phases include (Yong, 2000):

1. *Exchangeable Metals* (*exchangeable phase*): Metals extracted in this group are non-specifically adsorbed and ion exchangeable and can be replaced by competing cations. The soil fractions involved are mostly clay minerals, soil organics, and amorphous materials. Neutral salts such as $MgCl_2$, $CaCl_2$, KNO_3 and $NaNO_3$, are commonly used as ion-displacing extractants. These will promote release of metal ions attached by electrostatic attraction to the negatively charged sites on the soil particle surfaces. In the procedure used for example by Yong et al. (2001) for extraction of the HMs in the exchangeable phase from estuarine alluvium, 1 M KNO_3 was used, with pH adjusted to the natural pH of the soil.

   There is little evidence to suggest that dissolution of the soil solids occur because of the neutral electrolytes. Pickering (1986) showed that $MgCl_2$ sediment leachate contained only low levels of Al, Si, and organic carbon, confirming the weakness of the neutral salts in interaction with the clay surfaces, sulfides and organic matter. If the neutral salt solutions are applied at neutral pH, dissolution of Fe or Mn oxides is not expected, and we would only expect minimal dissolution of carbonates. Other types of salts such as NH_4Cl and NH_4OAc, may dissolve considerable amounts of compounds such as $CaCO_3$, $MgCO_3$, $BaCO_3$, and $MgSO_4$. If $CaSO_4$, and NH_4OAc are used, they can cause some dissolution of Mn-oxyhydrates and metal oxide coatings.

2. *Metals Associated with Carbonates* (*carbonate phase*): Metals precipitated or co-precipitated as natural carbonates can be released by application of acidified acetate as the extractant. A solution of 1°M HOAc-NaOAc (pH 5) is generally sufficient to dissolve calcite and dolomite to release the metals bound to them without dissolving organic matter, oxides, or clay mineral particle surfaces.
3. *Metal Contaminants Associated with Metal Oxides* (*hydroxide/oxide phase*): The contaminant metals released in this sequence of extractant treatment are those metals which are attached to amorphous or poorly crystallized Fe, Al, and Mn oxides. The metal oxides include ferromanganese nodules, ranging from completely crystalline to completely amorphous which occur as coatings on detrital particles, and as pure concretions. Their varying degree of crystallization result in several types of association with the heavy metals: exchangeable forms via surface complexation with functional groups (e.g., hydroxyls, carbonyls, carboxyls, amines, etc.) and interface solutes (electrolytes), moderately fixed via precipitation and co-precipitation (amorphous) and relatively strongly bound.

   The extractant selected for oxyhydrates should not attack either the silicate minerals or the organic matter. A good example is the one used by Chester and Hughes (1967)—i.e., a combination of an acid reducing agent (1°M hydroxylamine hydrochloride) with 25% (v/v) acetic acid for the extraction of ferromanganese oxides. Similarly, Yong et al. (2001) used 0.04 M $NH_2OH.HCl$ in 25% (v/v) HOAc with occasional agitation for 6 h at 96°C.
4. *Metals Associated with Organic Matter* (*organic phase*): The binding mechanisms for metals in association with organic matter include complexation, adsorption, and chelation. Because of the different types of binding mechanisms, some overlapping effects will be obtained with those methods designed to release exchangeable cations. However, the general technique used with respect to release of metal binding to organic matter is to obtain detachment of the metals as a result of oxidation of the organic matter. The oxidants are generally used at levels well below their (organic matter) solubilities.

   Yong et al. (2001) considered this as a three-phase procedure. In Phase (1), 3 mL of 0.02 M HNO_3 together with 5 mL of 30% H_2O_2 was adjusted to pH 2 with HNO_3 and agitated occasionally for 2 h at room temperature. For Phase (2), 3 mL of 30% H_2O_2 at pH 2 was used. This was intermittently agitated for 3 h at a temperature of 85°C. For Phase (3), 3 mL of NH_4OAc in 20% (v/v) HNO_3, diluted to 20 was used. This was continuously agitated at room temperature for 30 min.

5. *Metals Contained in the Residual Fraction*: This metal fraction is generally not considered to be significantly large. The metals are thought to be contained within the lattice of silicate minerals, and can become available only after digestion with strong acids at elevated temperatures. Determination of the metal associated with this fraction is important in completing mass balance calculations.

The information given in Table 4.3 indicates that as we progress from step 1 upward to step 5, the extraction procedures become very aggressive. This can be used as an indicator of the degree of retention capability, i.e., a qualitative measure of the irreversibility of sorption by the soil. The SSE results shown in Figure 4.24 refer to the shaded portion in Figure 4.18 indicating "Pb sorbed by illite." At pH values below 4, Pb is present in the solution as a free cation (Pb^{2+}), and the dominant mechanism for Pb retention is by cation exchange. These ions can be considered to be non-specifically adsorbed and are ion exchangeable, i.e., they can be replaced by competing cations. They are lesser in quantity when the precipitation pH of Pb is exceeded because other competing mechanisms begin to dominate. The amounts of Pb retained increases as pH increases. When the soil solution pH increases to a certain level, Pb begins to form hydroxy species—the beginning of Pb retention by the hydroxide fractions. Heavy metals tend to form hydroxy species in the presence of soil solids since the addition of an OH group on the metal reduces the free energy required for sorption. Sorbed meal ions hydrolyze more readily with deprotonation of coordinated water molecules—followed by hydroxy bridging and polymer formation onto soil (Farrah and Pickering, 1979). Hydroxy species of heavy metals form at different pH values, depending on the type of heavy metals—as shown for example in Figure 4.25 where the hydroxy species of Cd begin to form at about pH 6. Figure 4.26 shows the retention of Pb, Cu, and Zn by the estuarine alluvium tested by Yong et al. (2001). Since the pH of the combined test leachate and the soil was between pH 7 and pH 8, this demonstrates once again that the different hydroxy species form at different pH values—as witness retention as carbonates and hydroxides.

Most of the techniques for sequential extraction have been developed for cationic species. Since arsenic is mainly found in the anionic form, SSE procedures adopted follow those that have been used for studies of phosphorus retention. Arsenate, in particular, however does not exhibit the same stability as phosphorus does over a wide range of redox and pH values. Three procedures for sequential extraction were compared by Van Herreweghe et al. (2003). They included the BCR technique, scheme I (extractants of NH_4Cl, NH_4F, NaOH, diothionite-citrate-bicarbonate (DCB), and H_2SO_4 followed by concentrated HCl/.HNO_3/HF) and scheme II (water, anion exchange membranes, NH_4F, NaOH, DCB, and H_2O_2/HNO_3 followed by concentrated HCl/HNO_3/HF).

Schemes I and II, which were based on the phosphorus-like protocol, removed more arsenic than the BCR method. This was due mainly to the use of DCB for amorphous and crystalline oxide extraction as compared to hydroxylamine hydrochloride for amorphous oxide only extraction of arsenic. NaOH, used to remove the easily reducible Fe bound arsenic was found to extract most of the

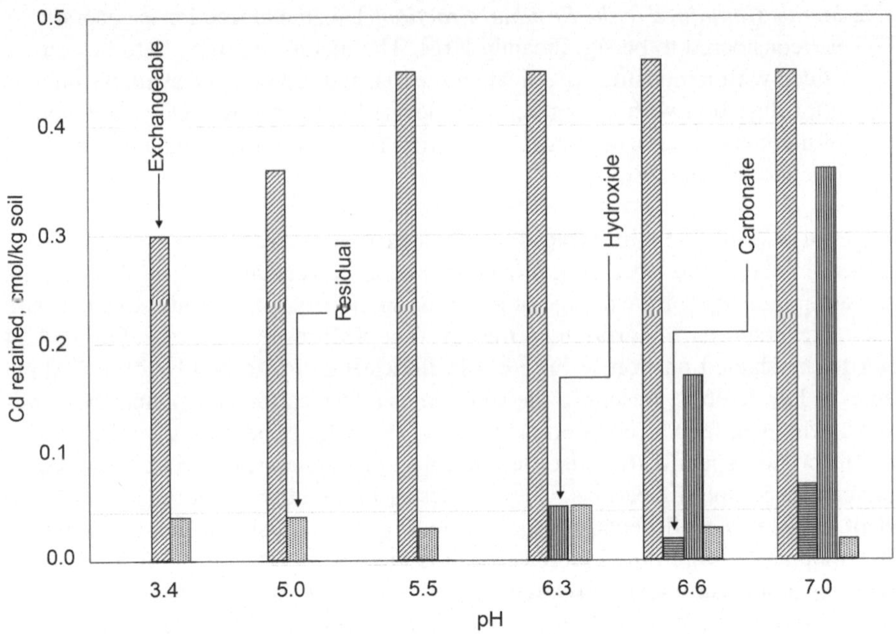

FIGURE 4.25 Distribution of Cd retained in a montmorillonite soil at various values of pH.

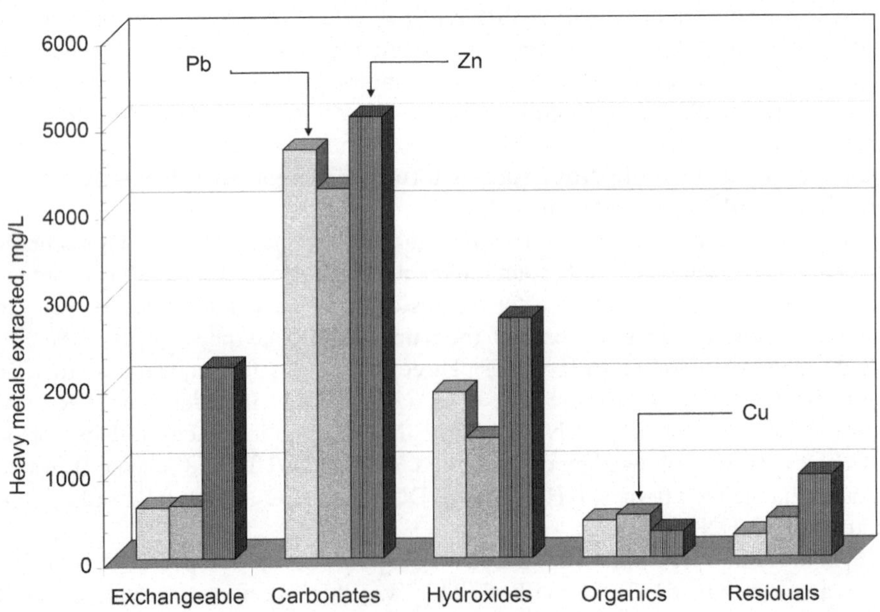

FIGURE 4.26 Estuarine alluvium retention of heavy metals. (Data from Yong, R.N. et al., *Eng. Geol.*, 60, 307–322, 2001.)

arsenic in highly contaminated samples. The anionic exchange membranes were not beneficial and should not be used.

Ashraf et al. (2012) using the Tessier (1979) SSE procedure to determine the sorption and retention of Pb, Cu, Cr, Zn, As, and Sn in ex-mining land soil consisting of sand-clay mixtures, determined that these were retained in greater quantities in the residue, and that the potential mobility of these contaminants, in decreasing order was Sn > Cu > Zn > Pb > Cr > As. What this means is that the environmental mobility for Cu for example was greater than Zn, or looking at it from the opposite view, retention of Zn was more robust than Cu. Although there is some resemblance to selectivity of metals for the ex-mining land soil, more work needs to be undertaken to fully determine such a process.

Yang et al. (2002) investigated the adsorption, sequestration, and bioavailability of arsenic in soils. They determined that at a pH of less than 6, more arsenic was adsorbed than at pH values above 6. In addition, it was established that iron oxide content had a significant effect on bioavailability. Dithionite-citrate-bicarbonate was used as the extracting agent to estimate bioavailability by simulating a child's digestive tract.

The results of the study on the influence of organic matter on the speciation of arsenic reported by Redman et al. (2002) showed that organic matter can influence both speciation and complexation of arsenic. The presence of organic matter decreased adsorption of both arsenate and arsenite by hematite (iron oxide). Furthermore, it was shown that organic matter was able to displace arsenic from hematite, thus enhancing mobility. Arsenite was desorbed more significantly than arsenate.

The use of the SSE techniques is not confined to conditions where contamination is solely by heavy metals. Mulligan et al. (2001) reported on the use of SSE technique on hydrocarbon and metal contaminated soils recovered from a site which contained leaking underground storage tanks from refineries and other associated oil products. A typical set of results is shown in Figure 4.27. It is not easy to distinguish between sorption-retention (physisorption and chemisorption) and precipitation-retention (i.e., retention by precipitation mechanisms). The Pb, Cu, Zn, or Cd precipitated or co-precipitated as natural carbonates can be released if the immediate environment is acidified.

4.7.2 Selective Sequential Fraction Addition or Removal

It is the source of the original sample that distinguishes between the SSE, SSFR, and SSFA types of test. The common thread throughout the extraction process begins with the soil solution and ends with the re-use of the cleaned residue for the next extraction step as shown in Figure 4.23. The SSE technique seeks to determine the retention characteristics of individual soil fractions from selective destruction of the bonds between the heavy metals and the target soil fraction. The SSFR technique selectively removes target soil fractions before proceeding with the extraction tests, and finally, the SSFA procedure selectively adds soil fractions for contamination and then proceeds forthwith with the extraction procedure.

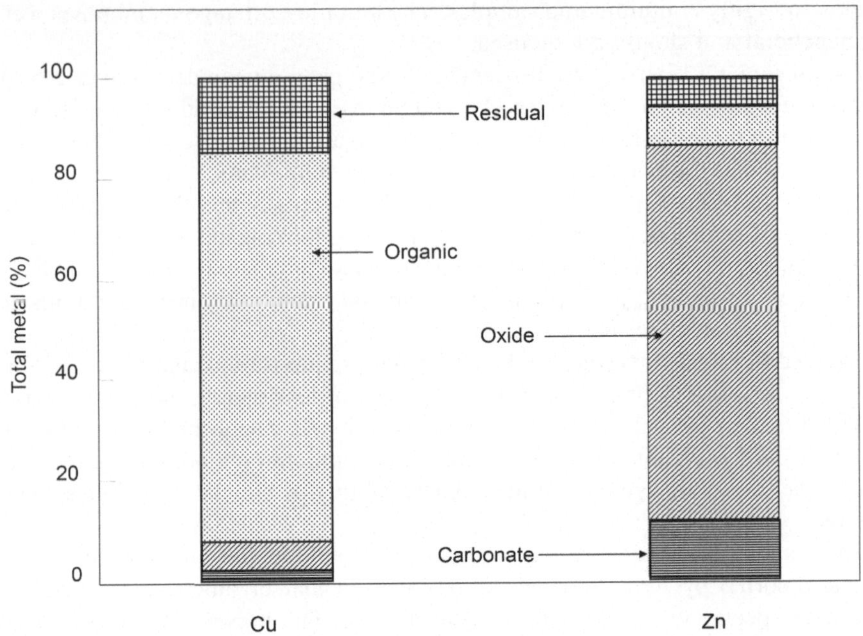

FIGURE 4.27 Selective sequential extraction (SSE) results for contaminated soil from site with leaking underground storage tanks. (Data from Mulligan, C.N. et al., The use of selective extraction procedures for soil remediation, in Adachi, K. and Fukue, M. (Eds.), *Clay Science for Engineering*, Balkema, Rotterdam, the Netherlands, 377–384, 2001.)

Yong and MacDonald (1998) used the selective sequential removal fraction technique for removal of carbonates and amorphous oxides from the soil. The procedure consisted of the following: (a) extraction of soil carbonates using 1 M NaOAc at pH°5 (adjusted with HOAc) at a soil:solution ratio of 40:1, and (b) extraction of amorphous oxides using 0.1 M oxalic acid, buffered to pH 3 by ammonium oxalate and mixed in the dark at the same soil:solution ratio. The procedure used by Xing et al. (1995) for removal of the various soil fractions was somewhat similar to that used by Yong and MacDonald (1998). For example, removal of carbonates was achieved by using 1 M NH_4Ac-HAc at pH 5. To distinguish between the amorphous Fe and the structural Fe (in the layer lattice), 0.2 M $(NH_4)_2C_2O_4$ + 0.2 M $H_2C_2O_4$ was used to extract the amorphous Fe whilst structural Fe was removed by the same extractant but with 0.1 M ascorbic acid added to the reagent. Extraction of the Fe was conducted in the dark. The results obtained by Xing et al. (1995) for retention of Cu, Cr, and Pb for a natural soil with selective fractions removed sequentially (SSFR technique) are shown in Figure 4.28. The results show that removal of the organics has a significant effect on the retention of Cr. The amount of Cr retained is considerably lessened. However, retention of Cu and Pb appear to be unaffected. When amorphous and crystal Fe oxides are removed from the soil, in addition to the removal of the exchange mechanisms, carbonates, Mn oxides and organics, we see a significant reduction in the retention of Cu. The amount of Pb does not seem to be as critically

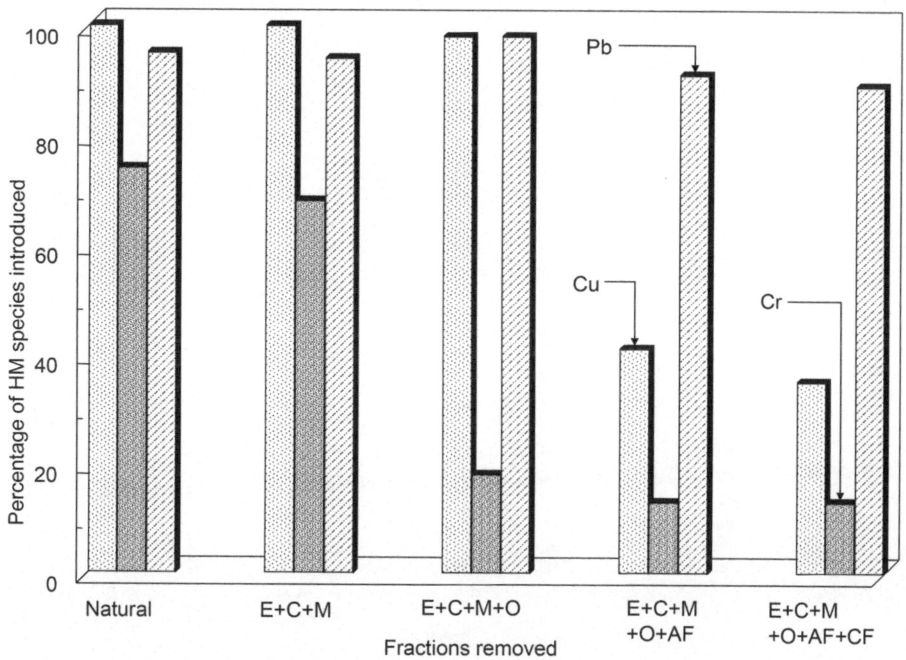

FIGURE 4.28 Results of selective soil fraction removal (SSFR) tests. The notations of E, C, M, O, AF, and CR refer to exchangeable, carbonates, Mn oxides, organics, amorphous Fe oxide, and crystal Fe oxide, respectively. (Data from Xing, G.X. et al., Role of amorphous Fe oxides in controlling retention of heavy metal elements in soils, in Huang, P.M. et al. (Eds.), *Environmental Impact of Soil Component Interactions: Metals, Other Inorganics, and Microbial Activities*, Lewis Publishers, Boca Raton, FL, 63–74, 1995.)

affected by the absence of these oxides. It is likely that precipitation of the metals, at this stage, is the primary mechanism available for their retention. Figure 4.29 shows the effect of pH on the retention of Pb where most of the precipitated Pb is associated with the carbonates.

Unlike the SSE and SSRF types of tests, the SSFA test constructs its soil sample from individual soil fractions, e.g., specified clay minerals, laboratory-produced amorphous oxides and carbonates, etc. In that sense, the SSFA technique is more suitable for studying the contribution of specific soil fractions to the retention characteristics of soil. SSFA tests could also be used as calibration or benchmark tests.

4.7.3 Influence of Particle Size on Retention of Metals

Up until now, we have been paying attention to the mechanisms involved in partitioning of heavy metals in soils where soil composition is the paramount factor. Much has been said about the available contact surface areas of the soil solids (particles) because these surfaces are the ones that in direct contact with the HM contaminants in the porewater. The results of the study reported by Liu et al. (2018) on four soils

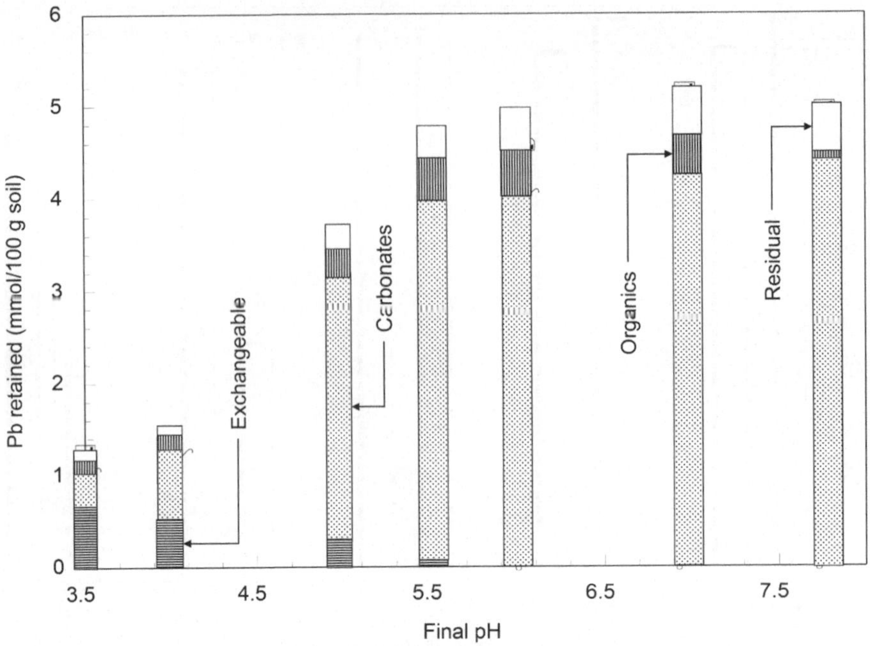

FIGURE 4.29 Retention of Pb by an illite soil—as determined from SSE (selective sequential extraction) test.

contaminated by heavy metals of geogenic and anthropogenic origin show that one could make a correlation between particle size distributions and retention or partitioning of HMs. Using the BCR SSE procedure, they conclude that partitioning of HMs in the different particle size fractions is closely related the composition of the particles, and that co-transport of HMs by HM-contaminated soil colloids could occur. The results of this study remind one that whilst one could use various sophisticated techniques to determine the kinds of HMs contaminating a candidate soil, care should be exercised to distinguish between anthropogenic and geogenic HMs. This means to say that one should have good information of the geogenic contaminants before embarking on detailed studies of assimilative capacities of soils and partitioning of contaminants.

4.8 CONCLUDING REMARKS

In this chapter, we have discussed the interactions between heavy metal contaminants and soil solids, together with the various reactions in contaminant-particle interactions that contribute to the assimilative capacity of soils. We consider these to be prime mechanisms and processes that contribute to the attenuation of these kinds of contaminants during their transport in soils. For natural attenuation to be used as a tool in managing ground contamination by heavy metals, we need to be concerned with the processes associated with partitioning and environmental mobility of these heavy metals. We have been primarily concerned with the interactions between soil particles and heavy metals, and particularly with the processes that lead to partitioning. Environmental mobility of these same metals will be discussed in Chapter 8 when we address the protocols and requirements for establishment of the lines of evidence and evidence of success of the monitored natural attenuation technique.

- At least five factors need to be considered in evaluating environmental mobility and bioavailability of heavy metal contaminants in the ground. These are: (1) hydrogeological setting, (2) changes in pH-regime of the affected ground, (3) changes in the ionic strength of the contaminated ground or soil-water system, (4) changes in the oxidation-reduction potential of the system, and (5) formation of complexes.
- The processes involved in the sorption and retention of heavy metals include ion exchange, precipitation as a solid phase (oxide, hydroxides, carbonates), and complexation reactions.
- With respect to clay soils, the two kinds of surface charge reactions responsible for most of the contaminant-particle interactions are (a) reactions associated with the net negative charge, and (b) pH-dependent reactions due to hydration of broken bonds.
- Not all of these processes result in retention, as for example, ion-exchange adsorption is, in actual fact, more of a retardation phenomenon than a retention phenomenon since desorption of contaminants sorbed by ion-exchange mechanisms can readily occur.
- The pH at which maximum adsorption of metals occurs can be expected to vary according to the first hydrolysis constant of the metal (cationic) ions.
- We note that metallic ions are generally coordinated (chemically bound) to water molecules, and that the water molecules which form the coordinating complex are the ligands.
- The characteristic feature of ligands is their free pairs of electrons.
- When the ionic activity of heavy metal solutes exceed their respective solubility products, precipitation of these metals as hydroxides and carbonates is likely.
- Alkaline conditions are favorable for precipitation of heavy metals. These will form new substances in the porewater or will be precipitated onto soil particles.
- Precipitation is a two-stage process: (1) nucleation, and (2) particle growth.
- Some of the key factors involved in precipitation of heavy metals include: (a) pH of the soil-water system, (b) concentration and species of heavy metals, (c) nature of inorganic and organic ligands, and (d) precipitation pH of the heavy metal contaminants.
- It is not easy to distinguish the various processes responsible for partitioning, since the various sorption mechanisms and precipitation all result in the removal of heavy metals from the porewater.
- Partitioning occurs as a result of mass transfer of the contaminants from the porewater of a soil-water system to the surfaces of the soil fractions.

- The two types of partitioning tests that appear to be the most popular are: (a) batch equilibrium adsorption isotherm tests, and (b) column or cell leaching tests.
- Batch equilibrium adsorption isotherm tests deal with soil solutions, and column or cell leaching tests deal with compact or intact (bulk) soil samples.
- Adsorption isotherms refer to characteristic curves relating adsorption of individual target contaminants by candidate soils to the available concentration of the target contaminant.
- The factors and processes involved in metal-soil interaction include: (a) mechanisms contributing to sorption of the HMs, (b) types of soil fractions involved in interaction with the HMs, (c) types and concentrations of the HMs, and (d) pH and redox environments.
- Selectivity in metal species sorption is different between soils. It is influenced by ionic size/activity, soil type and pH of the system.
- The amounts of HM retained is markedly affected by the presence of other metallic ions in the aqueous phase.
- Availability of reactive surfaces is a significant factor in determination of sorption capacity and selectivity.
- Speciation refers to the formation of complexes between heavy metals and ligands in the aqueous phase.
- The basic idea in application of the SSE procedure centers on the removal of sorbed HMs from individual soil fractions.
- Chemical reagents are chosen to selectively destroy the bonds established between heavy metal contaminants and specific individual soil fractions.
- The results obtained in SSE procedures are operationally defined.

REFERENCES

Agency for Toxic Substances and Disease Registry. (1995), *Division of Toxicology*, Atlanta, GA.

Ashraf, M.A., Maah, M.J., and Yusoff, I. (2012), Chemical speciation and potential mobility of heavy metals in the soil of former tin mining catchment, *Sci. World J.*, 2012, 125608.

Belzile, N., Lecomte, P., and Tessier, A. (1989), Testing readsorption of trace elements during partial chemical extractions of bottom sediments, *Environ. Sci. Technol.*, 23:1015–1020.

Benjamin, M.M., and Leckie, J.O. (1982), Effects of complexation by Cl, SO_4, S_2O_4, on adsorption behaviour of Cd on oxide surfaces, *Environ. Sci. Technol.*, 16:152–170.

Biddappa, C.C., Chino, M., and Kumazawa, K. (1981), Adsorption, desorption, potential and selective distribution of heavy metals in selected soils of Japan, *J. Environ. Sci. Health, Part B*, 156:511–528.

Bohn, H.L. (1979), *Soil Chemistry*, John Wiley & Sons, New York.

Bolt, G.H. (1979), *Soil Chemistry, Part B: Physico-Chemical Models*, Elsevier Scientific, Amsterdam, the Netherlands.

Brady, P.V., Cygan, R.T., and Nagy, K.L. (1998), Surface charge and metal sorption to kaolinite, in E.A. Jeanne (Ed.), *Adsorption of Metals by Geomedia*, Academic Press, San Diego, CA.

Butler, J.N. (1964), *Ionic Equilibrium: A Mathematical Approach*, Addison-Wesley, New York.

Cameron, R.E. (1992), Guide to site and soil description for hazardous waste site characterization, Volume 1: *Metals*. Environmental Protection Agency EPA/600/4-91/029.

Chang, A.C., Page, A.L., Warneke, J.E., and Grgurevic, E. (1984), Sequential extraction of soils heavy metals following a sludge applications, *J. Env. Qual.*, 13:33–38.

Chester, R., and Hughes, R.M. (1967), A chemical technique for the separation of ferromanganese minerals, carbonate minerals and adsorbed trace elements from Pelagic sediments, *Chem. Geol.*, 2:249–262.

Clevenger, T.E. (1990), Use of sequential extraction to evaluate the heavy metals in mining waste, *Water Air Soil Pollut.*, 50:241–254.

Coles, C.A., Rao, S.R., and Yong, R.N. (2000), Lead and cadmium interactions with Mackinawite: Retention mechanisms and role of pH, *Environ. Sci. Technol.*, 34:996–1000.

Darban, A.K. (1997), Multi-component transport of heavy metals in clay barriers, PhD thesis, McGill University, Montreal, Canada.

de Haas, F.A.M., and Zwerman, P.J. (1978), Pollution of soil, Chapter 10, in G.H. Bold and Bruggenwert, M.G.M. (Eds.), *Soil Chemistry: Basic Elements*, Elsevier Scientific Publishing, Amsterdam, the Netherlands.

de Matos, A.T., Fontes, M.P.F., da Costa, L.M., and Martinez, M.A. (2001), Mobility of heavy metals as related to soil chemical and mineralogical characteristics of Brazilian soils, *J. Environ. Pollut.*, 111:429–435.

Deng, B., Lan, L., Houston, K., and Brady, P.V. (2003), Effects of clay minerals on Cr(VI) reduction by organic compounds, *Environ. Monitor. Assess.*, 88:5–18.

Doner, H.E. (1978), Chloride as a factor in mobilities of Ni (II), Cu(II), and Cd(II) in soil, *Soil Sci. Soc. Am. J.*, 42:882–885.

Egozy, Y. (1980), Adsorption of cadmium and cobalt on montmorillonite as a function of solution composition, *Clays Clay Miner.*, 28:311–318.

Elliott, H.A., Liberati, M.R., and Huang, C.P. (1986), Competitive adsorption of heavy metals by soils, *J. Environ. Qual.*, 15:214–219.

Emmerich, W.E., Lund, L.J., Page, A.L., and Chang, A.C. (1982), Solid phase forms of heavy metals in sewage sludge-treated soils, *J. Environ. Qual.*, 11:178–181.

Engler, R.N., Brannon, J.M., Rose, J., and Bigham, G. (1977), A practical selective extraction procedure for sediment characterization, in T.F. Yen (Ed.), *Chemistry of Marine Sediments*, Ann Arbor Science Publisher, Ann Arbor, MI.

Evans, J.L. (1989), Chemistry of metal retention by soils, *Environ. Sci. Technol.*, 23:1046–1056.

Farrah, H., and Pickering, W.F. (1976a), The sorption of copper species by clays. I. Kaolinite, *Aust. J. Chem.*, 29:1167–1176.

Farrah, H., and Pickering, W.F. (1976b), The sorption of copper species by clays. II. Illite and montmorillonite, *Aust. J. Chem.*, 29:1177–1184.

Farrah, H., and Pickering, W.F. (1976c), The sorption of zinc species by clay minerals, *Aust. J. Chem.*, 29:1649–1656.

Farrah, H., and Pickering, W.F. (1977a), Influence of clay-solute interactions on aqueous heavy metal ion levels, *Water Air Soil Pollut.*, 8:189–197.

Farrah, H., and Pickering, W.F. (1977b), The sorption of lead and cadmium species by clay minerals, *Aust. J. Chem.*, 30:1417–1422.

Farrah, H., and Pickering, W.F. (1979), pH effects in the adsorption of heavy metal ions by clays, *Chem. Geol.*, 25:317–326.

Fassett, D.W. (1980), Cadmium, in H.A. Waldron (Ed.), *Metals in the Environment*, Academic Press, New York, pp. 61–110.

Fergusson, J.F. (1990), *The Heavy Elements: Chemistry, Environmental Impact and Health Effects*, Pergamon Press, Oxford, UK.

Fetter, C.W. (1993), *Contaminant Hydrogeology*, Macmillan Publishers, New York.

Forbes, E.A., Posner, A.M., and Quirk, J.P. (1974), The specific adsorption of inorganic Hg (II) species and Co (II) complex ions on goethite, *J. Colloid Interface Sci.*, 49:403–409.

Garcia-Miragaya, J., and Page, A.L. (1976), Influence of ionic strength and inorganic complex formation on the sorption of trace amounts of Cd by montmorillonite, *Soil Sci. Soc. Am. J.*, 40:658–663.

Gibson, M.J., and Farmer, J.G. (1986), Multi-step sequential chemical extraction of heavy metals from urban soils, *Environ. Pollut. B*, 11:117–135.

Guy, R.D., Chakrabarti, C.L., and McBain, D.C. (1978), An evaluation of extraction of copper and lead in model sediments, *Water Resour.*, 12:21–24.

Harter, R.D. (1979), Adsorption of copper and lead by Ap and B2 horizons of several northeastern United States soils, *Soil Sci. Soc. Am. J.*, 43:679–683.

Hester, R.E., and Plane, R.A. (1964), A Raman spectrophotometric comparison of interionic association in aqueous solutions of metal nitrates, sulphates and perchlorates, *Inorg. Chem.*, 3:769–770.

Ho, M.D., and Evans, G.J. (2000), Sequential extraction of metal contaminated soils with radiochemical assessment of readsorption effects, *Environ. Sci. Technol.*, 34:1030–1035.

Jacobs, L.W., Seyers, J.K., and Keeney, D.R. (1970), Arsenic sorption by soils. *Soil Sci. Soc. Am. Proc.*, 34, 750–754,

Jaworsky, J. (1978), *Effects of Lead in the Environment I & II*, National Research Council Canada.

Kinniburgh, D.G., Jackson, M.L., and Syers, J.K. (1976), Adsorption of alkaline earth, transition and heavy metal cations by hydrous oxide gels of iron and aluminium, *Soil Sci. Soc. Am. J.*, 40:796–799.

Kirk-Othmer (1992), *Arsenic and Arsenic Alloys. The Kirk-Othmer Encyclopaedia Technology*, Volume 3, John Wiley & Sons, New York.

Klanberg, F., Hunt, J.P., and Dodgen, H.W. (1963), Nuclear magnetic resonance studies on the resistance of perchlorate acid, *Inorg. Chem.*, 2:139–142.

Lewis, G.N. (1923), *Valences and the Structure of Atoms and Molecules*, The Chemical Catalogue, New York.

Lester, J.N. (1987), *Heavy Metals in Wastewater and Sludge Treatment Processes*, CRC Press, Boca Raton, FL.

Liu, G., Wang, J., Liu, X., Liu, X., Li, X., Ren, Y., Wang, J., and Dong, L. (2018), Partitioning and geochemical fractions of heavy metals from geogenic and anthropogenic sources in various soil particle size fractions, *Geoderma*, 312:104–113.

MacDonald, E. (1994), Aspects of Competitive adsorption and precipitation of heavy metals by a clay soil, ME thesis, McGill University, Montreal, Canada.

MacLean, J.E., and Bledsoe, B.E. (1993), Groundwater issue-behaviour of metals in soils EPA/540/S-92/018, October.

Maguire, M., Janece, S., Vimpany, I., Higginson, F.R., and Pickering, W.F. (1981), Influence of pH on copper and zinc uptake by soil clays, *Aust. J. Soil Res.*, 19:217–229.

Manahan, S.E. (1990), *Environmental Chemistry*, 4th ed., Lewis Publishers, Chelsea, MI.

Matthews, P.J. (1984), Control of metal application rates from sewage sludge utilization in agriculture. *Crit. Rev. Environ. Control.*, 14:199.

Mortland, M.M., and Raman, K.V. (1968), Surface acidity of smectites in relation to hydration, exchangeable cation, and structure, *Clays Clay Miner.*, 16:393–398.

Mulligan, C.N., Yong, R.N., and Gibbs, B.F. (2001), The use of selective extraction procedures for soil remediation, in K. Adachi and M. Fukue (Eds.), *Clay Science for Engineering*, Balkema, Rotterdam, the Netherlands, pp. 377–384.

Murray, J.W. (1975), The interaction of metal ions at the manganese dioxide solution interface. *Geochim. Cosmochim. Acta*, 39:505–519.

Nyffeler, U.P., Li, Y.H., and Santschi, P.H. (1984), A kinetic approach to describe trace-element distribution between particles and solution in natural aquatic systems, *Geochim. Cosmochim. Acta*, 48:1513–1522.

Pearson, R.G. (1963), Hard and soft acids and bases, *J. Am. Chem. Soc.*, 85:3533–3539.

Pearson, R.G. (1968), Hard and soft acid and bases, Part I, *J. Chem. Educ.*, 45:581–587.

Phadungchewit, Y. (1990), The Role of pH and Soil Buffer Capacity in Heavy Metal Retention in Clay Soils, Ph.D. Thesis, McGill University. Pickering,W.F. (1986), Metal ion speciation: Soils and sediments (a review), *Ore Geol. Rev.*, 1:83–146.

Pickering,W.F. (1986) Metal ion speciation: Soils and sediments (a review), *Ore Geol. Rev.* 1, 83–146.

Po, H.N., and Senozan, N.M. (2001), Henderson–hasselbalch equation: Its history and limitations, *J. Chem. Educ.*, 78:1499–1503.

Puls, R.W., and Bohn, H.L. (1988), Sorption of cadmium, nickel, and zinc by kaolinite and montmorillonite suspensions, *Soil. Sci. Soc. Am. J.*, 52:1289–1292.

Quevauviller, P., Rauret, G., Muntau, H., Ure, A.M., Rubio, R., Lopez-Sanchez, J.F., Fiedler, H.D., and Griepik, B. (1994), Evaluation of a sequential extraction procedure for the determination of extractable trace metal contents in sediments. *Fresenius J. Anal. Chem.*, 349(12):808–814.

Rai, D., Zachara, J.M., Eary, L.E., Girvin, D.C., Moore, D.A., Resch, C.T., Sass, B.M., and Schmidt, R.L. (1986), *Geochemical Behaviour of Chromium Species*, EPRI-4544, Batelle Pacific Northwest Laboratories, Washington, DC.

Ray, D., Sass, B.M., and Moore, D.A. (1987), Chromium hydrolysis constants and stability of chromium(III) hydroxide, *Inorg. Chem.*, 26:345–349.

Redman, A.D., Maclady, D.L., and Ahmann, D. (2002), Natural organic matter affects arsenic speciation and sorption onto hepatite, *Environ. Sci. Technol.*, 36:2889–2896.

Sauvé, S., Hendershot, W., and Allen, H.E. (2000), Soild-solution partitioning of metals in contaminated soils: Dependence on Ph, total metal burden, and organic matter, *Environ. Sci. Technol.*, 34:1126–1131.

Sawyer, C.N., McCarty, P.L., and Parkin, G.F. (1994), *Chemistry for Environmental Engineering*, 4th ed., McGraw-Hill, New York.

Schnitzer, M., and Skinner, S.I.M. (1967), Organo-Metallic interaction in soils: 7. Stability constants of Pb^{++}_, Ni^{++}_, CO^{++}_ and Mg^{++}_ fulvic acid complexes, *Soil Sci.*, 103:247–252.

Sposito, G., and Mattigold, S.V. (1979), *GEOCHEM: A Computer Programme for Calculation of Chemical Equilibria in Soil Solutions and Other Natural Water Systems*, University of California, Riverside, CA.

Stevenson, F.J. (1977), Stability constants of Cu^{2+}, Pb^{2+}, and Cd^{2+} complexes with humic acids, *Soil Sci. Soc. Am. J.*, 40:665–672.

Tessier, A., Campbell, P.G.C., and Bisson, M. (1979), Sequential extraction procedure for the speciation of particulate trace metals, *Anal. Chem.*, 51:844–851.

Ure, A.M., Quevauviller, P.H., Muntau, H., and Griepink, B. (1993), Speciation of heavy metal in soils and sediments. An account of the improvement and harmonisation of extraction techniques undertaken under the auspices of the BCR of the commission of the European Communities. *Int. J. Environ. Anal. Chem.*, 51:135–151.

U.S. EPA. (1984), Health assessment document for chromium final report, EPA Environment Criteria Assessment Office, Research Triangle Part, NC, p. 27711.

U.S. EPA. (2000). National primary drinking water regulations: Arsenic and clarifications to compliance and new source contaminants monitoring: Proposed rule. *Federal Register*, 65(121):38888. June 22. http://www.epa.gov/safewater/ars/arsenic.pdf.

Valocchi, A.J. (1984), Describing the transport of ion-exchanging contaminants using an effective k_d approach, *Water Resour. Res.*, 20:499–503.

Vance, D.B. (2001), *Arsenic—Chemical Behaviour and Treatment*, October 2001, http://2the4.net/arsenicart.htm.

Van Herreweghe, S., Swennen, R., Vandecasteele, C., and Cappuyns, V. (2003), Solid phase speciation of arsenic by sequential extraction in standard reference materials and industrially contaminated soil samples, *Environ. Pollut.*, 122:323–342.

Xing, G.X., Xu, L.U., and Hou, W.H. (1995), Role of amorphous Fe oxides in controlling retention of heavy metal elements in soils, in P.M. Huang, J. Berthelin, J.M. Bollag, W.B. McGill, and A.L. Page (Eds.), *Environmental Impact of Soil Component Interactions: Metals, Other Inorganics, and Microbial Activities*, Lewis Publishers, Boca Raton, FL, Chapter 6, pp. 63–74.

Yanful, E.K., Quigley, R.M., and Nesbitt, H.W. (1988), Heavy metal migration at a landfill site, Ontario, Canada—II: Metal partitioning and geotechnical implications, *Appl. Geochem.*, 3:623–629.

Yang, J.K., Barnett, M.O., Jardine, P.M., Basta, N.T., and Casteel, S.W., (2002), Adsorption, sequestration and bioaccessibility of As(V) in soils, *Environ. Sci. Technol.*, 36:4562–4569.

Yong, R.N., Warkentin, B.P., Phadungchewit, Y., and Galvez, R. (1990), Buffer capacity and lead retention in some clay materials, *Water Air Soil Pollut.*, 53:53–67.

Yong, R.N., and Sheremata, T.W. (1991), Effect of chloride ions on adsorption of cadmium from a landfill leachate, *Can. Geotech. J.*, 28:378–387.

Yong, R.N., Galvez-Cloutier, R., and Phadungchewit, Y. (1993), Selective sequential extraction analysis of heavy metal retention in soil, *Can. Geotech. J.*, 30:834–847.

Yong, R.N., and Phadungchewit, Y. (1993), pH influence on selectivity and retention of heavy metals in some clay soils, *Can. Geotech. J.*, 30:821–833.

Yong, R.N., and MacDonald, E.M. (1998), Influence of pH, metal concentration, and soil component removal on retention of Pb and Cu by an illitic soil, Chapter 10, in E.A. Jenne (Ed.), *Adsorption of Metals by Geomedia*, Academic Press, San Diego, CA.

Yong, R.N. (1999), Overview of partitioning and fate of contaminants: Retention, retardation and regulatory requirements, in R. Metcalfe and C.A. Rochelle (Eds.), *Chemical Containment of Waste in the Geosphere*, Geological Society, London, UK, pp. 1–20.

Yong, R.N. (2000), *Geoenvironmental Engineering: Contaminated Soils, Pollutant Fate and Mitigation*, CRC Press, Boca Raton, FL.

Yong, R.N., Yaacob, W.Z.W., Bentley, S.P., Harris, C., and Tan, B.K. (2001), Partitioning of heavy metals on soil samples from column tests, *Eng. Geol.*, 60:307–322.

Yong, R.N., Nakano, M., and Pusch, R. (2012), *Environmental Soil Properties and Behaviour*, CRC Press, Boca Raton, FL, 435 p.

5 Interactions and Fate of Organic Chemical Contaminants in Soils

5.1 INTRODUCTION

The discussions in Chapter 4 focused on the interactions between heavy metal contaminants and soil fractions. We have examined the various mechanisms of interactions and factors contributing to the assimilation of heavy metals (HMs) in soils, such as (a) retention and retardation of HMs, (b) sorption, (c) pE-pH environment, (d) precipitation and dissolution, (e) acid-base reactions, (f) oxidation-reduction processes, and (g) distribution and partitioning.

In this chapter we will extend this discussion to the interactions between organic chemicals and soil fractions—i.e., soil material making up the subsoil. Although some of the mechanisms and processes involved will be like those attributed to interactions between soil fractions and inorganic contaminants, others will be different due to the specific properties of organic chemical contaminants. Unlike heavy metal contaminants, organic chemical contaminants will degrade and can volatilize. We will be primarily concerned with the abiotic reactions between the organic chemicals and soil materials in this chapter. We recognize that biological reactions are major contributors to the processes which serve to attenuate organic chemical contaminants and have separated these for detailed discussion in the next chapter.

5.2 ORGANIC CHEMICAL CONTAMINANTS

Organic chemicals or compounds are defined generally as chemicals/compounds that contain carbon atoms with one or more of the carbon atoms bonded to other elements or radicals. It has often been said, for example, that compounds containing carbon and hydrogen number in the millions. They range from petroleum to unsaturated and aromatic hydrocarbons, organic halogen compounds, carbonyl compounds, phenols, etc.

5.2.1 Sources of Organic Chemical Contaminants

There are many sources of organic chemical contaminants found in the ground. Since the late 1970s, improved methods of detection have allowed us to measure or detect organic chemical contaminants in the range of parts per billion levels of contaminants in the soil and groundwater. This is particularly important since indiscriminate dumping of materials, bankrupt and abandoned manufacturing plants, insufficient waste storage, treatment and disposal facilities, spills and discharges from pipelines and other transport mechanisms, fertilizers and pest control aids, etc. have contributed to the contamination of many sites. It is reported that there are at least a million organic chemical compounds registered in the various Chemical Abstracts Services available, with many thousands of these in commercial use. The major organic chemicals found in the landscape include (a) those resulting from accidents and spills in the transport of petrochemicals; (b) agricultural fertilizers and pest control chemicals such as insecticides, pesticides, fungicides, etc.; and (c) inadvertent or deliberate discharge of organic aqueous wastes, organic liquids (solvents from dry cleaning), oils (lubricating oils, automotive oils, hydraulic oils, fuel oils) and organic sludges/solids (painting operations, tars from dyestuffs intermediates). Figure 5.1 and Table 5.1 show some of the sources of both inorganic and organic chemicals found in the ground.

Organic chemical wastes generated from manufacturing plants usually correlate with the amount of chemicals produced. Most soil-contaminating accidental spills and leaks originate from cleaning of equipment, residues left in used containers and outdated materials. Smaller generators of organic chemical contaminants include landfills that are improperly managed, automobile service establishments, maintenance shops, and photographic film processors. Household wastes that include

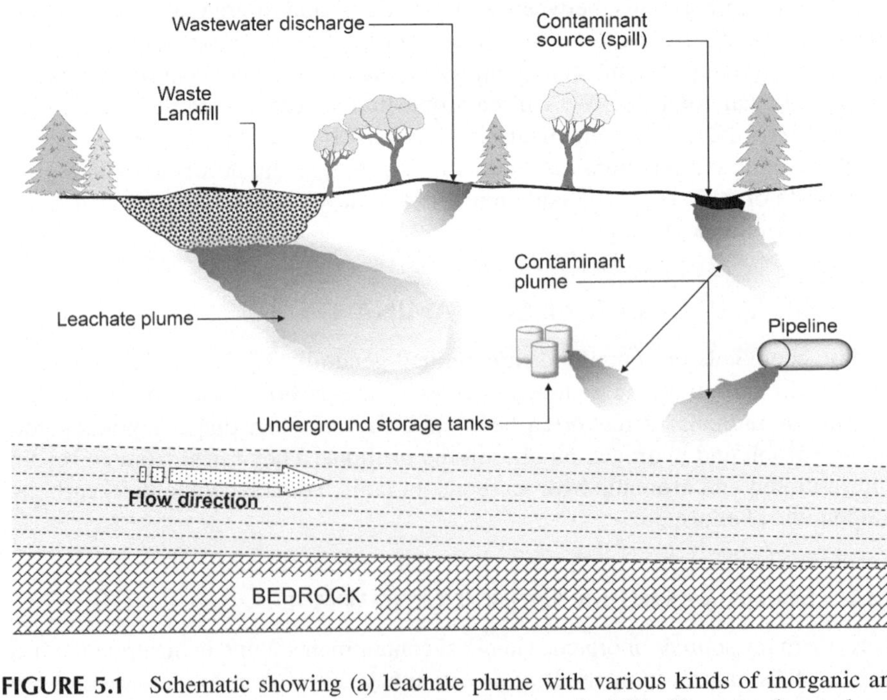

FIGURE 5.1 Schematic showing (a) leachate plume with various kinds of inorganic and organic chemical contaminants emanating from a waste landfill, (b) contaminant plumes from a leaking underground storage tanks and leaking pipeline, (c) a surface contaminant source (spill).

TABLE 5.1
Sources of DNAPL Contamination

Manufacturing	Disposal
Coal gas plants	Drum storage
Dry cleaning	Lagoon or ponds
Pesticide and herbicide manufacturing	Landfill disposal
Pipeline compressor stations	Metal degreasing
Steel industry	Paint stripping
Oil refining and transportation	Solvent loading and unloading
Wood preservation	Tool and die operations
	Underground tank storage

Source: Brar, G.S., Phytoremediation of chlorinated solvents: Progress and challenge, *IBC's 2nd Annual Conference on Innovative Technologies*, Boston, MA, 1997.

pesticides, paint products, household cleaners, and automotive products can also contribute significantly as sources of organic chemicals (LaGrega et al., 2001). For the next few sections, we will examine the natural attenuation processes that occur in soils contaminated by organic chemicals.

5.2.2 Functional Groups and Soil Particle Bonding

We can classify organic chemical compounds into families according to their reactivity, i.e., according to their functional groups, and since the chemical reactivities of individual members of a family are similar, we can study the reactions between chemical compounds and soil solids by studying the reactions of the functional groups of both the soil solids and the organic chemical compounds. This is a helpful because it would be virtually impossible to take into account the behavior and performance of each of the more than 10 million known organic chemicals when it comes to evaluation of their reactions in the soil-water system.

The functional groups of organic chemical compounds found as contaminants in soil are chemically reactive atoms or groups of atoms bound into the structure of the chemical compound. Some of the more common ones can be seen in Figure 5.2. Note that the functional groups shown in Figure 2.10 in Chapter 2 for natural organic matter in soils are like those found in synthetic organic compound contaminants shown in Figure 5.2. These functional groups are either acidic or basic. The factors that influence the adsorption of the organic chemicals by soil particles include the nature of the functional groups in the (organic) molecule, the shape and size of the molecule, configuration, polarity, polarizability, and water solubility. Organic molecules such as amines, alcohols, and carbonyl groups are positively charged by protonation (adding a proton or hydrogen). Accordingly, the surface acidity of the soil fractions becomes very important in the adsorption of these ionizable organic molecules.

It is useful to note that even without organic chemical contaminants present in the ground, we already have many of the same kinds of functional groups present in

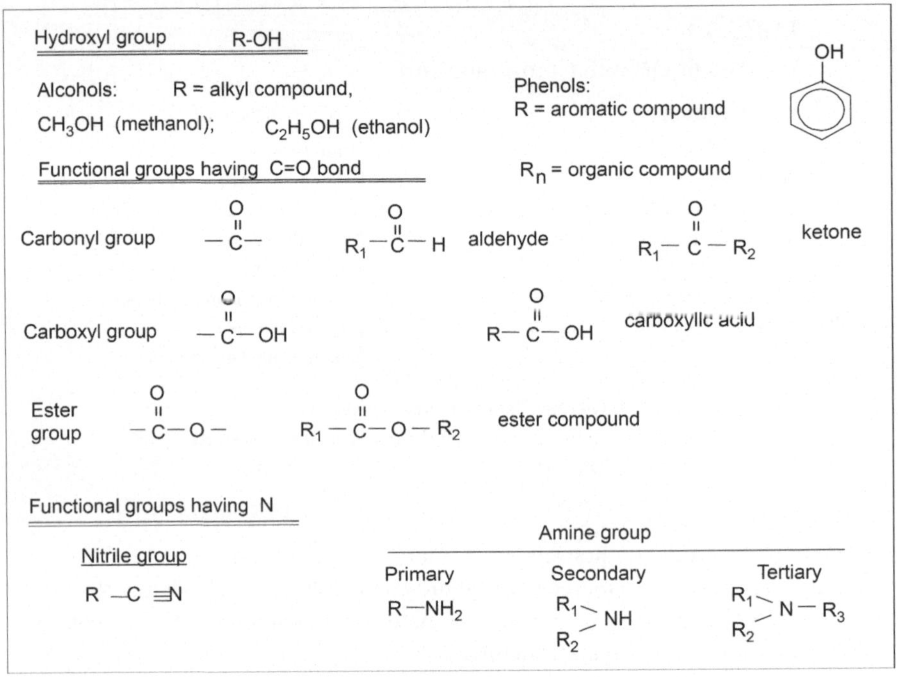

FIGURE 5.2 Some common functional groups for organic chemical contaminants.

the soil-water system without such chemicals. From Chapter 2, we have learnt that most of the soil fractions, such as clay minerals, amorphous silicate minerals, metal oxides, oxyhydroxides, and hydroxides have hydroxyl groups as their dominant functional group. However, we also learnt that it is the soil organic matter (SOM) that contains many of the same functional groups associated with organic chemicals, such as hydroxyls, carboxyls, carbonyls, amines, and phenols, as shown previously in Figure 2.10 in Chapter 2.

5.2.2.1 Hydroxyl Functional Group

For organic chemical contaminants, the hydroxyl functional group shown in the top left-hand corner of Figure 5.2 is present in two broad classes of compounds—the aliphatic hydroxy and the aromatic hydroxy compounds.

- *Alcohols* are aliphatic hydroxy compounds. These include methyl (CH_3 –), ethyl (C_2H_5 –), propyl (C_3H_7 –), and butyl (C_4H_9 –). Alcohols are hydroxyl alkyl compounds (R– OH), with a carbon atom bonded to the hydroxyl group. The more familiar ones are CH_3OH (methanol) and C_2H_5OH (ethanol)—as shown in Figure 5.2.
- *Phenols* are aromatic hydroxy compounds. They are compounds which possess a hydroxyl group directly attached to an aromatic ring. These compounds include monohydric (aerosols) and polyhydric (obtained by oxidation of acclimatized activated sludge such as pyrocatechol, trihydroxybenzene).

The summary account provided by Yong (2000) on the properties and characteristics of the main functional groups and chemicals shows that alcohols are considered to be neutral in reaction since the *OH* group does not ionize. In regard to interactions with soil fractions, adsorption of the hydroxyl groups of alcohol can be obtained through hydrogen bonding and cation-dipole interactions. Most primary aliphatic alcohols form single layer complexes on the negatively charged surfaces of the soil fractions, with their alkyl chain lying parallel to the surfaces of the soil fractions. Double-layer complexes are also possible with some short-chain alcohols such as ethanol. Alcohols act as acids when they lose their OH proton and will act as bases when their oxygen atom accepts a proton. The next section provides a more detailed discussion of the intermolecular interactions leading to bonding between organic compounds and soil fractions.

5.2.2.2 C=O Functional Groups

In the group of organic chemicals with carbon-oxygen double bonds (C=O, carbonyl functional group) such as aldehydes and ketones shown in the second group in Figure 5.2, we should note that the C=O bonds are polarized due to the high electro-negativity of the oxygen O relative to the carbon C. This is because of the greater electron density over the more electronegative oxygen atom. The C functions as an electrophilic site and the O is in essence a nucleophilic site. We could say that the electrophilic site is a Lewis acid and the nucleophilic site is a Lewis base. Reactions of compounds containing carbonyl functional groups will be nucleophilic additions. Aldehydes are considered to be more reactive than ketones in nucleophilic addition reaction. The addition of a nucleophile to the carbonyl group is defined as a nucleophilic addition reaction. Possible nucleophile candidates include the hydroxide ion (OH^-), hydride ion (H^-), water, ammonia (H_3N), carbon anions (R_3C^-) and alcohols (ROH), (McMurray, 1994).

Organic chemical contaminants with: (1) functional groups having a C=O bond, e.g., carboxyl, carbonyl, methoxyl, and ester groups, and (2) nitrogen-bonding functional groups, e.g., amine and nitrile groups, are fixed or variable-charged organic chemical compounds. They can acquire a positive or negative charge through dissociation of H^+ from or onto the functional groups—dependent on the dissociation constant of each functional group and the pH of the soil-water system. The fate of organic chemical contaminants can be significantly affected when a high pH regime replaces an original lower pH regime in the soil. As with the case of organic compounds with OH functional groups, a high pH regime will cause these functional groups (i.e., groups having a C=O bond) to dissociate. The release of H^+ (dissociation) would result in the development of negative charges for the organic chemical compounds, as shown for example by a carboxyl compound and an alcohol as follows:

$$R-COOH \leftrightarrow R-COO^- + H^+$$

$$R-OH \leftrightarrow R-O^- + H^+$$

where R represents any chemical structure (e.g., hydrocarbon moiety) and –COOH is the carboxyl functional group. If cation bonding is responsible for sorption between organic chemicals and the soil fractions, charge reversal (i.e., to negative charges) will result in the possible release of the organic chemical contaminant. When this happens, the released organic chemical contaminant could be sorbed by soil fractions possessing positive-charged surfaces—e.g., edges of kaolinites, oxides and soil organics. If such soil fractions are unavailable, the contaminants will be free to move. This situation is not desirable since it represents a classic case of environmental mobility of contaminants.

Carbonyl compounds (aldehydes, ketones, esters, amides, and carboxylic acids) are often obtained as products of photochemical oxidation of hydrocarbons. They most often possess dipole moments because the electrons in the double bond are unsymmetrically shared. Aldehydes have one hydrocarbon moiety (R) and a hydrogen atom (H) attached to the carbonyl C=O group as shown in Figure 5.2. They can be oxidized to form carboxylic acids. Ketones, on the other hand, have two hydrocarbon moieties (R and R_1) attached to the carbonyl group. While they can accept protons, the stability of complexes between carbonyl groups and protons is considered to be very weak.

The carboxyl group of organic acids (benzoic and acetic acids) can interact either directly with the interlayer cation or by forming a hydrogen bond with the water molecules coordinated to the exchangeable cation associated with the soil fractions. Adsorption of organic acids depends on the polarizing power of the cation. Because of their ability to donate hydrogen ions to form basic substances, most carboxyl compounds are acidic—weak acids as compared to inorganic acids. Carboxylic acids are essentially carbonyl compounds with the acyl group bonded to an electronegative atom. Oxygen, halogen, nitrogen, and sulfur are some of the electronegative atoms (McMurray, 1994). These acids are difficult to oxidize since they are end products and are also difficult to reduce.

5.2.2.3 Amino Functional Group

The amino functional group, NH_2, is found in primary amines. Much in common with alcohols, amines are highly polar and are more likely to be water soluble. Their chemistry is dominated by the lone-pair electrons on the nitrogen, rendering them nucleophilic. As shown in Figure 5.2, the amino group consists of primary, secondary, and tertiary amines depending on the nature of the organic compound R_n. They can be adsorbed with the hydrocarbon chain perpendicular or parallel to the reactive surfaces of the soil fractions, depending on their concentration. The phenolic functional group, which consists of a hydroxyl attached directly to a carbon atom of an aromatic ring, can combine with other components such as pesticides, alcohol, and hydrocarbons to form new compounds; e.g., anthranilic acid, cinnamic acid, ferulic acid, gallic acid, *p*-hydroxy benzoic acid.

5.2.3 Petroleum Hydrocarbons

The various petroleum fractions in petroleum hydrocarbons (PHCs) are primarily constituted by non-polar organics with low dipole moments (generally less than one), and dielectric constants less than three (see Figure 5.3 for some examples). Adsorption of non-ionic organic compounds by soil fractions is governed by the

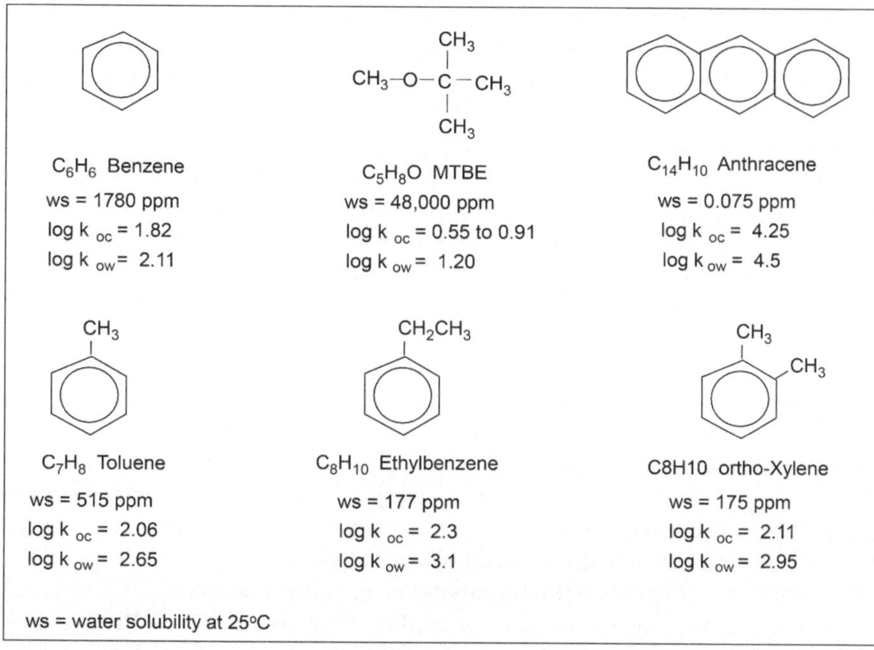

FIGURE 5.3 Chemical structure, water solubility, log k_{oc}, and log k_{ow} values of some petroleum hydrocarbons (PHC) and an additive methyl tert-butyl ether (MTBE). ws = water solubility at 25°C.

CH activity of the molecule; the *CH* activity arises from electrostatic activation of the methylene groups by neighboring electron-withdrawing structures, such as C=0 and C=N. Molecules possessing many C=0 or C=N groups adjacent to methylene groups would be more polar and hence more strongly adsorbed than those compounds in which such groups are few or absent.

The chemical structures of petroleum hydrocarbons such as monocyclic aromatic hydrocarbons (MAHs), and polycyclic aromatic hydrocarbons (PAHs), shown in Figure 5.3 for example, indicate the absence of electron-withdrawing units such as C=0 and C=N associated with the molecules. Accordingly, the PHC molecules would be weakly adsorbed (mainly by van der Waals adsorption) by the soil functional groups, and do not involve any strong ionic interaction with the various soil fractions.

Weakly polar (resin) to non-polar compounds (saturates and aromatic hydrocarbons) of PHCs develop different reactions and bonding relationships with the surfaces of soil fractions. Weakly polar compounds are more readily adsorbed onto soil surfaces in contrast to non-polar compounds. The adsorption of non-polar compounds onto soil surfaces is dominated by weak bonding (van der Waals attraction)—and is generally restricted to external soil surfaces primarily because of their low dipole moments (less than 1) and their low dielectric constants (less than 3) (Yong and Rao, 1991). Aqueous solubility and partition coefficients are important factors that control the interactions of organic compounds. Most hydrocarbon molecules are hydrophobic and have low aqueous solubilities. As shown in the next section, partitioning of PHCs onto soil surfaces occurs to a greater extent than in the aqueous phase. This results in lower environmental mobility and higher retention of the PHCs.

Studies on the desorption of PHCs using soil column leaching tests show that these can be desorbed as an aqueous phase or as a separate liquid phase (i.e., non-aqueous phase liquid—NAPL). Figure 5.4 shows the results of a leaching cell experiment with a clayey silt contaminated with 4% (by weight) PHC. The water solubility of the PHC is a significant controlling factor in determination of whether the PHC is desorbed as an aqueous phase or as a NAPL. As can be seen in Figure 5.3, the water solubility (ws) of the different PHC types varies considerably. The water solubility of an organic chemical contaminant is of significant importance in the control of the fate of the contaminant. Organic molecules, by and large, demonstrate less polar characteristics than water, and their varied nature (size, shape, molecular weight, etc.) render them as being considerably different than water. The water solubility of organic molecules will influence or control the partitioning of the organic contaminant, and the transformations occurring as a result of various processes associated with oxidation/reduction, hydrolysis and biodegradation.

The behavior of MTBE (methyl tert-butyl ether) differs substantially from other gasoline contaminants due to its water solubility. When the desorbed PHC remains as a NAPL, viscosity and surface wetting properties are critical. Light hydrocarbons are more likely to volatilize and be leached whereas heavier constituents will tend to be retained in the soil fractions.

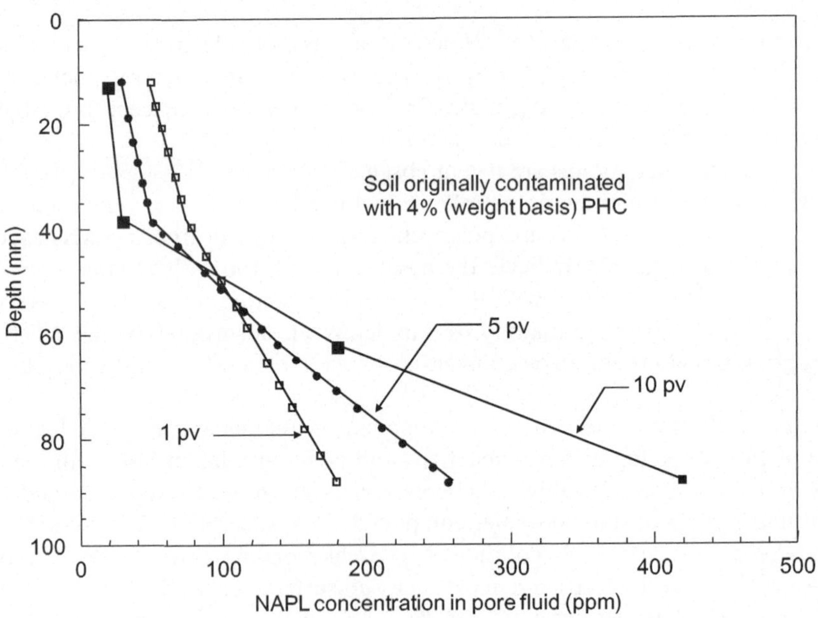

FIGURE 5.4 Results from leaching cell experiments on a clayey silt contaminated with petroleum hydrocarbons (PHC).

It is quite common to use a simple initial characterization scheme for organic chemical contaminants. This type of scheme would distinguish between organic acids/bases and non-aqueous phase liquids. The NAPLs are liquids that exist as a separate fluid phase in an aqueous environment and are not readily miscible with water. Because of the distinct differences in properties between the dense and less dense NAPLs, it is not uncommon to distinguish between NAPLs on the basis of densities greater than, or less than that of water. Thus, NAPLs with densities greater than water are called *Dense NAPLs* (DNAPLs) and those with densities less than water are called *Light NAPLs* (LNAPLs). Because the densities of DNAPLs are greater than water, they have a tendency to "plunge" all the way downward in the substrate until progress is impeded by an impermeable boundary (Figure 5.5).

The major organic chemicals that classify as DNAPLs found in contaminated soils include those associated with anthropogenic sources, e.g., chlorinated hydrocarbons such as polychlorinated biphenyls (PCBs), carbon tetrachloride, 1,1,1-trichloroethane, chlorophenols, chlorobenzenes, and tetrachloroethylene. The chemistry of the soil porewater plays a prominent role in the many processes that result in partitioning of the organic chemicals. The bonds established between organic chemical contaminants and soil fractions are controlled not only by the inorganic and organic ligands in the porewater, but also by the chemically reactive groups of the organic chemical contaminants and the soil fractions.

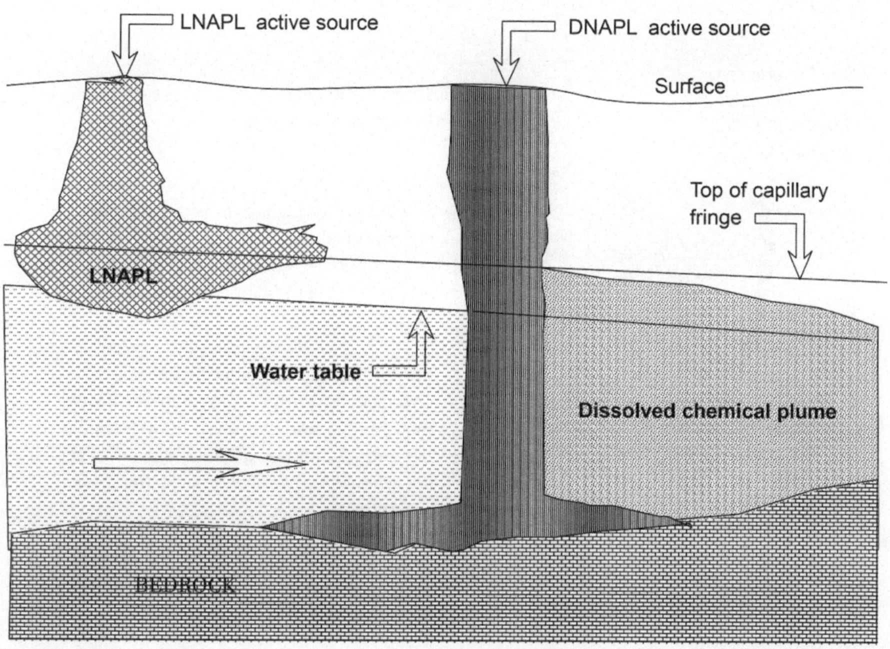

FIGURE 5.5 Comparison of LNAPL and DNAPL penetration from active sources. Note the influence of the water table on NAPL penetration. (From Yong, 2000). NAPL = non-aqueous phase liquid; LNAPL = light non-aqueous phase liquid; DNAPL = dense non-aqueous phase liquid; PHC = petroleum hydrocarbons.

Adsorption of organic cations is related to the molecular weight of the organic cations. Large organic cations tend to adsorb more strongly to clay soils in comparison to inorganic cations because of their greater lengths and higher molecular weights. Depending on how the functional groups are located and depending on the pH and chemistry of the soil-water system, these will influence the characteristics of their respective organic compounds and will thus contribute greatly in the development of the mechanisms which control accumulation, persistence and fate of these compounds in soil. These will be discussed in greater detail in the next section.

5.3 INTERMOLECULAR INTERACTIONS

Interactions at the intermolecular level contribute directly to the mechanisms that bind organic chemicals to soil fractions. These can be physical or chemical interactions and/or exchange-type mechanisms. Yong (2000) has provided a simple schematic (Figure 5.6) that summarizes these processes. The development in this and the next section follows the discussion presented by Yong (2000) who points out that although all of these are not included in the sketch, the basic sets of forces, reactions, and processes that constitute the major sets of interactions include three types of forces that classify as London-van der Waals forces. These are (a) Keesom forces developed as a result of instantaneous dipoles resulting from fluctuations in the electron distributions in the atoms and molecule, (b) Debye forces developed as a result of induction, and (c) London dispersion forces.

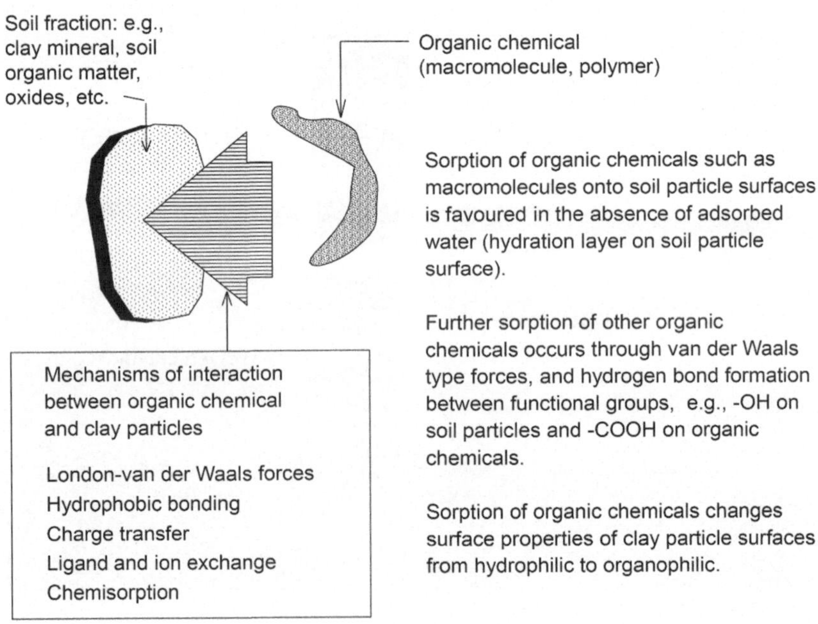

FIGURE 5.6 Some mechanisms of the interactions of organic contaminants and clay particles. (From Yong, R.N., *Geoenvironmental Engineering: Contaminated Soils, Pollutant Fate and Mitigation*, CRC Press, Boca Raton, FL, p. 307, 2000.)

While London-van der Waals influence decreases in proportion to the inverse of the sixth power of the separation distance R between molecules, i.e., proportional to $1/R^6$, the result of their interactions can lead directly to disruption of the liquid water structure immediately next to the soil solids. This leads to the development of entropy-generation hydrophobic bonding. Larger-sized organic molecules tend to be more favorably adsorbed because of the greater availability of London-van der Waals forces.

In regard to the reaction mechanisms depicted in Figure 5.6:

- Hydrophobic reactions contribute significantly to the bonding process between organic chemicals and soil fractions. The tendency for organic chemical molecules to bond onto hydrophobic soil particle surfaces such as soil organic matter, is in part because this will result in the least restructuring of the pre-existing water structure in the soil pores. This phenomenon allows for water in the vicinity of the organic chemical to continue its preference for association with itself (i.e., water-to-water attachment) as opposed to being in close proximity with the hydrophobic moiety of the organic chemical. This type of interaction results in the development of organic-soil particle bonding commonly called *hydrophobic bonding.*
- Charge transfers, or more specifically charge transfer complexes (of which hydrogen bonding is a special case) are formed between electron-donor and electron-acceptor molecules where some overlapping of molecular orbitals occurs together with some exchange of electron densities (Hamaker and Thomson, 1972). These transfer mechanisms appear to be involved in bonding between chemicals and soil organic matter because of the presence of aromatic groups in humic acids and humins. In the case of hydrogen bonding, the hydrogen atom provides the bridging between two electronegative atoms (Dragun, 1988) via covalent bonding to one and electrostatic bonding to the other (atom).
- For ligand exchange to occur as a sorption (binding) process, it is necessary for the organic chemical to have a higher chelating capacity than the replaced ligand. Humic acids, fulvic acids, and humins are important soil fractions in such exchanges and also in ion exchange phenomena. Since organic ions can be hydrophobic structure makers or breakers, the structure of water becomes an important factor in establishing the extent and rate of ion exchange sorption phenomena. As in the case of electrostatic interactions and chemical sorption between inorganic contaminants and soil fractions, the ionic properties of the organic ion are significant features that require proper characterization. This will be considered further when the influence of functional groups is examined.
- Ion exchange mechanisms involving organic ions are essentially similar to those that participate in the interaction between inorganic contaminants and soil fractions. Since molecular size is a factor, the structure of water immediately adjacent to soil particle surfaces becomes an

important issue in the determination of the rate and extent of sorption—similar to the processes associated with ligand exchange. Fulvic acids are generally hydrophilic and thus produce the least influence on the structuring of water. This contrasts considerably with humins which are highly hydrophobic, i.e., these play a highly important restructuring role in the water structure.

- The nature of chemical adsorption (chemisorption) is such that higher adsorption energies are obtained in reactions. These reactions can be either endothermic or exothermic, and generally require activation energies to surmount the energy barrier for reactions to occur between the molecules and the soil particles. A strong chemical bond is obtained when high exothermic heat of reaction is observed. The first layer is chemically bonded to the surface of the soil particles, and subsequent layers are held by van der Waals forces.

One should not assume or expect that bonding relationships between organic chemical contaminants and soil fractions at the intermolecular level are the result of any one single process. Recognizing that a soil mass consists of different types of soil fractions, it is clear that we will have different types of reactive surfaces. Accordingly, because of the variety in functional groups for both the organic chemical contaminants and the soil fractions, it is reasonable to expect that bonding between the organic chemical contaminants and soil will be comprised of more than one type of process—e.g., ion exchange and hydrophobic bonding.

5.4 ADSORPTION AND BONDING MECHANISMS

Adsorption reactions and/or processes involving organic chemicals and soil fractions are governed by: (a) the surface properties of the soil fractions, (b) the chemistry of the porewater, (c) microbial activity in the soil-water system, and (c) the chemistry and physical-chemistry of the contaminants. In general, organic chemical compounds develop mechanisms of interactions that are somewhat different from those discussed previously in Chapter 5 for inorganic contaminants. In respect to the transport of PHCs (petroleum hydrocarbons), in soils for example, it is important to develop an understanding of the interactions between oil and soil surfaces and the degree of retention of oil by the soil particle surfaces. This is not only because of our need to know how partitioning of the PHC occurs, but also to recognize the significant issue of bioavailability of the oil retained by the soil. By this, we mean that in regard to the possibility for biodegradation of the contaminant oil, it is important for us to determine the degree to which the contaminant is available for biologically-mediated degradation and transformations. The contaminant-soil interaction mechanisms are influenced by soil fractions, type (composition) of oil, micro-organisms, and presence of water. Because of their large surface areas, high surface charges and surface characteristics, surface active fractions in soil such as soil organic matter (SOM), amorphous materials and clays, can significantly enhance oil retention in soils.

Microorganisms can affect the sorption of hydrophobic components such as PAHs and pesticides by production of biosurfactants that can enhance desorption and increase bioavailability (Guerin and Boyd, 1992; Volkering et al., 1993; Mata-Sandoval et al., 2000; Voparil and Mayer, 2000; Sahnoun, et al., 2017). The latter parameter is highly difficult to quantify.

The studies reported by Noordman et al. (2002) on the effect of the biosurfactant from *Pseudomonas aeruginosa* on hexadecane degradation showed that the biosurfactant could enhance biodegradation if the process is rate-limited. For example, in the case of small soil pore sizes (6 nm), the hexadecane is entrapped and of limited availability. The rhamnolipid stimulates release of entrapped substrates (if mixing conditions are low such as in a column) and enhances uptake by cells (if the substrate is available). This could then become important in the stimulation of bacterial degradation under in situ conditions.

The problem of "*first wetting*" is most important in the case of organic chemical penetration into the subsurface soil. The nature of the liquid that first wets the soil solids or is made available to the dry surfaces of the soil fractions is critical. This has a considerable influence on subsequent bonding of contaminants to the soil solids—regardless of whether they are inorganic or organic. If we take alcohols for example, which have OH functional groups, these are directly coordinated to the exchangeable cations on soil mineral particle surfaces when these surfaces are dry. However, if the soil solids are wetted with water, since the cations are hydrated, the attachment of the alcohols to the soil particles surfaces will be through water bridging mechanisms.

If the surfaces of the soil solids are first wetted with water, the hydration layer will influence and affect soil-PHC bonding relationships. The amount of oil associated with the soil fractions will decrease in proportion to the amount of first wetting, i.e., in proportion to the thickness of the hydration layer. Because of the low aqueous solubilities and large molecular size of many organic chemicals, penetration into the outer Helmholtz plane and the Stern layer cannot be easily achieved—bearing in mind that the effective diameter of many different kinds of hydrocarbon molecules varies from 1 to 3 nm for a complex hydrocarbon type. In comparing this to a water molecule which has a diameter of approximately 0.3 nm, it can be clearly seen that the size difference is not in favor of the hydrocarbons in respect to attachment directly onto the soil particle surfaces when these surfaces are wetted with water. When determining the retention of hydrocarbons (HCs) and most NAPLs, it is prudent to examine or study the initial and subsequent wettings of the soil-engineered barriers and subsurface soil material.

Leaching tests with organic chemical contaminants and fluid conductivity experiments have often shown significant *shrinkage* in the soil samples tested. One explanation for this shrinkage has been suggested to be the inability of the diffuse double layers (DDL) to fully develop. The basis for this suggestion can be found in the nature of the counterions that form the DDL itself. By and large, interaction of clay minerals with organic chemicals with dielectric constants lower than water will result in the development of thinner interlayer spacing because of the *contraction* of the soil-water volume in the interlayer spaces in the soil-water system. Transport of organic molecules through the subsurface soil will be by diffusive means through the micropores, and by advection through the macropores. Partitioning between the pore-aqueous phase and soil fractions occurs in large measure in the micropores and to some extent in the macropores throughout the flow region. It would appear that in the macropores, the weakly adsorbed molecules can be re-mobilized if the advective velocities are sufficiently aggressive.

Hydrophobic substances such as heptane, xylene, and aniline which are well partitioned will develop resultant soil-organic chemical permeabilities that will be much lower than the corresponding soil-water permeability. By and large, organic fluid transport in soil is conditioned not only by the hydrophobicity or hydrophilic nature of the fluid, but also by other properties such as the dielectricity of the substance. This will be further evident from the examination of the partitioning of organic chemicals during, and as a result of, transport in the soil. For compounds such as MTBE that are difficult to biodegrade and do not sorb onto soil matter, dispersion and dilution could play a significant role in the attenuation of this compound (Anthony et al., 1999). Landmeyer et al. (1998) found that dilution and dispersion were much more significant for MTBE than for benzene. However, an analysis by Schirmer and Barker (1998) indicated that dispersion was not significant for a MTBE and benzene, toluene, ethylbenzene, and xylene (BTEX) contaminated site at Borden.

5.4.1 Partitioning and Organic Carbon Content

As considerable similarity exists between the composition of natural organic matter in soils and synthetic organic chemicals, it is therefore not surprising that the primary mechanism of organic adsorption is the hydrophobic bond between the synthetic chemical and natural organic matter. The amount of organic chemical sorbed can be estimated as follows—according to Karickhoff (1984):

$$C_S = k_d C_A = k_{oc} f_{oc} \tag{5.1}$$

where C_S and C_A represent the sorbed concentration of chemical and equilibrium aqueous concentration of the chemical respectively, $k_d = k_{oc} f_{oc}$ = partition coefficient, f_{oc} is the fractional organic carbon content of the soil and k_{oc} is the proportionality constant of the chemical. Good predictions are obtained when $f_{oc} > 0.001$.

Correlations relating k_{oc} to more commonly available chemical properties such as solubility and octanol-water partition coefficient, k_{ow} can be obtained (Chiou et al., 1982). The relationships obtained from various data and are usually expressed in a log-log form regression equation as follows:

$$\log k_{oc} = a + b \log S = m + n \log k_{ow} \tag{5.2}$$

where *a, b, m,* and *n* are constants. The octanol-water partition coefficient k_{ow} is defined as the ratio of the equilibrium concentration (*C*) of the dissolved candidate substance in two immiscible solvents (*n*-octanol and water) and is expressed as: $k_{ow} = C_{octanol}/C_{water}$. Chemicals with k_{ow} values of about less than 10 are considered to be relatively hydrophilic and tend to have high water solubilities and low soil adsorption coefficients. At the higher end of the scale, chemicals with k_{ow} values in excess of 10^4 are considered to be very hydrophobic.

Studies of the correlation of k_p with soil and contaminant factors have revealed good correlations between sorption and organic carbon content over a wide range of organic carbons content—from as low as 0.1% to nearly 20% of the soil. Typical values for k_{oc}, and k_{ow} are listed for a considerable variety of organic chemicals in the various handbooks on environmental data, e.g., Verscheuren (1983), Montgomery and Welkom (1991), Mackay et al. (1992), LaGrega et al. (2001).

5.4.2 Volatilization

Volatilization is an important attenuation process for volatile organic contaminants in subsurface as a free product, in the groundwater and in the vadose zone. Freshly spilled petroleum products such as gasoline can exhibit high rates of volatilization. Volatilization can occur from the free phase or dissolved phase. Henry's constant law describes volatilization from the dissolved phase. The rate of volatilization slows as the age of the spill increases. As a general guideline, a Henry's constant greater than 0.05 means that volatilization or off-gassing is likely while if it is less than 0.05, volatilization would be unlikely.

Volatilization involves first passing of the contaminant into the void space of the soil from the contaminated media. If the subsurface is shallow, then the volatile contaminant will be transported into the atmosphere. Free product in the vadose zone can be exposed to significant surface areas of soil air. The rate of volatilization will depend on Henry's constant, temperature, soil moisture, depth to the contaminated zone, sorption, soil porosity, and type of soil (Dragun, 1988; Larson and Weber, 1994) and is given by the following relationship:

$$C_a = HC_w \tag{5.3}$$

where C_a is the concentration in air (g/cm^3-vapor), C_w is the concentration in water (g/cm^3—water) and H is the Henry's law constant (dimensionless).

Constants for various compounds are shown in Table 5.2. Chlorinated compounds can have a wide range of constants, but many are similar to BTEX. It has been estimated by Chiang et al. (1989) that approximately 5% of benzene mass is lost due to volatilization. A fluctuating or shallow water table (<1 m) leads to higher rates of volatilization. It can also be seen that MTBE has a lower Henry's law constant compared to BTEX compounds (Squillace et al., 1997). Other oxygenates have similar values. However, equilibrium gas concentrations can still be higher than BTEX compounds because of the high-water solubility of MTBE (50,000 mg/L). PAHs have typically very low Henry's law constants. However, the lower the molecular weight, the higher is the value. Compounds such as naphthalene may volatilize from the capillary fringe or vadose zone (Rogers et al., 2002). Volatilization, however, is usually negligible, particularly compared to biodegradation as is the case for BTEX.

TABLE 5.2
Henry's Law Constants of Various Hydrocarbons

Chemical	Henry's Constant H (Dimensionless)
Benzene	0.22[a]
Toluene	0.26[a]
Ethyl benzene	0.32[a]
Xylene (isomers)	0.29[a]
o-xylene	0.2084[b]
m-xylene	0.3139[b]
p-xylene	0.3139[b]
Methyl tert-butyl ether (MTBE)	0.0240[b]
Ethyl tertiary butyl ether (ETBE)	0.11[c]
Tertiary amyl methyl ether (TAME)	0.081[c]
Diisopropyl ether (DIPE)	0.199[c]
Tetrachloroethylene (PCE)	1.21[d]
Trichloroethylene (TCE)	0.417[d]

Note: PAHs (LaGrega et al. 2001).
[a] Guide E 1979 Standard for Risk-Based Corrective Action (RBCA) at Petroleum Release Sites.
[b] Squillace et al. (1997).
[c] USEPA (1998).
[d] Wiedemeier et al. (1999).

5.5 PARTITIONING OF ORGANIC CHEMICAL CONTAMINANTS

Partitioning of organic chemical contaminants in a soil-water system is defined in the same manner as in the case of inorganic contaminants in Chapter 4. By this, we mean that partitioning of organic chemical contaminants occurs when a portion of the organic chemical contaminants in the porewater (aqueous phase) is removed from the aqueous phase and is assumed to be sorbed by the soil solids. The partitioning or division of organic chemical contaminants in soils is described by a coefficient identified as k_d, much similar to that used in the description of partitioning of HM contaminants in Chapter 4. As defined previously, this coefficient refers to the ratio of the concentration of contaminants held by the soil fractions to the concentration of contaminants remaining in the porewater (aqueous phase), i.e., $C_s = k_d C_w$, where C_s refers to concentration of the organic contaminants sorbed by the soil fractions and C_w refers to the concentration remaining in the aqueous phase (porewater), respectively.

5.5.1 Adsorption Isotherms

Partitioning of organic chemical contaminants in the subsurface soil is the result of a variety of interaction mechanisms and processes occurring between the contaminants and soil fractions in the soil-water system. Determination of the partition coefficient k_d for organic chemical contaminants from laboratory tests cannot fully replicate the processes that are associated with microbial activities. Only results from interacting physico-chemical mechanisms between the chemicals and soil solids in partitioning are obtained in laboratory tests. The partitioning coefficient k_d is generally obtained using procedures similar to those described in Chapter 3 in respect to adsorption isotherms. The soil-suspension tests utilize target contaminants and specified (or actual) soil fractions. Figure 5.7 shows 3 "classes" of adsorption isotherms describing the partitioning behavior of organic chemicals.

The general Freundlich isotherm given previously is used to characterize the three classes, i.e., $C_s = k_1 C_w^n$. To avoid confusion with the isotherms used previously in the inorganic contaminant sorption tests, we will use the relationship shown in the graph depicted in Figure 5.7. As before, denoting C_s and C_w as the organic chemical sorbed by the soil fractions and remaining in the aqueous phase respectively, the k_1 and n terms are better known as Freundlich constants. These were identified as k_1 and k_2 respectively in the partitioning tests for inorganic contaminants. The parameter n is associated with the nature of the slope of any of the curves shown in Figure 5.7. When $n = 1$, linearity is obtained, and one concludes that the sorption of the chemical contaminant by the soil fractions is a constant proportion of the available contaminant. When $n < 1$, the sorbed chemical contaminant decreases proportionately as the available contaminant increases, suggesting therefore, that all available mechanisms (processes) for sorption are being exhausted. However, when $n > 1$ we obtain the reverse situation. For such a situation to exist, enhancement of the sorption capacity of the soil must result from sorption of the chemical

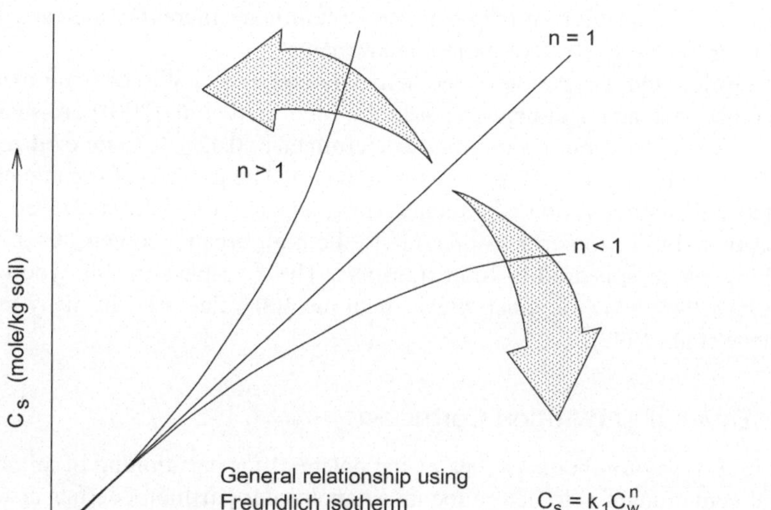

FIGURE 5.7 Influence of the n value from the Freundlich adsorption isotherms for organic chemicals sorbed onto soil fractions.

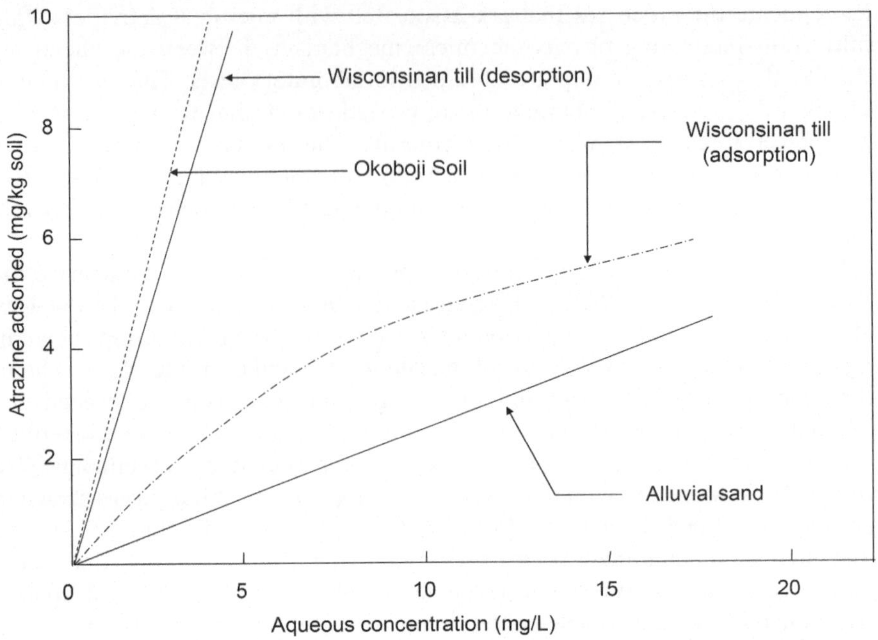

FIGURE 5.8 Adsorption of atrazine on various soils. Solid lines represent adsorption. The dashed line represents desorption from the Wisconsinan oxidized till. (Moorman, T.B. et al., *Soil Science*, 166, 921–929, 2001.)

contaminant, i.e., sorption of the chemical contaminant increases the capacity of the soil to proportionately sorb more contaminants.

The sorption and desorption of the herbicide atrazine in Wisconsinan oxidized till and other soils and sediments reported by Moorman et al. (2001) are shown in Figure 5.8. The k_d adsorption values were determined as 0.43 ± 0.25 for oxidized till, 0.51 ± 0.02 for loess and 0.55 ± 0.24 for alluvium. The amount of organic carbon correlated well with sorption of atrazine. However, unoxidized tills showed levels of adsorption similar to surface soils due to the high organic content (0.63%) and mineralogy which appeared to limit transport. The k_{oc} values in this type of soil were 1198 ± 943, which is much higher than the 100 value used in many models (Wauchope et al., 1992).

5.5.2 Equilibrium Partition Coefficients

A very useful way to obtain an idea of the potential for partitioning of an organic chemical contaminant is to determine its octanol-water partition coefficient k_{ow}. To better appreciate what this means, we need to consider the *equilibrium partition coefficient*, i.e., coefficient pertaining to the ratio of the concentration of a specific organic contaminant in other solvents to that in water. The partitioning process is studied in terms of the relative fugacity of the organic compound in an organic solvent and in water. This says that the ratio or coefficient k_{os} will be obtained as: $k_{os} = C_{os}/C_w$, where C_{os} is the concentration of the organic chemical in the organic solvent, and C_w is the concentration of the organic chemical in water. A popular organic solvent used for determination of this coefficient is *n*-octanol. One of the reasons for this is its very low solubility in water—as will be discussed later. Hence, the term octanol-water partition coefficient k_{ow}. This coefficient is considered to be well correlated to water solubilities of most organic chemicals. Chiou et al. (1977), for example have reported good correlations between solubilities of organic compounds and their *n*-octanol-water partition coefficient k_{ow}. Since *n*-octanol is part lipophilic and part hydrophilic (i.e., it is amphiphilic), it can accommodate organic chemicals with various kinds of functional groups such as those shown in Figure 5.2. As stated earlier, one of the reasons for the choice of *n*-octanol as the organic solvent for these tests is because of its very low solubility in water. The dissolution of *n*-octanol in water is roughly eight octanol molecules to 100,000 water molecules in an aqueous phase—a ratio of about one to twelve thousand (Schwarzenbach et al., 1993). Since water-saturated *n*-octanol has a molar volume of 0.121 L/mol as compared to 0.16 L/mol for pure *n*-octanol, this close similarity allows us to ignore the effect of the water volume on the molar volume of the organic phase. This is particularly important in experiments conducted to determine the octanol-water equilibrium partition coefficient.

The relationship for the *n*-octanol-water partition coefficient k_{ow} given in terms of the solubility S (Chiou et al., 1982) is seen in Equation (5.4):

$$\log k_{ow} = 4.5 - 0.75 \log S \text{ (ppm)} \tag{5.4}$$

The k_{ow} octanol-water partition coefficient has been widely adopted as a significant parameter in studies of the environmental fate of organic chemicals. Concerns have been raised over the accuracy of k_{ow} values for Dichlorodiphenyltrichloroethane (DDT) and 1,1-dichloro-2,2-bis(p-chlorophenyl)ethylene (DDE) and Pontolillo and Eganhouse (2001) have cautioned that one should carefully consider the data source. Log values range from 4.9 to 6.9 for DDT and from 4.3 to 7.0 for DDE. It has been found to be sufficiently correlated not only to water solubility, but also to soil sorption coefficients. In the experimental measurements reported, the octanol is considered to be the surrogate for soil organic matter. Organic chemicals with low k_{ow} (e.g., less than 10) may be considered to be relatively hydrophilic. They tend to have high water solubilities and small soil adsorption coefficients. Conversely, chemicals with a high k_{ow} value (e.g., greater than 10^4) are very hydrophobic and are not very water soluble (i.e., they have low water solubilities). Solvent systems which are almost completely immiscible (e.g., alkanes-water) are fairly well behaved and if the departures from ideal behavior exhibited by the more polar solvent systems are not too large, a thermodynamic treatment of partitioning can be applied to determine the distribution of the organic chemical without serious loss of accuracy.

Studies on the aqueous concentrations of hydrophobic organics such as PAHs, and compounds such as nitrogen and sulfur heterocyclic PAHs, and some substituted aromatic compounds indicate that the accumulation of the hydrophobic chemical compounds is directly correlated to the organic content (SOM) of a soil. This is not surprising since a large proportion (by weight) of SOM is carbon. This has been shown in Figure 2.10 in Chapter 2—i.e., the SOM functional groups are similar to most of the organic chemicals. They (SOM) occupy a position in between water and hydrocarbons insofar as polarity is concerned. Because of their composition and structure, they are well suited for hydrophobic bonding with organic chemical contaminants.

Studies have shown that whereas the variability in sorption coefficients between different soils may be due to the attributes of soil fractions (surface area, cation exchange capacity, pH, etc.), and the amount and nature of the organic matter present, a good correlation of sorption can be obtained with the proportion of organic carbon in the soil. The partition coefficient k_{ow} can be related to the organic content coefficient k_{oc}. The organic carbon content in soil organic matter can be used to characterize the k_d performance. Amongst the relationships commonly used are (Olsen and Davis, 1990):

$$k_{oc} = \frac{k_d}{f_{oc}} = 1.724 k_{om} \tag{5.5}$$

where f_{oc} refers to the organic carbon content (dimensionless) in the SOM, and k_{om} refers to the partition coefficient expression using the (soil) organic matter content. Values for k_d and k_{oc} for numerous organic chemicals can be found in the various handbooks detailing the environmental data for such chemicals. A sampling of these (log k_{oc}) can be seen in Figure 5.9. Soils with very low organic matter content

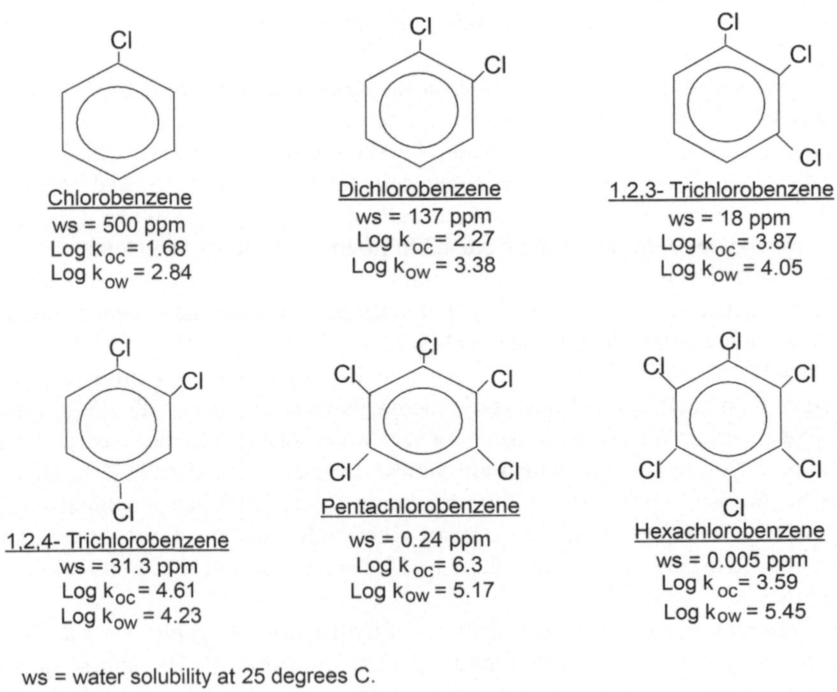

FIGURE 5.9 Water solubility, log k_{oc} and log k_{ow} values for some chlorinated benzenes. ws = water solubility at 25°C.

(less than one percent by weight) will tend to give high values for k_{oc} because of the competing sorption processes offered by the other soil fractions in the soil. Because of that, Equation (5.6) should not be used when $f_{oc} < 1$. McCarty et al. (1981) gives a critical minimum level for the organic carbon content f_{oc-cr} as:

$$f_{oc-cr} = \frac{SSA}{200(k_{ow})^{0.84}} \tag{5.6}$$

where SSA denotes the specific surface area of the soil.

Benzene, toluene, ethylbenzene, and xylene (BTEX) compounds have different batch equilibrium isotherms (Stuart et al., 1991; Kemblowski et al., 1987; Zytner, 1994) and different mobilities in column studies (Stuart et al., 1991, Angley et al., 1992; Larsen et al., 1992) and even in some groundwater aquifers (Odermatt 1994) different mobilities for each of these compounds have been obtained. Retardation factors determined based on k_{oc} values show values that can be predicted as benzene > toluene > ethylbenzene and xylenes. This order has been postulated by Ptacek et al. (1987) in a sandy aquifer with a low fraction of organic carbon content (<0.02%). Whilst higher organic contents can enhance sorption significantly (Zytner, 1994), sorption is also possible on mineral surfaces (Rogers et al., 1980). The retardation R (see Equation [5.10]) values for MTBE have also been estimated by field studies (Schirmer and Barker, 1998) and have been found to be not as significant as for BTEX. Mobility of MTBE is much higher than for BTEX since sorption is insignificant. Mechanisms of adsorption by MTBE are by adsorption onto mineral surfaces such as clays, and then onto organic matter. Equations normally used to determine typical adsorptions cannot be used since the adsorption is not only a surface phenomenon but also a solvation process or dissolving into the organic matter (Conklin, 2001).

The graphical relationship shown in Figure 5.10 uses some representative values reported in the various handbooks (e.g., Verscheuren, 1983, Montgomery and Welkom, 1991) for log k_{ow} and log k_{oc}. The PHC compounds shown in Figure 5.7 and the chlorinated benzenes shown in Figure 5.12 are identified in the chart. The values used for log k_{ow} are considered to be the mid-range results reported from many studies. Not all log k_{oc} values are obtained as measured values. Many of these have been obtained through application of the various log k_{oc}-log k_{ow} relationships reported in the literature, e.g., Kenaga and Goring (1980), Karickhoff et al. (1979). The approximate relationship shown by the solid line Figure 5.10 is given as:

$$\log k_{oc} = 1.06 \log k_{ow} - 0.68 \tag{5.7}$$

Equation (5.7) covers chemicals ranging from PAHs to pesticides and PCBs. We can compare this to other relationships shown in Equation (5.8) which were obtained for certain classes of chemical compounds.

$$\log k_{oc} = \log k_{ow} - 0.21$$

$$\log k_{oc} = 1.029 \log k_{ow} - 0.18 \tag{5.8}$$

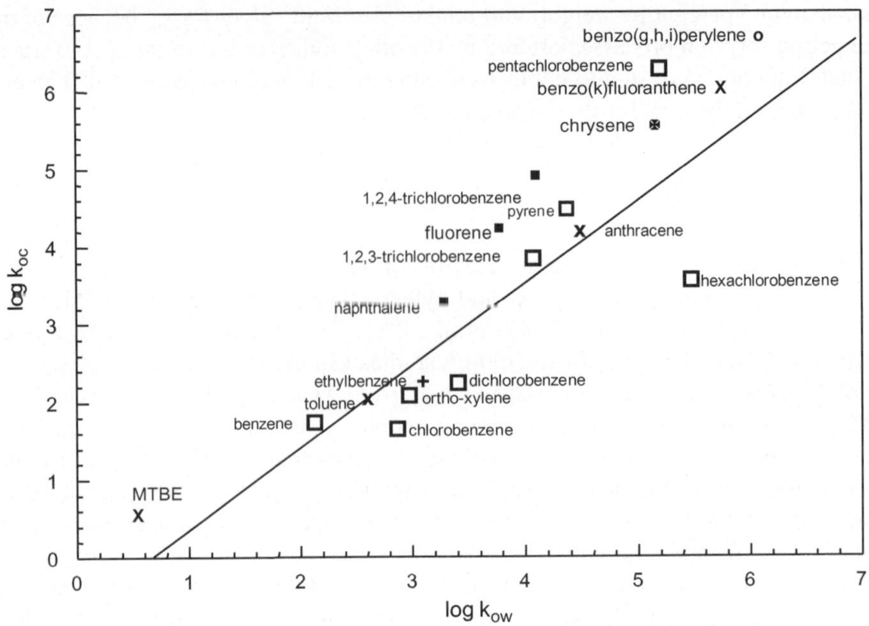

FIGURE 5.10 Relationship between log k_{oc} and k_{ow} for several organic compounds.

$$\log k_{oc} = 0.72 \log k_{ow} + 0.49$$

The first relationship shown in Equation (5.8) was reported by Karickhoff et al. (1979) in respect to 10 PAHs, whilst the second one referring to pesticides was reported by Rao and Davidson (1980). The relationship describing the group containing chlorinated benzenes, which also includes methylated benzenes, has been reported by Schwarzenbach and Westall (1981). The graphical diagram shown in Figure 5.10 provides us with an opportunity to anticipate partitioning of the organic chemical compounds shown in the diagram. Thus, for example, the k_{oc} value shown in Figure 5.9 for dichlorobenzene indicates that it partitions well to sediments, and particularly to the organic fractions (SOM). Because it is quite resistant to anaerobic degradation, it is very persistent. Although there exist three isomers of trichlorobenzene ($C_6H_3Cl_3$), 1,2,4-, 1,2,3- and 1,3,5-, the isomer 1,2,4- is most common. The low water solubilities, high log k_{ow} and log k_{oc} values indicate that it (i.e., the 1,2,4-trichlorobenzene) partitions well to the soil fractions. As in the case of the dichlorobenzene, the trichlorobenzene is well adsorbed by the SOM. The similarly high values of k_{oc} for pentachlorobenzene (C_6HCl_5) and hexachlorobenzene (C_6Cl_6) are also indicative of the ability of these chemicals to partition to soil fractions, in common with the trichlorobenzenes.

Pentachlorobenzene which has been identified in waste streams from pulp and paper mills, iron and steel mills, inorganic and organic chemical plants, petroleum refineries, and activated sludge waste water treatment plants (Meyers and Quinn, 1973; Laflamme and Hites, 1978) appears to have the highest k_{oc} value of the various chlorobenzenes. The low water solubilities of the dichloro-, trichloro-, pentachloro- and hexachlorobenzenes combined with their respective high k_{oc} values indicate that they can be well adsorbed by the soil fractions. Desorption of chlorobenzenes from soil fractions can occur (Oliver, 1984; Oliver et al., 1989).

5.5.3 Organic Chemicals—Partitioning and Adsorption Isotherms

Studies on adsorption of the petroleum hydrocarbons (PHCs) by the active soil fractions' surfaces show that adsorption occurs only when the water solubility of the PHCs is exceeded and the hydrocarbons are accommodated in the micellar form (Yong, 2000). Instead of using the k_{ow} and k_{oc} partition coefficients, the accommodation concentration of hydrocarbons in water is sometimes used to reflect the partitioning tendency of organic substances between the aqueous and soil solids. Hydrocarbon molecules with lower accommodation concentrations in water (i.e., higher k_{oc} values) would be partitioned to a greater extent onto the soil fractions than in the aqueous phase. Meyers et al. (1973, 1978) have shown that we obtain a general inverse relationship between the accommodation concentration of the hydrocarbons and the proportion (percent) adsorbed; i.e., the lower the accommodation concentration of the hydrocarbon in water, the greater is the tendency of the organic compound to be associated with the reactive surfaces of the sediment fractions. The important consequence of such a relationship is that the aromatic fraction of petroleum products which are the most toxic, would have the least affinity for the reactive surfaces associated with the soil fractions. As might be expected, a study of adsorption data of hydrocarbons shows that anthracene is substantially adsorbed. This is confirmed by the high k_{oc} value and the very low solubility of the organic compound in water (Figure 5.3). The higher accommodation concentrations of the aromatic hydrocarbons inhibit their association with the clay particles.

Other methods have been devised to predict soil-water partitioning coefficients. Krauss and Wilcke (2001) used a sequential extraction technique for four temperature/methanol fractions, based on the "solvophobic" approach of Rao et al. (1985). They determined the partitioning coefficients for 20 PAHs and 12 PCBs using 11 urban soils. The co-solvent fractions used ranged from 0.1 to 0.9 and temperatures varied from 20°C to 80°C.

Modelling of sorption and desorption of PAHs can be difficult due to nonlinearities and hysteresis as the age of the contamination increases (Hatzinger and Alexander 1995; MacLeod and Sample, 2000). Redistribution of previously sorbed PAHs can become covalently bound to humic acids or coupled to metal-oxide surfaces (Bollag, 1992). Diffusion of the PAHs through micropores and organic matter between soil particles can also affect desorption from the soil to the water phase (Brusseau et al., 1991; Luthy et al., 1997).

An analysis of veterinary pharmaceutical sorption in soils was recently performed by Tolls (2001). The behavior of these compounds is quite similar to pesticides due to their functional groups. Attempts made to correlate k_{oc} and k_{ow} were not successful. Data for log k_{ow} < 3 lie above the regression line of Equation (5.7).

However, for these pharmaceuticals, hydrophobic partitioning did not appear to be a major contributor to sorption. Mechanisms of hydrogen bonding, surface complexation, bridging to clay surfaces, and cation exchange were significant and should be incorporated into the calculations of k_d (Schwarzenbach et al., 1993). For example, quinolone carboxylic acid, enrofloxacin, and tetracycline sorption increase with surface area (Nowara et al., 1997). Langmuir-type isotherms were found for tetracycline. Another pharmaceutical, enrofloxacin, shows sorption increasing as one progresses from kaolinite to illite to vermiculite. The performance with vermiculite is about the same as for montmorillonite (Sithole and Guy, 1987). The adsorption and desorption of VOCs (1,2-dichloroethane, 1,1,2 trichloroethane an 1,1,2,2-tetrachloroethane) by a clayey soil at a Superfund site reported by Kommalapati et al. (2002), showed that the adsorption isotherms fitted well to Freundlich isotherms. Only 20%–70% of the compounds could be desorbed.

Sorption can play a significant role for Perchloroethylene (PCE), trichloroethylene (TCE), and 1,2-dichloroethylene (DCE). In the calculations made by Witt et al. (2002) to evaluate their movement at Dover Air Force Base, Delaware, it was determined that by assuming that the first contamination was in the late 1940s to early 1950s, significant PCE levels were expected near a particular well location. However, some were found only 900 m from that well, leading them to conclude that sorption was definitely not the only attenuation mechanism.

The results of sorption tests of aniline and TCE in a 6% bentonite soil barrier material using batch equilibrium and column samples (Khandelwal et al., 1998) and a lithium chloride tracer indicated that isotherms obtained for 7 and 10 days for both components were similar and could be fitted to linear and Freundlich curves. Since sorption did not appear to be significant, the advective-dispersive reactive equation could be used to predict movement of the organic chemical contaminants in the column samples. Sorption of volatile organic compounds (VOCs) from the gas phase to soils in columns studies reported by Ruiz et al. (1998) indicated that polar compounds (water and methyl ethyl ketone) were sorbed more than aliphatic (n-hexane, n-heptane, and n-octane) or aromatic compounds (toluene, xylenes, and ethylbenzene). The effect of increased humidity appeared to be limited to low VOC concentrations. At low concentrations of VOC, a decrease in sorption was obtained with increased humidity. At high VOC concentrations, the effect of increased humidity appeared to be insignificant. The sorption results for low VOC concentrations appear to fit the BET (Brunauer-Emmett-Teller) adsorption isotherm model.

The effects of clay content (17%–50%) and organic matter (0.3%–9.8%) on the sorption of the pesticide carbendazim were examined for some Vietnamese soils by Berglof et al. (2002). Carbendazim sorbed more strongly to acid sulfate soils in comparison to other soils. Values of k_d for this soil were 127 mL/mg compared to less than 15 mL/g for the others since the organic and clay content was the highest. Sorption was also found to decrease as the pH decreased for the sandy soil.

In the studies on carbendazim and other pesticides performed by Nemeth-Konda et al. (2002), values of k_{oc} for carbendazim were calculated from the Freundlich equation as 2805. This compared well with the 2700 k_{oc} value obtained for a sandy soil by Berglof et al. (2002). This value was higher than the other pesticides, acetochlor, atrazine, diazonin, imidacloprid, and isoproturon. Attempts made to correlate k_{ow} values with the Freundlich adsorption constants showed that the correlations were slight. As might be expected, other components in the soil and soil water appear to have contributed some significant influence on the development of partitioning of these chemicals—as shown in the discussion on partitioning and adsorption isotherms in this and the previous chapters.

Paddy soils were studied in Japan to determine the adsorption of herbicides after runoff (Kibe et al., 2000). The tests on five herbicides, esprocarb, mefenacet, pretilachlor, simetryn, and thiobencarb indicated that the Freundlich-type relationship, with n coefficients ranging from 1.0 to 1.6 could model the adsorption isotherms obtained. The k_d partition coefficients however failed to correlate with k_{ow}, cation exchange capacities (CECs), organic carbon contents, specific surface areas, and the minerals in the paddy soils. The range in values for the various properties and characteristics of the paddy soils were: (a) organic contents ranging from 1.2% to 6.8%, (b) specific surface areas from 6.2 to 34 m^2/g-dry soil, and (c) CEC from 12 to 45 meq/100 g dry. Considering the significant range in values for CEC and SSA, and the narrow range in n coefficients for the Freundlich model, it is not surprising that calculations of k_d from k_{oc} differed significantly from experimental values. Nevertheless, the results indicate that the herbicide concentrations from runoffs calculated from the Freundlich relationship could serve as preliminary equilibrium estimates.

In the batch equilibrium studies on a soil-tar-water system material consisting of solid coal tar particles, pure crystals of naphthalene on the soil, crystal particles of fluoranthene and solubilized organic molecules from coal tar were conducted by Bayard et al. (2000) to provide partitioning information on naphthalene for determination of transport in a soil tar water system equilibrium periods of 24–60 h were used. While the presence of other PAHs did not influence the rate of naphthalene sorption, it was established that high concentrations (120 mg/kg) of phenanthrene did. Sorption was also very strong for the coal tar (k_{oc} of 2000 cm^3/g) compared to the soluble organic matter (k_{oc} of 300 cm^3/g). Almost 90% of the naphthalene was adsorbed in the coal tar compared to only one third of the naphthalene in the soluble organic matter. The batch equilibrium test results showed that sorption could be modelled by linear adsorption isotherms. In commenting on the work by Luthy et al. (1997) and their postulate that sorption will be in different domains of geosorbents (in soils, sediments, and aquifer solids), the authors noted that: (a) the xenobiotic organic matter consisted of hard and softly condensed organic matter, and (b) the naphthalene must have adsorbed onto the soft matter because of the short equilibrium time. Longer times would lead to diffusion in and out of the dense organic matter and soil aggregates (Luthy et al., 1997). A double domain model was used to simulate column studies and was applicable to short-term, abiotic studies. The two domains consisted of: (a) an external domain which involved local adsorption, and (b) an internal domain based on first-order kinetics with local adsorption and resistance to mass transfer.

5.6 INTERACTIONS AND FATE

5.6.1 Abiotic Reactions

Abiotic reactions and transformations of organic chemical contaminants are sensitive to at least two factors: (a) the physico-chemical properties of the contaminant itself, and (b) the physico-chemical properties of the soil (i.e., soil fractions comprising the soil). Similar to inorganic contaminants, abiotic chemical reactions with organic compounds occur and include (a) hydrolysis, (b) formation of a double bond by removal of adjacent groups, and (c) oxidation/reduction reactions.

5.6.2 Hydrolysis

Hydrolysis reactions, for example, for PAHs are not known since they are very stable (Radding et al., 1976; Howard et al., 1991). Benzene, toluene, ethylbenzene, and xylene (BTEX) compounds are also unlikely to undergo chemical reactions despite the high degree of unsaturation of benzene-like alkenes (McAllister and Chiang, 1994). This is also supported by column experiments by Angley et al. (1992) and Larsen et al. (1992) who performed experiments under abiotic conditions and found insignificant losses of BTEX. Hydrolysis reactions for PCE and TCE do not seem to play a major role as hydrolysis half-lives for 9.9×10^8 and 1.3×10^6 years, respectively, have been reported (Jeffers et al., 1989). However, halogenated alkanes such as chloroethane and 1,2 dibromoethane are known to undergo hydrolysis when there is insignificant biodegradation. These rates would be typically much slower in comparison to most biodegradation processes. For pH values between 2 and 11, methyl fluoride, methyl chloride, and allyl chloride have hydrolysis half-lives of 10 years, 100 days and 80 h, respectively (USEPA, 1989). Clays can also increase the rates of hydrolysis by catalysis (Vogel et al., 1987). Additionally, dissolved organic matter and metal ions can also influence hydrolysis rates.

McCarty (1996) found that 20% and 80% of 1,1,1 TCA were transformed to 1,1 DCE and acetic acid, respectively. This is an important consideration since 1,1 DCE is more toxic than the parent compound. Reductive dehalogenation of 1,1,1, TCA progresses towards 1,1 DCA and then to chloroethane (CA). The latter can be hydrolyzed to ethanol (Vogel and McCarty, 1987) or dehydrohalogenated to vinyl chloride (Jeffers et al., 1989). Chlorinated ethenes have hydrolysis half-lives in the order of thousands of years. As the number of chlorines increase, the importance of dehydrohalogenation seems to increase (Jeffers et al., 1989).

Methyl tert-butyl ether (MTBE), also, does not typically undergo chemical reactions in groundwater (Squillace et al., 1997). This is expected as ethers are not very reactive since the carbon-oxygen-carbon bond is very strong. Strong acids and high temperatures would be required to break this bond. In various batch studies, abiotic reactions of MTBE have not been detected (Yeh and Novak, 1995; Borden et al., 1995; Bradley et al., 1999). Although alcohols such as ethanol can form strong hydrogen bonds with their surroundings, ethers have lone electron pairs that do not form bonds. The oxygen electrons of the ethers do not exhibit strong bonds with the mineral components of the soil. This is because the negative charges of the clays repel the oxygen in the ether (Conklin, 2001). However, interactions can occur with the organic matter and can provide an important sink for the MTBE. Higher concentrations of soil organic matter lead to more ether adsorption.

5.6.3 Dehydrohalogenation

Dehydrohalogenation is the removal from a halogenated alkane of a halogen atom from a carbon atom followed by removal of a hydrogen atom from an adjacent carbon. An alkene is thus produced. Although chloroethane can be dehydrohalogenated (Jeffers et al., 1989), monohalogenated aliphatics, in general, do not undergo this reaction. Vogel et al. (1987) have found that polychlorinated alkanes naturally undergo this reaction. Bromines are more reactive than chlorines. Dehydrohalogenation reactions reported by Jeffers et al. (1989) include 1,1,1 TCA and 1,1,2 TCA, that go to 1,1 DCE, in addition to tetrachloroethanes and pentachloroethane which go to TCE and PCE.

5.6.4 Oxidation-Reduction Reactions

Oxidation-reduction (redox) reactions can occur in interactions between organic chemical contaminants and soil fractions under abiotic and biotic conditions. In contrast to transformations occurring through nucleophilic replacement reactions where no net transfer of electrons occurs, electron transfer occurs in redox reactions. A general brief discussion of redox reactions and the redox potential can be found in Chapter 3. At this time, we need to understand how these affect the fate of organic chemical contaminants in a soil-water system. The significant points that require attention are the nature and result of electron transfer between the interacting participants (contaminants, microorganisms, and soil fractions). We recall that: (a) the chemical reaction process defined as *oxidation* refers to a removal of electrons from the subject of interest, and (b) reduction refers to the process where the "subject (electron acceptor or *oxidant*)" gains electrons from an electron donor (*reductant*). By gaining electrons, a loss in positive valence by the subject of interest results and the process is called a reduction.

It is not often easy to distinguish between redox reactions that occur abiotically and those that occur under biotic conditions since direct involvement of any (or some) microbial activity cannot be readily ruled out. It is not clear that insofar as organic chemical contaminants are concerned whether there is a critical requirement to distinguish between the two—since redox conditions are more likely than not to be the product of factors which include microbiological processes. The number of functional groups of organic chemical contaminants that can be oxidized or reduced under abiotic conditions is considerably smaller than those under biotic conditions (Schwarzenbach et al., 1993). Quantification of reaction rates is difficult because the interactions between the contaminants occur with both microorganisms and the many different soil fractions, thus making determination of reaction pathways almost impossible. While the scarcity of kinetic data makes it difficult to provide for quantitative calculations of redox reactions, it is nevertheless instructive and informative to obtain a qualitative or descriptive appreciation of these reactions.

Abiotic redox reactions of organic chemical contaminants in soil systems occur when electron acceptors such as those described above are present. Clay soils function well as electron acceptors (oxidizing agents or oxidants), i.e., they are electrophiles. The structural elements of clay minerals such as Al, Fe, Zn, and Cu can transfer electrons to the surface-adsorbed oxygen of the clay minerals. These can be released as hydroperoxyl radicals (–OOH) which can function as electron acceptors, i.e., these radicals can abstract electrons from the organic chemical contaminants. An example of this type of reaction is shown in Figure 5.11 using the results reported by Yong et al. (1997).

The possible mechanisms for oxidizing the two kinds of phenols shown in the figure include: (a) the structural elements of the montmorillonite clay (Fe and Al), and (b) the partially coordinated aluminum on the edges of the clay minerals. These function as Lewis acids which can accept electrons from aromatic compounds. In addition, the exchangeable cations such as Fe(III) and Cu(III) contribute to phenol polymerization through coupling of radical cations with phenols. We see from the results given in Figure 5.12 that more effective oxidation of the 2,6-dimethylphenol is obtained by the Fe-clay—presumably because of the greater oxidizing capability of the Fe(III). The intermediate product formed is a 2,6-dimethylphenol dimer of mass 242, as shown by the degree of "abundance" on the ordinate of the graph in Figure 5.12. Other intermediate products such as trimers and traces of oligomers of the 2,6-dimethylphenol have also been obtained (Desjardins, 1996). The presence of zero valent iron can lead to the dehalogenation of chlorinated aliphatics, but this is unlikely to occur naturally (Gillham and O'Hannesin, 1994).

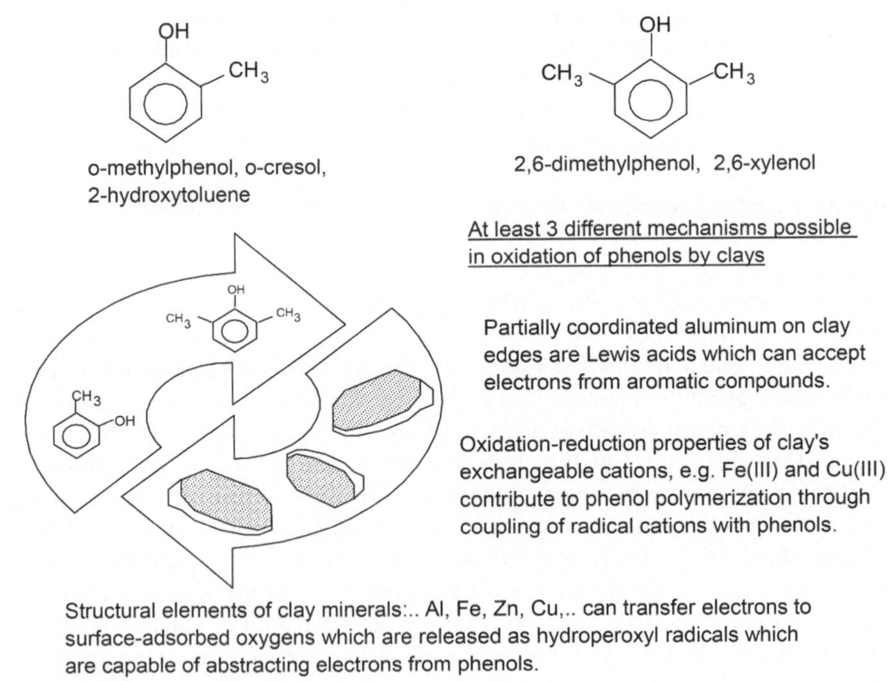

FIGURE 5.11 Different kinds of mechanisms of possible phenol oxidation by clay minerals.

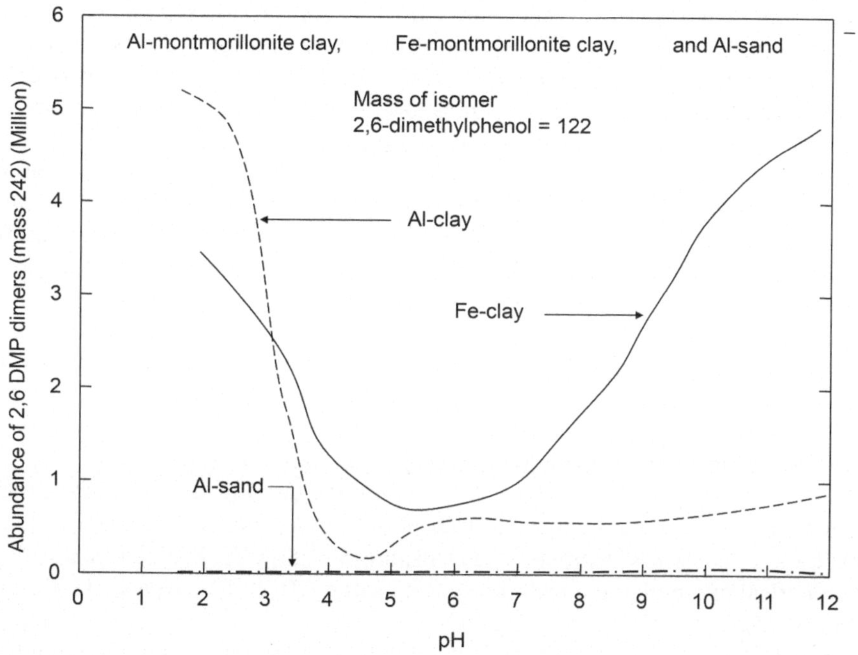

FIGURE 5.12 Oxidation of 2,6-dimethylphenol by Al-clay, Fe-clay and Al-sand. (Data from Yong, R.N. et al., *Appl. Clay Sci.*, 12, 93–110, 1997.)

5.7 MOBILITY OF ORGANIC CHEMICAL CONTAMINANTS

A key element in the application of monitored natural attenuation (MNA) as a tool in remediation of contaminated sites is the *tracking requirement* for detection and monitoring of the advance of the contaminant plume associated with the contaminants in the contaminated site. In addition to continuous monitoring of contaminant plume advance, we need to know whether the reductions in concentration and toxicity of the target contaminant, in a definite time frame, are consistent with the outcome of natural attenuation processes. To do this, we need to rely on predictive models of transport and fate of the contaminants of interest. In the summary discussion of fate and transport models in this section, we will note that the discussion applies to both inorganic and organic chemical contaminants.

Transport and fate models come in many forms, depending on the end-use of the models and the goals of the developers of the models. They are required elements in planning and design of containment systems, and in assessment of effectiveness of remediation technologies. As predictive tools, these models are essential tools in the decision-making process for regulatory agencies and practitioners in: (a) assessment and evaluation of attenuation competence of soil-engineered barriers or soil substrate, (b) prediction of continued progress of contaminant plumes, and (c) risk assessment and management. Transport and fate models can be designed to provide analyses and evaluation of performance characteristics of contaminant plumes in transport through the soil medium. The particular sets of issues addressed include those dealing with evaluation of the capability of: (a) engineered clay barriers to function as natural attenuation barrier systems over a period of time, (b) affected subsoils to attenuate contaminants in transport through the subsoils, and (c) tracking progress of contaminants in transport in subsoils. Other significant roles for transport and fate models include risk evaluation as, for example, in determination of the potential for groundwater and environmental contamination. Issues related to health threats are of particular importance. Transport and fate models are also very useful forensic tools. Determination of the source of pollution requires determination of both transport history and fate behavior of the contaminants in question. It stands to reason that the reliability of transport and fate models to properly evaluate and predict performance is of paramount importance. This means that the models must be able to properly represent the many interacting relationships which govern transport of contaminants in the soil-water system. Figure 5.13 illustrates the requirements for proper development and application of transport and fate models.

It is not the purpose of this book to deal with the development of transport and fate models. For the development of transport and fate models, the reader should consult more specialized textbooks on transport modelling for detailed treatment of the analytical and numerical techniques. e.g., Little (2012); Kinzelbach (1986); and Crank (1975).

Instead, the focus on transport and fate models is in the direction of the capability of the models to fully accommodate or account for the phenomena (results) resulting from the various interactions between contaminants and soil fractions. It is not

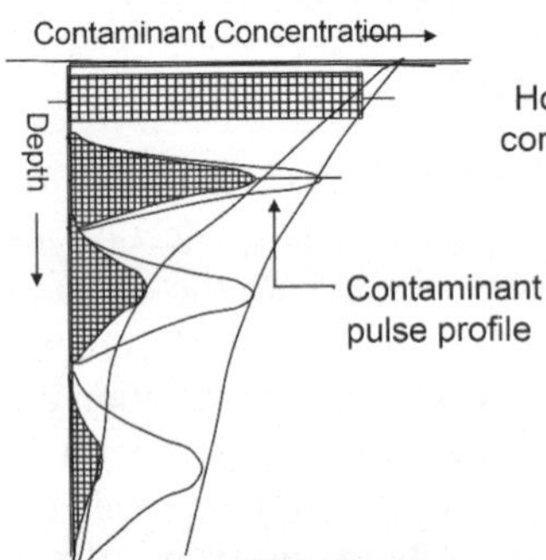

FIGURE 5.13 Requirements and issues for transport and fate modelling.

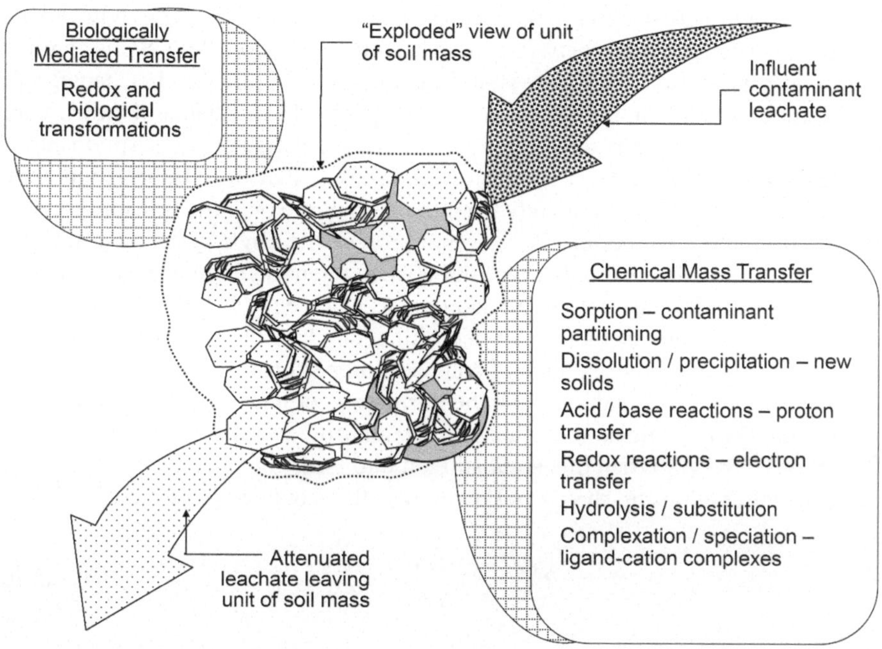

FIGURE 5.14 Interactions and processes involved in attenuation of contaminants in transport through a soil element. Advective transport and molecular diffusion are not included in the schematic illustration.

always clear that: (a) the various processes contributing to the physical environment, such as those represented in Figure 5.14 have been properly recognized, and (b) the various driving forces responsible have been well incorporated in to the models.

Prediction of the transport of contaminants are most often considered in terms of saturated transport, i.e., transport in fully saturated soil media. Without considering storage, the commonly used "advection-diffusion" relationship is given as follows:

$$\frac{\partial c}{\partial t} = D_L \frac{\partial^2 c}{\partial x^2} - v \frac{\partial c}{\partial x} - \frac{\rho}{n\rho_w} \frac{\partial c^*}{\partial t} \tag{5.9}$$

where:

c = Concentration of contaminant of concern
t = Time
D_L = Diffusion-dispersion coefficient (most often called "*diffusion* coefficient")
v = Advective velocity
x = Spatial coordinate
ρ = Bulk density of soil media
ρ_w = Density of water
n = Porosity of soil media
c^* = Concentration of contaminants adsorbed by soil fractions.

The c^* term in Equation (5.9) is of particular interest since this deals directly with the partitioning of contaminants between the soil fractions and pore fluid. Some common types of adsorption isotherms were shown previously as Figure 3.10 in Chapter 3 for high affinity, Freundlich, etc. isotherms, and Figure 4.13 in Chapter 4 which compares the batch equilibrium adsorption isotherm with characteristic sorption curves from leaching column tests. It should be emphasized that little information about the transfer process of contaminants from the fluid phase onto the soil solids can be obtained from these isotherms. They do not provide the kinds of detailed information which would allow us to distinguish between retention and retardation mechanisms responsible for partitioning. In addition, as batch equilibrium tests are used to generate the adsorption isotherms for determination of k_d, the soil suspensions used for such tests are not representative of compact soil performance.

To determine c^* in Equation (5.1), the partition coefficient k_d defined previously will be used. With the linear (constant) adsorption isotherm shown in Figure 3.9, we will obtain: $c^* = k_d\, c$. When this is substituted into Equation (5.9), the readily recognized popular relationship shown as Equation (5.10) is obtained as follows:

$$R\frac{\partial c}{\partial t} = D_L\frac{\partial^2 c}{\partial x^2} - v\frac{\partial c}{\partial x} \tag{5.10}$$

with $R = \left[1 + \frac{\rho}{n\rho_w} k_d\right]$

If a non-linear adsorption isotherm is used instead of the linear isotherm, k_d will not be constant. It will then be necessary to determine the variation of c^* with the equilibrium concentration of contaminants in the pore water. The proper k_d must be specified as a function of the concentration of the species under consideration, ionic strength, presence of other contaminants, temperature, etc. The predictions made from the relationships will tell us about the concentration of the specific target contaminants used in the model in relation to time and spatial distance. We must remember that the predicted distribution of target contaminant in relation to time and space can only be considered to be as accurate as the source and quality of the inputs provided for determination of c^*. This will be discussed when the mechanisms of partitioning and the distribution of partitioned contaminants are examined.

If we take the *Freundlich* relationship shown in Figure 3.10 and express it in logarithmic format, the following is obtained:

$$\log c^* = \log k_1 + k_2 \log c \tag{5.11}$$

A plot of the above equation with $\log c^*$ as the ordinate and $\log c$ as the abscissa will show an intercept on the ordinate of $\log k_1$ and a straight line with a slope of k_2. It is this slope k_2 that has been defined as the *distribution coefficient* k_d (see Figure 5.15) and is used to evaluate and/or predict the transport of contaminants as seen later.

As discussed in Chapter 4, the distribution coefficient k_d is used to describe contaminant partitioning between liquid and solids. We can develop various k_d values for different kinds of soils in relation to different contaminants and also for various conditions. The primary requirement is the availability of batch

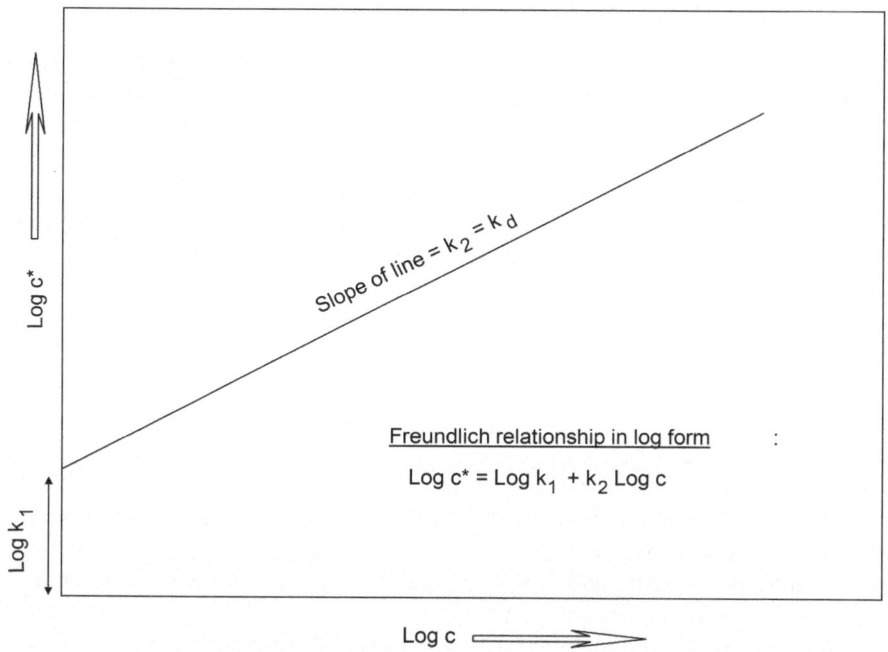

FIGURE 5.15 Determination of k_d from a Freundlich relationship.

equilibrium test data. The use of batch equilibrium testing and leaching column testing to obtain partitioning and distribution coefficients has been the subject of considerable discussion and debate by those concerned with the accuracy or reliability of predictions of models using this method of attenuation assessment. The essence of the debate focuses around the nature of the test sample used to determine partitioning of the contaminant under investigation. The question that needs to be addressed is whether soil solutions used for batch equilibrium tests can properly represent partitioning behavior that occurs with intact soil samples such as those used for leaching column tests.

However, regardless of the nature of the soil samples used, the contaminant partitioning tests assume that chemical equilibrium conditions are attained within a 24-h period. This is perhaps acceptable for many inorganic contaminants in interaction with soil particles but will most likely not be applicable to many organic chemical compounds. PCBs for example, can take up to several years to reach equilibrium conditions (Pignatello, 1989). We need to emphasize that the use of k_d in transport models developed along the lines shown previously, should only be undertaken for those circumstances where reactions that cause the partitioning are fast, i.e., equilibrium conditions are reached almost instantaneously. We will note that the distribution coefficient has been found to be useful in evaluation of partitioning of inorganic contaminants and some organic chemical contaminants. Various studies have reported linear adsorption characteristics for low concentrations of hydrophobic organic chemicals such as PAHs and some substituted aromatic compounds.

5.8 CONCLUDING REMARKS

Chemical mass transfer is responsible for partitioning of contaminants in the fate and transport of contaminants. Oxidation-reduction reactions can also play important roles in the fate of the contaminants. Assessment of whether the processes lead to retention or retardation is necessary if one wishes to determine the mobility of the contaminants. If potential contaminant hazards and threats to public health and the environment are to be minimized or avoided, we must ensure that the processes for contaminant attenuation are irreversible or below the levels of allowable limits or levels. We need to know how partitioning of the organic chemical compounds occur, and also to recognize the degree to which an organic chemical contaminant is available for biologically-mediated transformations.

Some interesting and useful points of note in partitioning include:

- Partitioning of PHCs onto soil particle surfaces occurs to a greater extent than in the aqueous phase, resulting in lower environmental mobility and higher retention of the PHCs.
- Weakly polar compounds are more readily adsorbed onto soil surfaces in contrast to non-polar compounds.
- Adsorption of non-ionic organic compounds by soil fractions is governed by the *CH* activity of the molecule.
- The water solubility of organic molecules will influence or control the partitioning of the organic contaminant, and the transformations occurring because of various processes associated with oxidation/reduction, hydrolysis and biodegradation.
- The k_{ow} octanol-water partition coefficient has been widely adopted as a significant parameter in studies of the environmental fate of organic chemicals.
- Organic chemicals with low k_{ow} (e.g., less than 10) may be considered to be relatively hydrophilic. They tend to have high water solubilities and small soil adsorption coefficients.
- Chemicals with a high k_{ow} value (e.g., greater than 10^4) are very hydrophobic and are not very water-soluble (i.e., they have low water solubilities).
- The constant k_d, partition coefficient, used in the computation of the retardation coefficient R means that infinite adsorption by the soil fractions cannot be accepted. As the concentration of contaminants increases, sorption will also increase. The relationship cannot be linear since a maximum level must be defined. Non-linear adsorption isotherms are more representative of field situations and should be used.

The various sorption processes that contribute to bonding between organic contaminants and soil fractions include partitioning (hydrophobic bonding) and accumulation through adsorption mechanisms involving the clay minerals and other soil particulates such as carbonates, organic and amorphous materials.

The more prominent properties affecting the fate of organic molecules by soil fractions include the following:

Soil fractions—Surface area, nature of surfaces (composition of surface fractions), configuration of reactive surfaces.
Organic chemical molecules—Functional groups, structure, charge, size, shapes, flexibility, polarity, volatility, water solubility, hydrophobicity, polarizability, partitioning and equilibrium constants,
Soil environment—Temperature, inorganic/organic ligands available, pH, *pE*, salinity, physical gradients.

Desorption must also be considered with regards to the fate of organic contaminants. It is the result of physical and chemical transformations. Desorption is very difficult to predict since characterization and quantification is not fully achieved. Very little field data are available, and most of the current values are based on experimental laboratory equilibrium tests or column test data, making the transport of organic chemicals very difficult to predict accurately. In addition, the effect of sorption and desorption on biodegradability has not been adequately taken into account. Components produced by microorganisms in situ also have the potential to significantly affect the fate of organic contaminants.

Evaluation of the ability of contaminants to absorb and desorb is an important aspect in determining attenuation of contaminants in the soil matrix. Many of the studies, as shown here, have been done with single or a few contaminants, not the mixtures currently found at many sites. In addition, contaminants such as antibiotics are now being recognized as problematic and their transport and fate in soils have yet to be fully determined or understood. At present, we have very little information on their sorption properties. Contaminants must be irreversibly sorbed to meet the requirements of protecting potential receptors in the environment and satisfy the regulatory agencies. It is not clear if we have the proper tools to determine whether sorption is irreversible. Unless sorption of contaminants is irreversible, threats to the environment cannot be minimized. Methods to determine this potential must be developed. One approach may be to use sequential extraction as is performed for metals to determine their potential for desorption. Use of more severe extractants would correlate with more strongly adsorbed contaminants.

We have also noted that other physical mechanisms including volatilization, and chemical reactions such as hydrolysis, dehydrohalogenation, and oxidation-reduction occur in the subsurface. The role of these attenuating mechanisms has not been thoroughly investigated and can be significantly underestimated and hard to distinguish from biotic reactions, especially on a long-term basis. Hydrolysis and dehydrohalogenation reactions can be quite significant for chlorinated solvents. Even the role of soil components as catalysts for contaminant transformations has not been fully understood. These types of reactions can produce contaminants that do not sorb as strongly and can be released into the environment. We have, therefore, many processes that still require characterization and quantification. These types of information need to be incorporated into natural attenuation models so that they can be used to more accurately assess and predict the capability of natural attenuation as a remedial technology.

REFERENCES

Angley, J.T., Brusseau, M.L., Miller, W.L., and Delfino, J.J. (1992), Nonequilibrium sorption and aerobic biodegradation of dissolved alkylbenzenes during transport in aquifer material during transport in aquifer material. Columns experiments and evaluation of a coupled-process model, *Environ. Sci. Technol.*, 26(7):1404–1410.

Anthony, J.W. et al. (1999), Methodology to evaluate natural attenuation of methyl tertiary-butyl ether, in B.C. Alleman and A. Leeson (Eds.), *Natural Attenuation of Chlorinated Solvent, Petroleum Hydrocarbons, and Other Organic Compounds*, Batelle, Columbus, OH, pp. 121–133.

Bayard, R., Barna, L., Mahojoub, B., and Gourdon, R. (2000), Influence of the presence of PAHs and coal on naphthalene sorption in soil, *J. Contam. Hydrol.*, 46:61–80.

Berglof, T., Van Dung, T., Kylin, H., and Nilsson, I. (2002), Carbendazim sorption-desorption in Vietnames soils, *Chemosphere*, 48(3):267–273.

Bollag, J.M. (1992), Decontaminating soil with enzymes, *Environ. Sci. Technol.*, 26(10):1876–1881.

Borden, R.C., Gomez, C.A., and Becker, M.T. (1995), Geochemical indicators of intrinsic bioremediation, *Ground Water*, 33(2):180–189.

Bradley, P.M., Landmeyer, J.E., and Chapelle, F.H. (1999), Aerobic mineralization of MTBE and tert-butyl alcohol by stream-bed sediment microorganisms, *Environ. Sci. Technol.*, 33(11):1877–1879.

Brar, G.S. (1997), Phytoremediation of chlorinated solvents: Progress and challenge, *IBC's 2nd Annual Conference on Innovative Technologies*, Boston, MA.

Brusseau, M.L., Jessup, R.E., and Rao, P.S.C. (1991), Nonequilibrium sorption of organic chemicals: Elucidation of rate-limiting process, *Environ. Sci. Technol.*, 25(1):134–142.

Chiang, C.Y., Salanitro, J.P., Chai, E.Y., Colthart, J.D., and Klein, C.L. (1989), Aerobic biodegradation of benzene, toluene, and xylene in a sandy aquifer- Data analysis and computer modelling, *Ground Water*, 27:823–834.

Chiou, G.T., Freed, V.H., Schmedding, D.W., and Kohnert, R.L. (1977), Partition coefficient and bioaccumulation of selected organic chemicals, *Environ. Sci. Technol.*, 11:5.

Chiou, G.T., Schmedding, D.W., and Manes, M. (1982), Partition of organic compounds on octanol-water system, *Environ. Sci. Technol.*, 16:4–10.

Conklin, Jr., A.R. (2001), Oxygenates in soil, *Contam. Soil Sed. Water Special Issue*, Spring, 10–11.

Crank, J. (1975), *The Mathematics of Diffusion*, 2nd ed., Oxford University Press.

Desjardins, S. (1996), Investigation on montmorillonite-phenol interactions, PhD thesis, McGill University, Montreal, Canada.

Dragun, J. (1988), *The Soil Chemistry of Hazardous Materials*, The Hazardous Materials Control Research Institute, Silver Spring, MD.

Elliott, H.A., Liberati, M.R., and Huang, C.P. (1986), Competitive adsorption of heavy metals by soils, *J. Environ. Qual.*, 15:214–219.

Gillham, R.W., and O'Hannesin, S.F. (1994), Enhanced degradation of halogenated aliphatics by zero-valent iron, *Ground Water*, 32:958–967.

Guerin, W.F., and Boyd, S.A. (1992), Differential bioavailability of soil-sorbed naphthalene to two bacterial species, *Appl. Environ. Microbiol.*, 58(4):1142–1152.

Hamaker, J.W., and Thompson, J.M. (1972), Adsorption, Chapter 3, in C.A.I. Goring and J.W. Hamaker (Eds.), *Organic Chemicals in the Soil Environment*, Marcel Dekker, New York, Vol. 1, pp. 49–144.

Hatzinger, P.B., and Alexander, M. (1995), Effect of aging of chemicals in soil on their biodegradability and extractability, *Environ. Sci. Technol.*, 29(2):537–545.

Howard, P.H., Boethling, R.S., Jarvis, W.F., Meylan, W.M., and Michalenko, E.M. (1991), *Handbook of Environmental Degradation Rates*, Lewis, Chelsea, MI.

Jeffers, P.M., Ward, L.M., Woytowitch, L.M., and Wolfe, N.L. (1989), Homogeneous hydrolysis rate constants for selected chlorinated methanes, ethanes, ethenes, and propanes. *Environ. Sci. Technol.*, 23:965–969.

Karickhoff, S.W., Brown, D.S., and Scott, T.A. (1979), Sorption of hydrophobic contaminants on natural sediments, *Water Res.*, 13:241–248.

Karickhoff, S.W. (1984), Organic contaminants sorption in aquatic system, *J. Hydraul. Eng.*, 110:707–735.

Kemblowski, M.W., Salanitro, J.P., Deeley, G.M., and Stanely, C.C. (1987), Fate and transport of residual hydrocarbon in groundwater: A case study, *Proceeding NWWA/API Conference on Petroleum Hydrocarbons and Organic Chemicals in Ground Water: Prevention, Detection and Restoration*, Water Well Journal Publishing, Dublin, OH, pp. 207–231.

Kenaga, E.E., and Goring, C.A.I. (1980), Relationship between water solubility, soil sorption, octanol-water partitioning and concentration of chemicals in biota, *ASTM-STP*, 707:78–115.

Khandelwal, A., Rabideau, A.J., and Shen, P. (1998), Analysis of diffusion and sorption of organic solutes in soil-bentonite barrier materials, *Environ. Sci.Technol.*, 32:1333–1339.

Kibe, K., Takahashi, M., Kameya, T., and Urano, K. (2000), Adsorption equilibriums of principal herbicides on paddy soils in Japan, *Sci. Environ.*, 263:115–125.

Kinzelbach, W. (1986), *Groundwater Modelling: An Introduction with Sample Programs in BASIC*, Elsevier, Amsterdam, the Netherlands.

Kommalapati, R.R., Valsaraj, K.T., and Constant, W.D. (2002), Soil-water partitioning and desorption hysteresis of volatile organic compounds from a Louisiana superfund site soil, *Environ. Monit. Assess.*, 73:275–290.

Krauss, M., and Wilcke, W. (2001), Predicting soil-water partitioning of polycycloc aromatic hydrocarbons and polychlorinated biphenyls by desorption with methanol-water mixtures at different temperatures, *Environ. Sci. Technol.*, 35:219–2325.

Laflamme, R.E., and Hites, R.A. (1978), The global distribution of polycyclic aromatic hydrocarbons in recent sediments, *Geochim. Cosmochim. Acta*, 42:289–303.

LaGrega, M.D., Buckingham, P.L., and Evans, J.C. (2001), *Hazardous Waste Management*, McGraw Hill, Boston, MA, 1202 p.

Landmeyer, J., Chapelle, F., Bradley, P., Pankow, J., Church, C., and Tratmyek, P. (1998), Fate of MTBE relative to benzene in a gasoline-contaminated aquifer (1993–98), *Ground Water Monit. Rev.*, 18(4):93–102.

Larsen, T., Kjeldsen, P., and Christensen, T.H. (1992), Sorption of hydrophobic hydrocarbons on three aquifer materials in a flow through system, *Chemosphere*, 24(4):429–451.

Larson, R.A., and Weber, W.J. (1994), *Reaction Mechanisms in Environmental Organic Chemistry*, Lewis Publishers, Boca Raton, FL, p. 433.

Little, K.W. (2012), *Environmental Fate and Transport Analysis with Compartment Modelling*, CRC Press, Boca Raton, FL, 244p.

Luthy, R.G., Aiken, G.R., Brusseau, M.L., Cunnigham, S.C., Gschwend, P.M., Pignatello, J.J., Reinhard, M., Traina, S.J., Weber, W.J., and Westall, J.C. (1997), Sequestration of hydrophobic organic contaminants by geosorbants, *Environ. Sci. Technol.*, 31:3341–3347.

MacLeod, C.J.A., and Semple, K.T. (2000), Influence of contact time on extractability and degradation of pyrene in soils, *Environ. Sci. Technol.*, 34(23):4952–4957.

Mackay, D., Shiu, W.Y., and Ma, K.C. (1992), *Illustrated Handbook of Physical-Chemical Properties and Environmental Fate for Organic Chemicals*, Vols. 1, 2, and 3, Lewis Publishers, Boca Raton, FL.

Mata-Sandoval, J.C., Karns, J., and Torrents, A. (2000), Effect of rhamnolipids produced by *Pseudomonas aeruginosa* UG2 on the solubilization of pesticides, *Environ. Sci. Technol.*, 34(23):4923–4930.

McAllister, P.M., and Chiang, C.Y. (1994), A practical approach to evaluating natural attenuation of contaminants in ground water, *Ground Water Monit. Rev.*, 14(2):161–173.

McCarty, P.L., Reinhard, M., and Rittman, B.E. (1981), Trace organics in groundwater, *Environ. Sci. Technol.*, 15(1):40–51.

McCarty, P.L. (1996), Biotic and abiotic transformations of chlorinated solvents in groundwater, *Proceedings of the EPA Symposium on Natural Attenuation of Chlorinated Organics*, EPA/540/R-96/509. U.S. EPA, Washington, DC.

McMurray, J., *Fundamentals of Organic Chemistry*, 3rd ed., Brooks/Cole Publishing, Pacific Grove, Belmont, CA, 536p.

Meyers, P.A., and Oas, T.G. (1978), Comparison of associations of different hydrocarbons with clay particles in simulated seawater, *Environ. Sci. Technol.*, 132:934–937.

Meyers, P.A., and Quinn, J.G. (1973), Association of hydrocarbons and mineral particles in saline solution, *Nature*, 244:23–24.

Montgomery, J.H., and Welkom, L.M. (1991), *Groundwater Chemicals Desk Reference*, Lewis Publishers, Chelsea, MI, 640p.

Moorman, T.B., Jayachandran, K., and Reungsang, A. (2001), Adsorption and desorption of atrazine in soils and subsurface sediments, *Soil Sci.*, 166:921–929.

Nemeth-Konda, L., Fuleky, G., Morovian, G., and Csokan, P. (2002), Sorption behaviour of acetochlor, atrazine, carbendazim, diazonin, imidachloprid and isoproturon on Hungariam agricultural soil, *Chemosphere*, 48(5):545–552.

Noordman, W.H., Wachter, J.J.J., de Boer, G.J., and Janssen, D.B. (2002), The enhancement by biosurfactants of hexadecane degradation by *Pseudomonas aeruginosa* varies with substrate availability, *J. Biotechnol.*, 94:195–212.

Nowara, A., Burhenne, J., and Spiteller, M. (1997), Binding of fluoroquinoline carboxylic acid derivatives to clay minerals, *J. Agric. Food Chem.*, 45:1459–1463.

Oliver, B.G. (1984), Distribution and pathways of some chlorinated benzenes in the Niagara river and Lake Ontario, *Water Pollut. Res. J. Can.*, 19:47–59.

Oliver, B.G., Charlton, M.N., and Durham, R.W. (1989), Distribution, redistribution, and geochronology of polychlorinated biphenyl congeners and other chlorinated hydrocarbons in Lake Ontario sediments, *Environ. Sci. Technol.*, 23:200–208.

Olsen, R.I., and Davis, A. (1990), Predicting the fate and transport of organic compounds in groundwater, Part 1, *Hazard. Mater. Control*, 3:38–64.

Pignatello, J.J. (1989), Sorption dynamics of organic compounds in soils and sediments, in B.L. Sawhney and K. Brown (Eds.), *Reactions and Movement of Organic Chemicals*, Soil Science Society of America, Madison, WI.

Pontilillo, I., and Eganhouse, R.P. (2001), *The Search for Reliable Aqueous Solubility (Sw) and Octanol-Water Partition Coefficient* (k_{ow}) *Data for Hydrophobic Organic Compounds DDT and DDE as a Case Study*, U.S. Geological Survey, Water-Resources Investigations Report 01-421, USGS, Reston, VA.

Ptacek, C.J., Cherry, J.A., and Gillham, R.W. (1987), Mobility of dissolved petroleum-derived hydrocarbon in sand aquifers, in J.H. Vandermeulen and S.E. Hrudey (Eds.), *Oil in Freshwater: Chemistry, Biology, Countermeasure Technology*, Pergamon, New York, pp. 195–214.

Radding, S.B. (1976), The environmental fate of selected polynuclear aromatic compounds, EPA-560/5-75-009, U.S. Environmental Protection Agency, Cincinnati, OH.

Rao, P.S.C., and Davidson, J.M. (1980), Estimation of pesticide retention and transformation parameters required in nonpoint source pollution models, in M.R. Overcash and J.M. Davidson (Eds.), *Environmental Impact of Nonpoint Source Pollution*, Ann Arbor Science Publishers, Ann Arbor, MI, pp. 23–27.

Rao, P.S.C., Hornsby, A.G., Kilcrease, D.P., and Nkedi-Kizzi, P.J. (1985), Sorption and transport of hydrophobic organic chemicals in aqueous and mixed solvent systems: Model development and preliminary evaluation, *Environ. Qual.*, 14:376–383.

Rogers, R.D., McFarlane, J.C., and Cross, A.J. (1980), Absorption and desorption of benzene in two soils and montmorillonite clay, *Environ. Sci. Technol.*, 14(4):457–460.

Rogers, S.W., Ong, S.K., Kjartanson, B.H., Golchin, J., and Stenback, G.A. (2002), Natural attenuation of polycyclic aromatic hydrocarbon-contaminated sites: Review, *Pract. Period. Hazard., Toxic, Radioact. Waste Manag.*, 6:141–155.

Ruiz, J., Bilbao, R., and Murillo, M.B. (1998), Adsorption of different VOCs on to soil minerals from gas phase: Influence of mineral, type of VOC and air humidity, *Environ. Sci. Technol.*, 32:1079–1084.

Sahnoun, R., Ellouz-Chaabouni, S., and Ghribi, D. (2017), Application of bacterial biosurfactants for removal and biodegradation of diesel oil in soil using a newly isolated consortium, *Process. Saf. Environ.*, 109:72–81.

Schirmer, M., and Barker, J.F. (1998), A study of long-term MTBE attenuation in the Borden Aquifer, Ontario, Canada, *Groundwater Monit. Remed.*, 18(2):113–122.

Schwarzenbach, R.P., and Westall, J. (1981), Transport of non-polar organic polar compounds from surface water to groundwater: Laboratory sorption studies, *Environ. Sci. Technol.*, 15(11):1360–1367.

Schwarzenbach, R.P., Gschwend, P.M., and Imboden, D.M. (1993), *Environmental Organic Chemistry*, John Wiley & Sons, New York.

Sithole, B.B., and Guy, R.D. (1987), Models for tetracycline in aquatic environments: I Interactions with bentonite clay systems, *Water, Air, Soil. Pollut.*, 32:303–314.

Squillace, P.J., Pankow, J.F., Korte, N.E., and Zogorski, J.S. (1997), Review of the environmental behavior and fate of methyl tert-butyl ether, *Environ. Toxicol. Chem.*, 16:1836–1844.

Stuart, B.J., Bowlen, G.F., and Kossen, D.S. (1991), Competitive sorption of benzene, toluene, and the xylenes in soil, *Environ. Prog.*, 10(2):104–109.

Tolls, J. (2001), Sorption of veterinary pharmaceuticals in soils: A review, *Environ. Sci. Technol.*, 35:3397–3406.

U.S. Environmental Protection Agency (USEPA) (1989), *Transport and Fate of Contaminants in the Environment in the Subsurface*, EPA/625/4-89/019, Center of Environmental Research Information, Cincinnati, OH.

U.S. Environmental Protection Agency (USEPA) (1998), *Oxygenates in Water: Critical Information and Research Needs*, EPA/600/R-98/048. Office of Research and Development, Washington, DC.

Verscheuren, K. (1983), *Handbook of Environmental Data on Organic Chemicals*, 2nd ed., Van Nostrand Reinhold, New York.

Vogel, T.M., and McCarty, P.L. (1987), Abiotic and biotic transformations of 1,1,1-trichloroethane under methanogenic conditions, *Environ. Sci. Technol.*, 21:1208–1213.

Vogel, T.M., Criddle, C.S., and McCarty, P.L. (1987), Transformation of halogenated aliphatic compounds, *Environ. Sci. Technol.*, 21:722–736.

Volkering, F., Bruere, A.M., and van Andel, J.G. (1993), Effect of micro-organisms on the bioavailability and biodegradation of crystalline naphthalene, *Appl. Microbiol. Biotechnol.*, 40:535–540.

Voparil, I.M., and Mayer, L.M. (2000), Dissolution of sedimentary polycyclic aromatic hydrocarbons in to the lugworm's (Arenicola marina) digestive fluids, *Environ. Sci. Technol.*, 34(7):1221–1228.

Wauchope, R.D., Buttler, T.M., Hornsby, A.G., Augustijin-Beckers, P.W.M., and Burt, J.P. (1992), The environmental decision making, *Rev. Environ. Contam. Toxicol.*, 123:1–164.

Wiedemeier, T.H., Rifai, H.S., Newell, C.J., and Wilson, J.T. (1999), *Natural Attenuation of Fuels and Chlorinated Solvents in the Subsurface*, John Wiley & Sons, New York.

Witt, M.E., Klecka, G.M., Lutz, E.J., Ei, T.A., Grosso, N.R., and Chapelle, F.H. (2002), Natural attenuation of chlorinated solvents at area 6, dover air force base: Groundwater biogeochemistry, *J. Contam. Hydrol.*, 57:61–80.

Yeh, C.K., and Novak, J.T. (1995), The effect of hydrogen peroxide on the degradation of methyl and ethyl tert-butyl ether in soils, *Water Environ. Res.*, 67(5):828–834.

Yong, R.N., and Rao, S.M. (1991), Mechanistic evaluation of mitigation of petroleum hydrocarbon contamination in soil medium, Can. *Geotech. J.*, 28:84–91.

Yong, R.N., Desjardins, S., Farant, J.P., and Simon, P. (1997), Influence of pH and exchangeable cation on oxidation of methylphenols by a montmorillonite clay, *Appl. Clay Sci.*, 12:93–110.

Yong, R.N. (2000), *Geoenvironmental Engineering: Contaminated Soils, Pollutant Fate and Mitigation*, CRC Press, Boca Raton, FL, p. 307.

Zytner, R.G. (1994), Sorption of benzene, toluene ethylbenzene and xylenes to various media, *J. Haz. Mat.*, 38(11):113–126.

6 Biological Transformation of Contaminants

6.1 INTRODUCTION

The main emphasis in this chapter is on the biological treatment of contaminants. Various types of organisms involved will be described, and growth and metabolism of bacteria, in particular, will be emphasized. The types of chemicals that can be biodegraded or transformed, and the pathways of conversion will be discussed as well as the toxicity and availability of several chemicals. A proper understanding of these aspects is very important because they will serve as the foundation of knowledge required for determining the potential for natural attenuation, and especially those facets dealing with natural bioremediation.

Microorganisms are the key to the biological treatment of contaminants. These include bacteria, protozoa, fungi, algae, and viruses. Microbes are smaller than plant and animal cells and are divided into two groups depending on their cell structures:

- Prokaryotes (simple, single cells less than 5 microns), and
- Eukaryotes (single or more complex multi-cells which are greater than 20 microns).

Prokaryotes have a nuclear region not encompassed in a membrane, with only a single strand of deoxyribonucleic acid (DNA). On the other hand, eukaryotes have a nucleus surrounded by a membrane containing DNA molecules. Eukaryotes are subdivided into unicellular organisms that have multipurpose cells and multicellular organisms (plants and animals) with special-purpose cells.

6.2 TYPES OF ORGANISMS

The various types of organisms and microorganisms included in the discussion in this section fall within the Whittaker (1969) five-kingdom classification. Figure 6.1 shows the five-kingdom classification on three levels of organization. Viruses are acellular and are not included in the listing in Figure 6.1, even though they are classified as micro-organisms. For bookkeeping purposes, they are included in this discussion although we must caution that their use in bioremediation purposes has yet to be determined. In large measure, because of the threats they pose, studies regarding their applicability in remediation have yet to be seriously mounted.

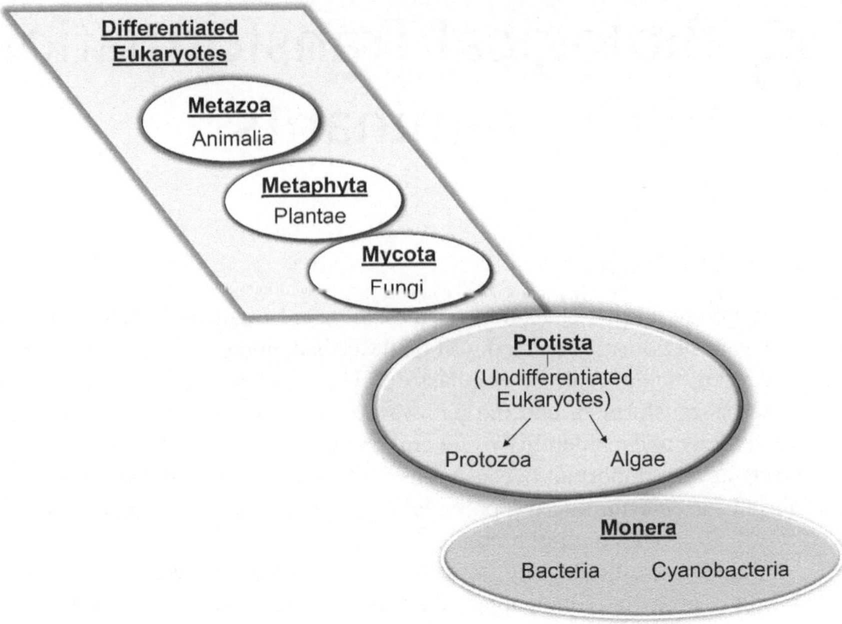

FIGURE 6.1 Types of organisms and microorganisms grouped according to the Whittaker 5-kingdom hierarchical system.

6.2.1 Protozoa

Protozoa include pseudopods, flagellates, amoebas, ciliates, and parasitic protozoa. Their sizes can vary from 1 to 2000 mm. They are aerobic, single-celled chemoheterotrophs, and are eukaryotes with no cell walls. They are divided into four main groups: (a) the Mastigophora which are flagellate protozoans, (b) the Sarcodina which are amoeboid, (c) the Ciliophone which are ciliated, and (d) the Sporozoa which are parasites of vertebrates and invertebrates.

In aerobic biological treatment processes, protozoa scavenge particles such as bacteria, yeasts, fungal spores, and other protozoa. Some protozoa such as *Entamoeba histolytica* and *Giardia lambia* cause diseases such as amoebic dysentery and giardiasis, respectively. Protozoa are found in water and soil. Water is an absolute requirement for their survival. In soil, protozoa are useful in reducing bacterial numbers near injection wells that become clogged due to excessive growth (Sinclair et al., 1993). In general, though, they do not biodegrade contaminants.

6.2.2 Fungi

Fungi are aerobic, multicellular, eukaryotes and chemoheterotrophs that require organic compounds for energy and carbon. They reproduce by formation of asexual spores. In comparison to bacteria, they: (a) do not require as much nitrogen, (b) are more sensitive to changes in moisture levels, (c) are larger, (d) grow more slowly, and

(e) can grow in a more acidic pH range (less than pH 5). Fungi mainly live in the soil or on dead plants and are sometimes found in fresh water.

Slime molds, yeasts, and mushrooms are all generally classified as fungi. Slime molds (filamentous fungus) form mycelium for adsorbing nutrients from a substrate. Since the term "substrate" is quite often used in the soil literature to mean "subsurface," the use of this term in that particular context has been diligently avoided in this book up until now. Our present use of the term substrate is in the context of biological activities. In this context, we define *substrate* to mean the food for microorganisms, i.e., substrate serves as a nutrient source and a source of carbon, or energy for microorganisms. Slime molds reproduce via formation of spore sacs or spores. *Mucor, rhizopus* and *penicillium* are examples of different species.

Yeasts are unicellular organisms that are larger than bacteria. They are shaped like eggs, spheres, or ellipsoids and reproduce by formation of buds or fission. *Neurospora* and *Candida* are two examples. Yeasts have been used industrially for many years for the production of beer, wine, cheese, antibiotics, and many other chemicals. They may or may not require oxygen.

The two main components of mushrooms are: (a) the basidia—the fruiting body above the ground, and (b) the mycelium, which is below the ground. The common mushroom is a well-known representative in this family of fungi. Appropriate aeration of the soil is required for effective growth of the fungus. *Phaenaerochaete chrysosporium* (also known as white rot fungus) produces an enzyme, peroxidase, under nitrogen limitation, which has the potential for attacking chlorinated compounds such as dioxins (Aust et al., 1988).

6.2.3 Algae

Algae are single-celled and multi-cellular micro-organisms that are green, greenish tan to golden brown, yellow to golden brown (marine), or red (marine). They grow in the soil and on trees or in fresh or salt water. Those that grow with fungi are called lichens. Seaweeds and kelps are examples of algae. Since they are photosynthetic, they can produce oxygen, new cells from carbon dioxide or bicarbonate (HCO_3^-), and dissolved nutrients including nitrogen and phosphorus. They use light of wavelengths between 300 and 700 nm. Red tides are indicative of excessive growth of dinoflagellates in the sea. The green color in a body of lakes and rivers is eutrophication due to the accumulation of nutrients such as fertilizers in the water. Algae can also bioaccumulate hydrophobic compounds (Matsumura and Essaac, 1979) or metals (Schiewer and Volesky, 1998).

6.2.4 Viruses

Although viruses are smaller than bacteria and require a living cell to reproduce, their relationship to other organisms is not clear. In order for them to replicate, they have to invade various kinds of cells. They consist of one strand of DNA and one strand of ribonucleic acid (RNA) within a protein coat. A virus can only attack a specific host. For example, those that attack bacteria are called bacteriophages. Whilst it would seem that introducing genes in bacteria by viruses would be a way for exploiting the potential of virus as a remediation treatment technique, we need to be mindful that viruses cause various diseases such as HIV, herpes, Ebola, meningitis, gastroenteritis, hepatitis, and poliomyelitis.

6.2.5 Animals (Worms)

The most significant animals in the soil are millimeter-sized worms. Nematodes are cylindrical in shape and are able to move within bacterial flocs. Flatworms such as tapeworms, eel worms, roundworms, and threadworms, which are nematodes, can cause diseases such as roundworm, hookworm, and filariasis.

6.2.6 Plants

We consider plants in this present category of organisms because they contribute significantly to remediation technology. Plants have a very evolved cell structure and are classified as eukaryotes. Under this category, we include aquatic plants, crop plants, macrophytes, seed plants, grasses, ferns and mosses. Recently, plants have been used to remove, degrade, or accumulate chemical contaminants in the soil, sediment, groundwater, surface water, or air. Plants can be used to treat petroleum hydrocarbons, chlorinated solvents, pesticides, metals, radionuclides, explosives, and excessive nutrients (Chappell, 1997). The mechanisms for contaminant treatment by the plants include rhizofiltration, phytotransformation, phytoextraction, phytostabilization, and phytovolatization. Genetic engineering or selective breeding of the plants can lead to faster uptake and higher yields of contaminants. Plant species that accumulate metals include *Thlaspi calaminare* (zinc accumulation), *Alyssum* species (nickel accumulation), *Brassica juncea* (lead accumulation) (PRC Environmental, 1997) and *Populus* (trichloroethylene [TCE] remediation) (Chappell, 1997). Bacteria can also be associated with plants particularly at the roots to assist in the natural attenuation of contaminants such as polychlorinated biphenyl (PCBs) (Vergani et al., 2017).

6.2.7 Bacteria

Bacteria are prokaryotes that reproduce by binary fission by dividing into two cells, in about 20 minutes. The time it takes for one cell to double, however, depends on the temperature and species. For example, the optimal doubling time for *Bacillus subtilis* (37°C) is 24 minutes and for *Nitrobacter agilis* (27°C) is 20 h. Classification is by shape, such as the rod-shaped bacillus, the spherical-shaped coccus and the spiral-shaped spirillum. Rods usually have diameters of one-half to one micron and lengths of three to five microns. The diameter of spherical cells varies from 0.2 to 2 microns. Spiral-shaped cells range from 0.3 to 5 microns in diameter and 6 to 15 microns in length. The cells grow in clusters, chains, or in single form and may or may not be motile. The substrate of the bacteria must be soluble. In most cases, classification is according to the genus and species (e.g., *Pseudomonas aeruginosa* and *Bacillus subtilis*). Some of the most common species are *Pseudomonas, Arthrobacter, Bacillus, Acinetobacter, Micrococcus, Vibrio, Achromobacter, Brevibacterium, Flavobacterium*, and *Corynebacterium*. Within each species, we will have various strains. Each of these can behave differently. Some strains can survive in certain conditions that others cannot. The ones that are better adapted will survive. For survival, strains, called mutants, originate due to problems in the genetic copying mechanisms. Some species are dependent on other species, though, for survival. Degradation of chemicals to an intermediate stage by one species of bacteria may be required for the growth of another species that utilizes the intermediate.

The maximum rate of growth and metabolism occurs during the log growth phase in which the log of the number of cells versus time is linear according to Monod (1949). The following equation represents the growth:

$$\frac{dN}{dT} = kN \tag{6.1}$$

where:

N = Number of cells per volume

t = Time and k is the growth rate in 1/time. Integration of this equation for time $t = 0$, when $N = N_o$ to time t gives the equation:

$$N = N_o e^{kt} \tag{6.2}$$

At the doubling time, t_d, when $N = 2N_o$, the specific growth rate, k, can be determined during exponential growth as:

$$k = \frac{\ln 2}{t_d} \tag{6.3}$$

Although the doubling time is from minutes to days, it is usually in the range of 10 to 60 minutes. This model can be used in fate and transport models since only the constant needs to be determined. However, site specific data cannot be incorporated. Laboratory data cannot be easily transferred to the field and it is assumed that biodegradation only occurs away from the source.

As the substrate, oxygen, or nutrient becomes limiting or is depleted (which takes place as the treatment process is nearing completion), pH shifts or toxic components start to accumulate as the declining growth phase occurs (Figure 6.2). Those micro-organisms capable of storing organics earlier in the growth phase, such as glycogen or poly-β-hydroxybutyrate (PHB), will survive very well in this phase. The rate of growth decreases and microorganisms begin to die. The stationary phase occurs as the numbers of cells produced and dying are equal. This occurs usually over a 12–36 h period.

When the number of microorganisms dying exceeds the amount growing, this signifies the death phase. The cells may either be inactivated or may start to degrade and break apart (lyse). An equation like the growth phase can be used as follows:

$$\frac{dN}{dt} = -bN \tag{6.4}$$

where b is the rate constant for the decrease in the cell number.

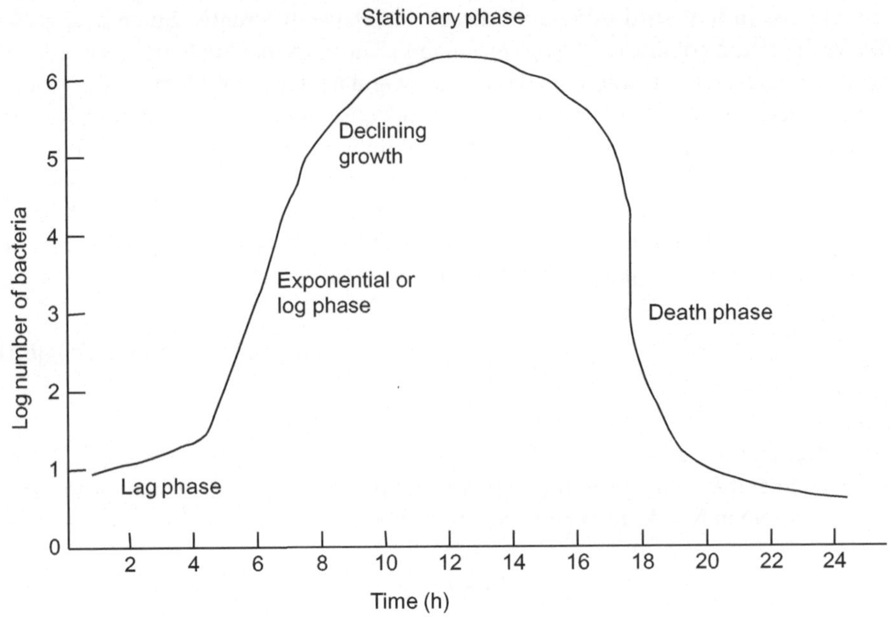

FIGURE 6.2 Bacterial growth as a function of time.

Bacterial growth rate is directly dependent on substrate concentration. The higher the substrate concentration, the higher is the growth rate. The substrate can be a single compound or a mixture. When it is a mixture, the parameters such as biochemical oxygen demand (BOD), chemical oxygen demand (COD), total organic carbon (TOC) or total petroleum hydrocarbons (TPH) can be used on a mass or liquid basis to indicate the total substrate concentration. The Monod equation (1949) is represented as:

$$\mu = \frac{\mu_{max}}{K_S + S} \quad (6.5)$$

where:

μ = Specific growth rate, time^{-1}
μ_{max} = The maximum specific growth rate, time^{-1}
S = Substrate concentration, mass/volume
K_S = Substrate concentration when the growth rate is half of the maximum growth rate, mass/volume.

The relationship between growth rate and substrate concentration can be seen in Figure 6.3. We note that maximum growth, μ_{max}, occurs at the highest substrate concentration. As the substrate concentration decreases, the growth rate also decreases. It can also be seen that S is equal to K_S at one half of the μ_{max} value. This value is dependent on pH, temperature, nutrient composition, and bacterial species. Lower K_S values mean that the microorganism has a higher affinity for the substrates.

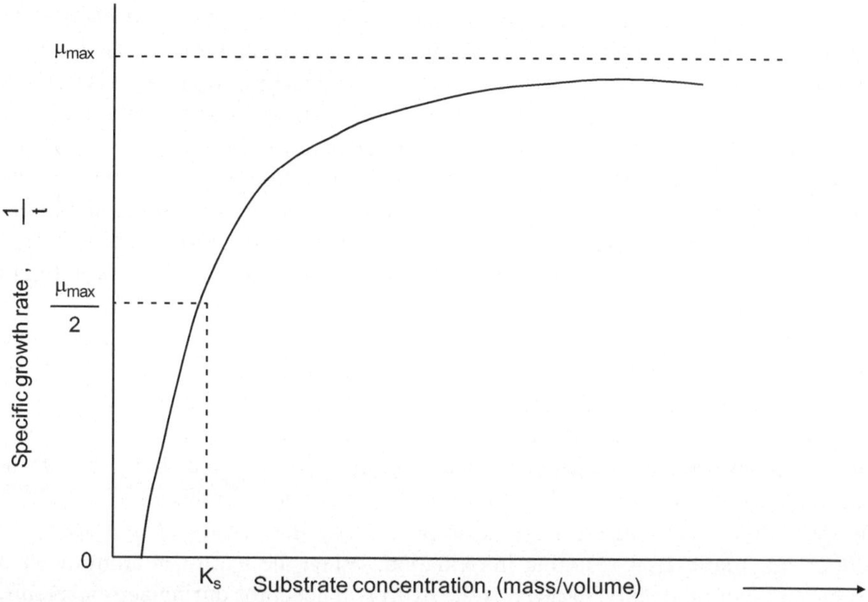

FIGURE 6.3 Effect of substrate concentration on the rate of growth according to the Monod model. K_S = substrate concentration when the growth rate is half of the maximum growth rate.

For example, K_S values for *Escherichia* for glucose and lactose are 4.0 and 20.0 mg/L, respectively. This means that this micro-organism has a higher affinity for glucose. In biological treatment processes, the substrate is the contaminant and the contaminant concentrations can be represented as:

$$\Delta C = \frac{M_t \mu_{max} C}{\left(K_c + C\right)} \Delta t \tag{6.6}$$

where:

ΔC = Contaminant concentration
M_t = Total microbe concentration
μ_{max} = Maximum rate of contaminant utilization per mass of micro-organisms
K_c = Contaminant half-saturation constant
Δt = Time interval.

Both the substrate and microbial concentrations are taken into account in this model. The values are difficult to estimate and have only been done for a few compounds.

6.3 CLASSIFICATION OF MICRO-ORGANISMS

Micro-organisms are classified by energy or carbon requirements. Most bacteria used for treatment that require organic substrates for energy are called chemo-organotrophs. If they use the organics as a carbon source, they are also called heterotrophs.

Those that use inorganic compounds as an energy source are named chemolithotrophs. Nitrifying bacteria (*Nitrosomonas* and *Nitrobacter*) that use carbon dioxide as a carbon source instead of organic compounds are called autotrophs. This is a class of chemolithotrophs. Nitrifying bacteria produce nitrite from ammonium ion which is then followed by conversion to nitrate. Photolithotrophs use light as an energy source and inorganic compounds for respiration. Photo-organotrophs use light with organic compounds for respiration. These organisms have chlorophyll to adsorb the light at specific wavelengths. In plants, blue (430 nm) and red light (680 nm) are absorbed by chlorophyll *a*. Purple and green bacteria with bacteriochlorophyll absorb light at wavelengths of 720–780 nm, 850 or 1020 nm. Photoautotrophic organisms convert light to chemical energy via adenosine triphosphate (ATP) and then reduce carbon dioxide to cellular materials by carbon dioxide fixation. Electron donors include sulfur, hydrogen sulfide, thiosulphate, hydrogen or organic compounds.

Bacteria can also be classified in respect to their requirements for oxygen. Those that require molecular oxygen for respiration are aerobes, and those that do not require molecular oxygen for respiration are anaerobes (Frobisher et al., 1974). Energy is obtained from the respiration process by the release of hydrogen from a donor enzymatically—resulting in oxidation. When the hydrogen atom meets an acceptor, reduction occurs. Oxygen is the hydrogen acceptor during aerobic respiration. It is the best one since the freest energy is released this way. Aerobic microbes thus grow faster than anaerobic strains. Water is produced as the end product. Carbonate, iron, nitrate, or sulfate ions serve as electron acceptors in anaerobic respiration. Methane, ammonia, hydrogen sulfide, or a reduced organic compound are the final products depending on the hydrogen acceptor. Facultative organisms use molecular oxygen when it is present, and then utilize nitrate or other oxides at low levels of oxygen concentrations.

For nutrition, microorganisms have specific needs. Carbon, oxygen, nitrogen, hydrogen, and phosphorus are the major elements required. The most often used empirical formula for the cell is $C_5H_7NO_2$ (Porgues et al., 1953). Incorporation of phosphate leads to the formula $C_{42}H_{100}N_{11}O_{13}P$ (McCarty, 1965). Sulphur, potassium, calcium, sodium, magnesium and chlorine are some of the minor elements. Other substances such as iron, cobalt, copper, zinc, molybdenum, boron, and aluminum are required in trace amounts. Vitamins, essential amino acids, and other precursors may also be required for growth since these are not produced by the cells. Some of these components may have to be added to the soil because deficiencies in the soil will not permit the optimal growth and performance of the microorganisms.

6.3.1 Actinomycetes

Actinomycetes are mainly aerobic prokaryotes that are multicellular, filamentous, rod-shaped bacteria. Like fungus, they can produce spores or spore chains called conidia. They are commonly found in soil and water and can degrade lignocellulosics, phenols, aromatics, steroids, and chlorinated aromatics (US EPA, 1983). Actinomycetes are very resilient since they resist desiccation and are able to withstand large changes or variations in pH and temperature.

6.4 PHYSICAL AND CHEMICAL ENVIRONMENTAL EFFECTS ON MICRO-ORGANISMS

Favorable environmental factors are important if optimal performance of microorganisms is to be obtained. The most important environmental physical factors are temperature, oxygen availability, presence of moisture, and osmotic pressure. There are three main temperature ranges in which the various classes of microorganisms can grow optimally. Psychrophiles grow at 0°C–10°C, mesophiles grow from 10°C to 45°C and thermophiles from 45°C to 75°C. Mesophiles are the most common, particularly for biological treatment processes. Few micro-organisms are able to survive exposure at temperatures above 75°C for extended periods. Capsule formation enables micro-organisms to survive at reduced temperatures and grow as the temperature increases to more favorable conditions (Sims et al., 1990). In general, the guidelines are that a 10°C increase in temperature will double the growth rate.

Moisture availability is essential for the survival of micro-organisms since it is the principal component of cell protoplasm, and is required for nutrient transport into the cell. Obviously, this can become a problem in the vadose zone where moisture contents can be relatively very low. In general, moisture levels of 50%–75% are optimal (US EPA, 1985). Too much moisture can also be a problem since this can cause oxygen limitations because of water saturation of soil pores. Transport of oxygen to bacteria in the soil can be a problem since oxygen solubility in water is low (9 mg/L at 20°C) and if the soil is very heterogeneous. Sufficient oxygen must be available for aerobic and facultative anaerobic bacteria. In general, a concentration of approximately 2.0 mg/L is considered a good design value (Reynolds and Richards, 1996).

Oxygen-releasing compounds (ORC) constitute a very useful means for delivery of oxygen to the microorganisms, and patented "designer" chemicals exist to provide such delivery. One such patented chemical is magnesium peroxide which releases oxygen over a time period when water is added—thus ensuring that oxygen supply is continuous. They are commercially available as powders or pellets and have been used to remediate petroleum and organic compounds, in addition to some chlorinated solvents, vinyl chloride, pentachlorophenol (PCP), polycyclic aromatic hydrocarbons (PAHs), and methyl tert-butyl ether (MTBE). A slurry mixture can be injected into a well as a powder or in boreholes as a slurry. In wells or trenches, they can be added as filter socks to the groundwater. Continuous mechanical operation is not required after injection. As a biostimulation technique, the availability of these kinds of designer chemicals in combination with their application technology can contribute significantly to a reduction in remediation costs (Fulton, 1996).

Another factor that can influence microbial activity is osmotic pressure since salt concentration affects the intake of nutrients by osmosis. A range of 500–35,000 mg/L is acceptable for salt concentrations.

6.4.1 Contaminant Availability

Availability of contaminants as a source material for metabolic activities by microorganism is an important factor, particularly when the contaminants are in a free form in the soil–water system. Because heavy petroleum components and others can sorb strongly onto soil, surfactants can be added in soil flushing situations to enhance the desorption of the contaminants and release of the organic chemicals (Aronstein et al., 1991). Production of microbial surfactants in the aquifer enhances desorption of hydrocarbons from the soils. However, these can be interpreted as contaminant levels higher than expected and can be seen as a spike. Rhamnolipids produced by *Pseudomonas aeruginosa* and surfactin produced by *Bacillus subtilis* can remove arsenic, chromium, copper, zinc, and lead from contaminated soils, sediments and mining residues (Mulligan et al., 1999; Arab and Mulligan, 2018). Other substances called chelants are produced by micro-organisms. These can bind insoluble metals which then can be solubilized and mobilized into the groundwater.

Organic chemicals can penetrate the micropores of the microstructural units described in Chapter 3. Since these micropores are less than one micron in diameter, bacteria cannot easily grow and neither can their enzymes penetrate. Molecules can also form complexes with humic substances in soils. These complexes can be resistant to micro-organisms and thus remain undegraded for extended periods of time (Alexander, 1994).

6.4.2 Chemical Factors

Some of the major chemical factors which affect microbial activity include pH, toxicity, heavy metals, molecular structure, and cometabolism. Most microbes prefer a pH range of 6.5–9.0. However, some micro-organisms can function at high or low pH values. For example, fungi grow at pH values below 5 while sulfur-oxidizing bacteria produce sulfuric acid which lowers the pH below 1 (e.g., $H_2S + 2O_2 \rightarrow H_2SO_4$). Adjustment of the pH may be necessary before biological treatment to obtain the preferred pH. However, once a neutral pH is established, generation of carbon dioxide by the micro-organisms can help to buffer the system and maintain the neutral pH. In soil or water, pH can be raised through addition of lime, calcium hydroxide, calcium carbonate, magnesium carbonate, etc. To lower soil pH, sulfur compounds such as sulfuric acid can be used (Dupont et al., 1988).

6.4.3 Toxicity

Toxicity inhibits microbial growth or substrate utilization. The process of degradation of substances which are normally biodegradable can be slowed or stopped by the presence of a toxic component. Several mechanisms are possible. Inhibition of a single enzyme in a metabolic pathway can adversely affect microbial growth or substrate utilization. Sulfonylurea is known to specifically inhibit acetolactate synthase which is required for the synthesis of amino acids which is the basis of proteins (LaRossa and Schloss, 1984). Mercuric, cupric or uranyl metal ions can non-specifically inhibit the functioning of the cell membrane by binding to the cell surface and breaking the cell membrane (Khovrychev et al., 1974). This can cause many problems in maintaining concentration gradients across the cell. Cyanide and azide are toxic in a different manner since they can bind to the heme iron present in cytochromes which are required for many electron transport reactions (Brock et al., 1984). Binding eliminates the cytochromes from taking part in these reactions and subsequently inhibits cell energy-generating processes. Organic acids, ethers, aldehydes, phenol, chlorophenol, cyanide, ammonia compounds, antibiotics, and dyes can also be toxic, depending on their concentrations. Greases can inhibit respiration by coating the micro-organism, inhibiting transport across the membrane of nutrients and oxygen.

6.4.4 Heavy Metals and Other Compounds

Numerous contaminant components can be toxic to micro-organisms at low or high concentrations. Contact time, temperature, and concentration of these agents will influence their effect on the microbes. In general, lower concentrations at short contact times are less toxic. For example, 2,4-D will degrade much faster at 1–100 ppm than at higher concentrations where it can remain undegraded for years. Toxic compounds include acids such as benzoic acid, bases such as ammonium hydroxide, oxidizing and reducing agents, and halogens (chlorine, iodine, bromine, fluorine). Heavy metals such as mercury, arsenic, and lead are particularly toxic. Other toxic heavy metals in wastewaters include chromium, zinc, cadmium, copper, barium, and nickel.

6.4.5 Molecular Structure

Molecules which are too large to enter the microbial cell (e.g., with high molecular weights or long chain lengths), or which have low water solubility, or containing structural factors inhibiting metabolism can be difficult to degrade. Amine, methoxy, sulfonate and nitro groups, benzene with meta substitution, branched carbon chains, and ether linkages are some examples of structural factors that affect biodegradability. The biodegradability of a substance is dependent on the degree of saturation or branching, and nature and amount of substitution. Water solubility is important since components enter the cell by water transportation. Increasing the saturation in a compound tends to decrease its solubility. The substrate to be degraded can become unavailable to the microorganism by sorption onto soil, or it may have limited water solubility.

A high level of branching of the organic chemical compound can also decrease biodegradability (Evans and Fuchs, 1988). N-octane, a straight chain compound, is easier to biodegrade than its branched isomer, 3-propyl pentane. Replacement of carbon atoms with sulfur, phosphorus or chlorine can decrease biodegradability. Substitution with nitrogen atoms can also adversely affect degradability—with the exception of amino acids and proteins which are vital for metabolism.

6.4.6 Halorespiration

The use of chlorinated compounds as electron acceptors by micro-organisms has been recently called dehalorespiration (Hollinger and Schumacher, 1994). The term *halorespiration* is preferred by Weidemeier et al. (1999). It seems to be a major mechanism in the natural attenuation of perchloroethylene (PCE), trichloroethylene (TCE), Trichloroacetic acid (TCA) and carbon tetrachloride and in the biodegradation of 1,2-dichloroethylene (DCE) and vinyl chloride which have fewer chlorine atoms. Michaelis-Menten kinetics are followed. Hydrogen is produced by fermentation of organic compounds in the first step of the two-step dechlorination process. Hydrogen next serves as the electron donor during the second step of halorespiration. The fermented organic compounds include benzene, toluene, ethylbenzene, and xylene (BTEX), acetone, or other organic non-chlorinated materials.

6.4.7 Cometabolism

Cometabolism is the degradation of a substrate in the presence of another substrate (a co-substrate). The co-substrate which is not required for energy or nutrient purposes is converted when the enzymes are induced for utilization of the principal substrate. However, the concentration of the co-substrate must not be too high if one wants to avoid competition for enzymes with the substance required for growth. Otherwise, degradation will be inhibited. Pulsing of the co-substrate into the soil to assist biodegradation of a contaminant is one technique that has been used (McCarty, 1988). Co-metabolism of recalcitrant organic compounds such as trichloroethene, dioxin, and polychlorinated biphenyls (PCBs) has been performed only on a small scale (Eweis et al., 1998). Another situation that can occur is diauxic growth, i.e., the substrate that can be more easily degraded will be used before any other (more difficult to degrade) substrate.

The injection of natural gas or methane with air can influence the metabolism of microorganisms called methanotrophs. These bacteria possess mono-oxygenase enzymes that oxidize methane, alkanes, alkenes, and halogenated alkanes (Thomas and Ward, 1989). Some compounds that are recalcitrant under normal conditions can be degraded when methane is used as a carbon and energy source. Some of the more common recalcitrant chemical compounds include the chlorinated solvents such as tetrachloroethylene, trichloroethylene, 1,1,1-trichloroethane. 1,1,2-trichloroethane, 1,1 dichloroethane, cis-and trans-1,2 dichloroethylene, 1,2 dichloroethene, vinyl chloride, dichloromethane, carbon tetrachloride, and chloroform. Alternating pulses of methane and oxygen into groundwater could enhance the biodegradation of trichloroethylene, cis-dichloroethylene, trans-dichloroethylene and vinyl chloride (Semprini et al., 1990).

The acclimation and subsequent growth of methanotrophs can take considerable time. A major factor is the breakthrough of methane. Fluid injection should be stopped once methane is detected to allow the bacteria to grow. It should then start when the population is well established. Residence time should be approximately 100 days (McCarty et al., 1991).

Degradation of TCE has been performed at field scale (Hazen, 1991). Competition of methane with TCE is a problem inasmuch as methane has a higher affinity for methane monooxygenase than TCE. Other problems include loss of enzyme activity upon TCE oxidation and the requirement for external energy sources to complete the oxidation.

6.5 BIODEGRADATION OF ORGANIC CHEMICALS

6.5.1 Petroleum Hydrocarbons

Petroleum contains a wide range of compounds from very biodegradable to recalcitrant due to their varying degrees of branching, chain lengths, molecular sizes, substitution with nitrogen, oxygen or sulfur atoms. A variety of microbial species would be required for efficient degradation of these compounds. Initially aerobic bacteria are used (Atlas, 1991), followed by anaerobic ones that utilize nitrogen or sulfur compounds as electron acceptors (Bartha, 1986).

6.5.2 Alkanes

The various components of petroleum hydrocarbons are alkanes, cycloalkanes, aromatics, polycyclic aromatic hydrocarbons, asphaltenes, and resins. Although alkanes, represented by the formula $C_2H_{2n+2,}$ (where n is the number of carbons and 2_{n+2} is the number of hydrogens), can have many isomers as the number of carbons increase, relatively few exist in petroleum. They are sometimes referred to as aliphatic compounds. Low molecular weight alkanes are the most easily degraded by micro-organisms despite the harsh conditions required for chemical degradation (strong acids, high temperatures, ultraviolet light, etc. [Morrison and Boyd, 1973]). As the chain length increases from C_{20} to $C_{40,}$ hydrophobicity increases while solubility and biodegradation rates decrease. Conversion of alkanes leads to the formation of an alcohol using a mono-oxygenase enzyme, followed by oxidation to an aldehyde and then to a fatty acid (Pitter and Chudoba, 1990). Further oxidation of the fatty acid yields products less volatile than the original contaminants. The oxidation of the fatty acids is called β-oxidation and is required for growth by many aerobic bacteria. Anaerobic bacteria such as sulfate-reducing bacteria are capable of degrading fatty acids via this step (Widdel, 1988).

Many of the alkanes found in petroleum are branched. Although bacteria that are capable of degrading n-alkanes cannot degrade branched ones (Higgins and Gilbert, 1978), *Brevibacterium ethrogenes, Corynebacteria* sp., *Mycobacterium fortuitum, Mycobacterium smegmatis*, and *Nocardia* sp. have been shown to grow on branched alkanes. The first degradation step is the same as for the unbranched alkanes. However, the β-oxidation is more difficult and less efficient (Pirnik, 1977). In addition, in the presence of n-alkanes, the metabolism of the branched alkanes will be repressed, which will cause difficulties during the degradation of mixtures such as petroleum.

6.5.3 Alkenes

Alkenes with a double bond between carbons and have not been extensively studied for biodegradation. Those containing the double bond on the first carbon may be more easily degradable than those alkenes with the double bond at other positions (Pitter and Chudoba, 1990). The products of oxidation of 1-alkenes can be either diols or the methyl group.

6.5.4 Cycloalkanes

Cycloalkanes are not as degradable as alkanes due to their cyclic structure. Their biodegradability also decreases as the number of rings increases, partly because of decreasing solubility (Pitter and Chudoba, 1990). Bacteria growing on cycloalkanes are not found frequently and can be difficult to isolate. Specific enzymes are needed to cleave the ring. Species of *Nocardia* and *Pseudomonas* are able to use cyclohexane as a carbon source as shown in Figure 6.4. Oxidation of the cycloalkanes with the oxidase enzyme leads to production of a cyclic alcohol and then a ketone (Bartha, 1986). The ketone forms to a lactone by insertion of an oxygen in the ring. In the case of cyclohexane, the lactone (e-caprolactone) is hydrolyzed to adipic acid, a substrate for β-oxidation (Perry, 1984).

6.5.5 BTEX

Aromatic compounds containing benzene structures are more difficult to degrade than cycloalkanes. Benzene, toluene, ethylbenzene and xylene, commonly called BTEX, are volatile, water soluble, hazardous components of gasoline. Benzene is a known carcinogen. Degradation is catalyzed by oxygenases from fungi and other eukaryotes to form trans-dihydrodiols. Dihydrodiols, formed by dioxygenases from bacteria, have been produced from naphthalene, anthracene, biphenyl, benzene, toluene, xylene, benzo(a)pyrene, and phenanthrene (Gibson, 1988). The dihydrodiol is

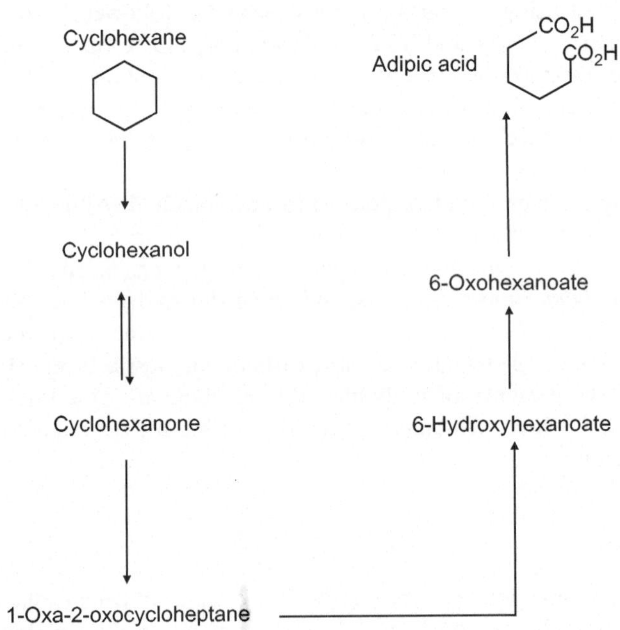

FIGURE 6.4 Pathway of microbial degradation of cyclohexane. (From Verschueran, K., *Handbook of Environmental Data on Organic Chemicals*, 2nd ed., Van Nostrand Reinhold, New York, 1983.)

then converted to catechol by dehydrogenation. Catechol is either converted by an ortho pathway to produce muconic acid (splitting of the ring between carbons with the hydroxyl groups) or the meta pathway to produce 2-hydroxymuconic semialdehyde (splitting of the ring at the carbon with a hydroxyl group) (Cerniglia, 1984). The descriptive diagram in Figure 6.5 shows these reactions and Figure 6.6 shows

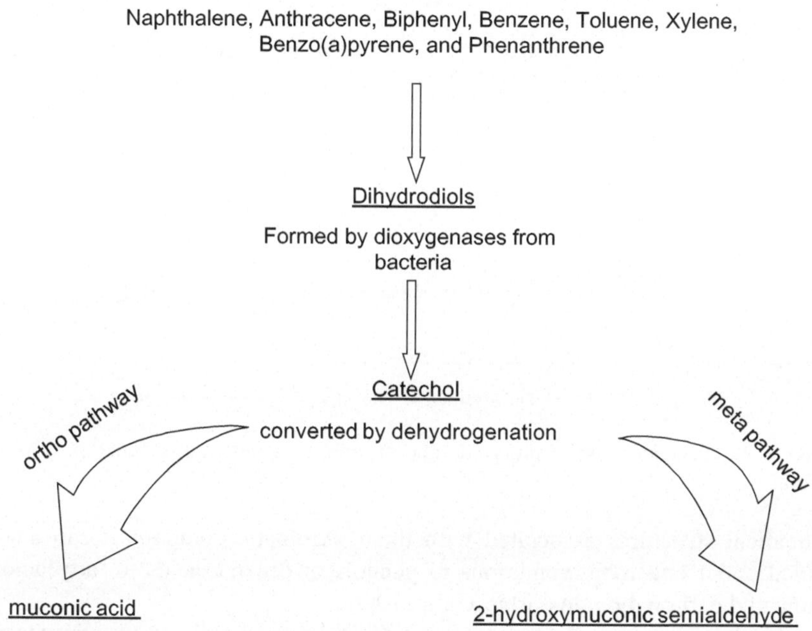

FIGURE 6.5 Schematic diagram showing oxidation of aromatic compounds to produce muconic acid or 2-hydroxymuconic semialdehyde.

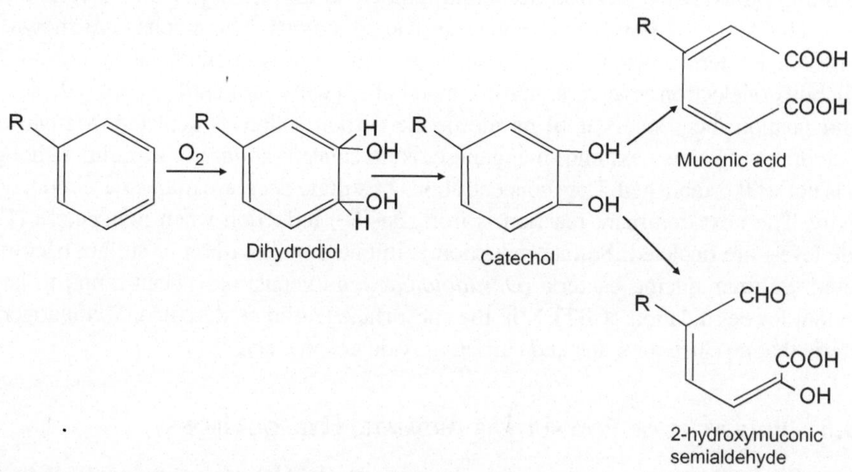

FIGURE 6.6 Oxidation of aromatic compounds to produce muconic acid or 2-hydroxymuconic semialdehyde—chemical reactions.

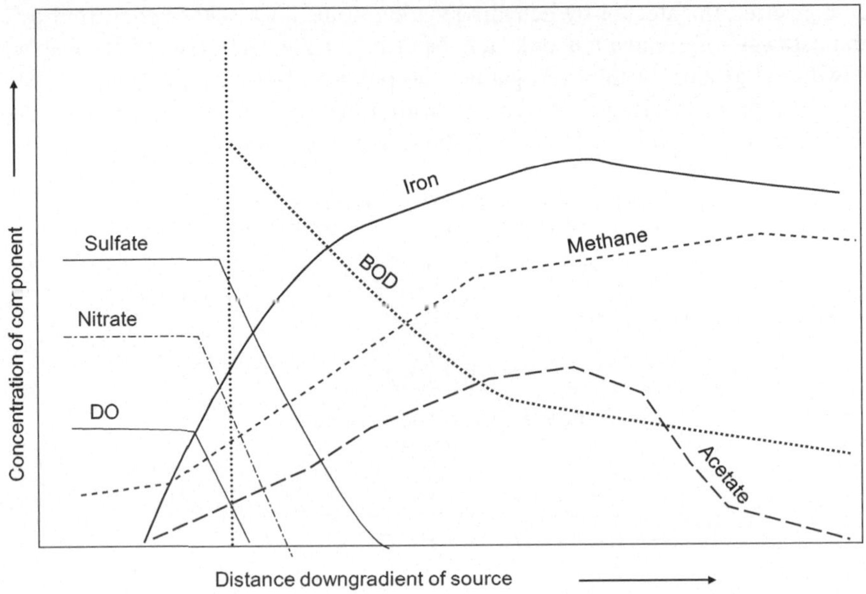

FIGURE 6.7 Concentration of inorganic and organic parameters as a function of oxygen depletion.

the chemical structures associated with them. Aromatic compounds can also be degraded under anaerobic conditions to phenols or organic acids to fatty acids to methane and carbon dioxide (Grbic-Galic, 1990).

Aerobic degradation of all components of BTEX occurs rapidly when oxygen is present at a concentration of 2 mg/L in water or 5% in gaseous vadose zone (Salinitro, 1993; Brown et al., 1995). Under anaerobic conditions, the formation of up to five zones occurs, depending on available electron acceptors (Anderson and Lovley, 1997). Degradation is less reliable and is slower than under aerobic conditions. These zones are shown in Figure 6.7. Bacterial metabolism proceeds through a series of steps depending on the availability of electron acceptors. Aerobic metabolism will occur until oxygen is depleted. Denitrification then occurs until nitrate or the carbon source is depleted. Manganese (IV) reduction occurs next until manganese oxide concentrations are limiting, although this is not well established. Low concentrations of sulfate decrease manganese reduction activity. The next dominant reaction is iron (Fe(III)) reduction when manganese (IV) oxide levels are depleted. Sulfate reduction is initiated until carbon or sulfate becomes limited. Sulfur reducing bacteria (*Desulfotomaculum* genus) have been found to have potential for degradation of BTEX in the subsurface (Aüllo et al., 2016). Methanogenic bacteria that produce methane and carbon dioxide become active.

6.5.6 Polycyclic or Polynuclear Aromatic Hydrocarbons

Polycyclic or polynuclear aromatic hydrocarbons (PAHs) are components of creosote and are produced during petroleum refining, coke production and wood preservation (Park et al., 1990). Many are suspected to be carcinogens. A general form for PAHs is $C_{4n+2}H_{2n+4}$, where n is the number of rings. A four-ring PAH would be $C_{18}H_{12}$. Similar to the case of cycloalkanes, as the number of rings increases, the compounds become more difficult to degrade due to decreasing volatility and solubility and increased sorption. Some examples are shown in Table 6.1. They are also degraded in a manner similar to single ring aromatics since PAHs are degraded one

TABLE 6.1
The Molecular Weight, Structure, Vapor Pressure, and Solubility of Various PAHs

Compound	Molecular Formula (Molecular Weight)	Molecular Structure	Vapor Pressure, mm Hg	Solubility (mg/L at 25°C)
2-ring: Naphthalene	$C_{10}H_8$ (128)	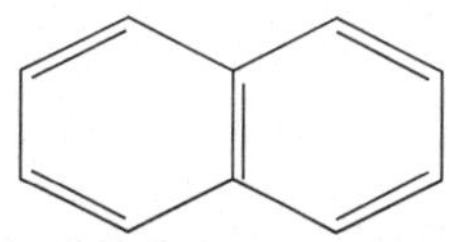	0.082 (25°C)	31.7
3-ring: Anthracene	$C_{14}H_{10}$ (178)	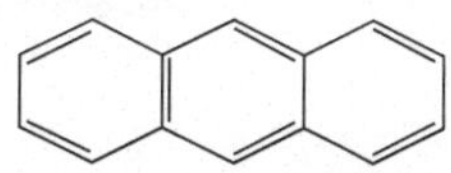	1.9×10^{-4} (20°C)	0.045
Phenanthrene	$C_{14}H_{10}$ (178)	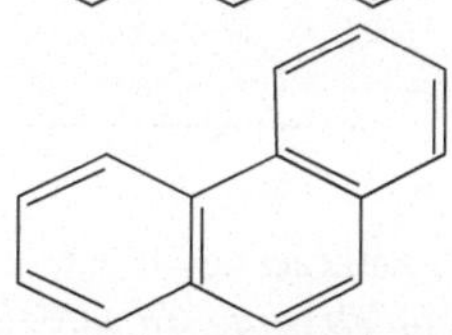	6.8×10^{-4} (20°C)	1.0
4-ring: Pyrene	$C_{16}H_{10}$ (202)	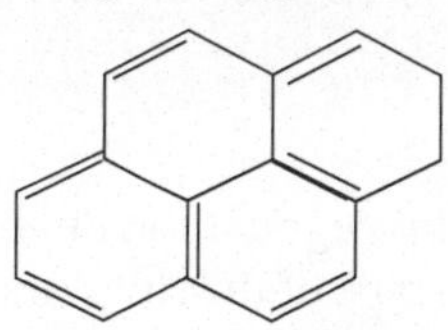	2.5×10^{-6}(25°C)	0.132
Benzo(a) anthracene	$C_{18}H_{12}$ (228)	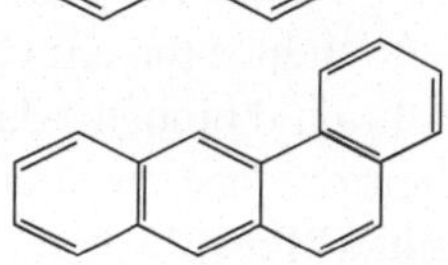	1.0×10^{-8}(20°C)	9.4×10^{-3}
Chrysene	$C_{18}H_{12}$ (228)	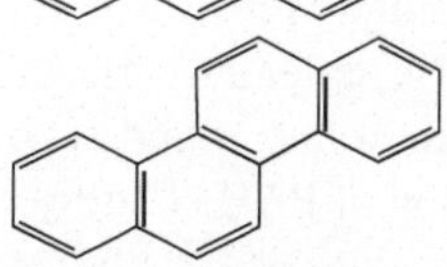	6.3×10^{-9}(25°C)	1.8×10^{-3}
5-ring: Benzo(a) pyrene	$C_{20}H_{12}$ (252)	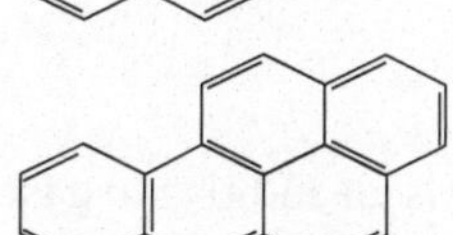	5.6×10^{-9} (25°C)	1.2×10^{-3}
6-ring: Benzo(g,h,i) perylene	$C_{22}H_{12}$ (276)	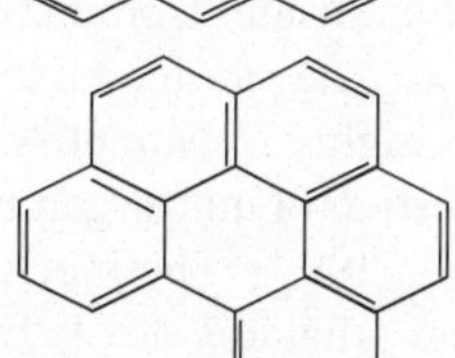	1.0×10^{-10} (25°C)	0.7×10^{-3}

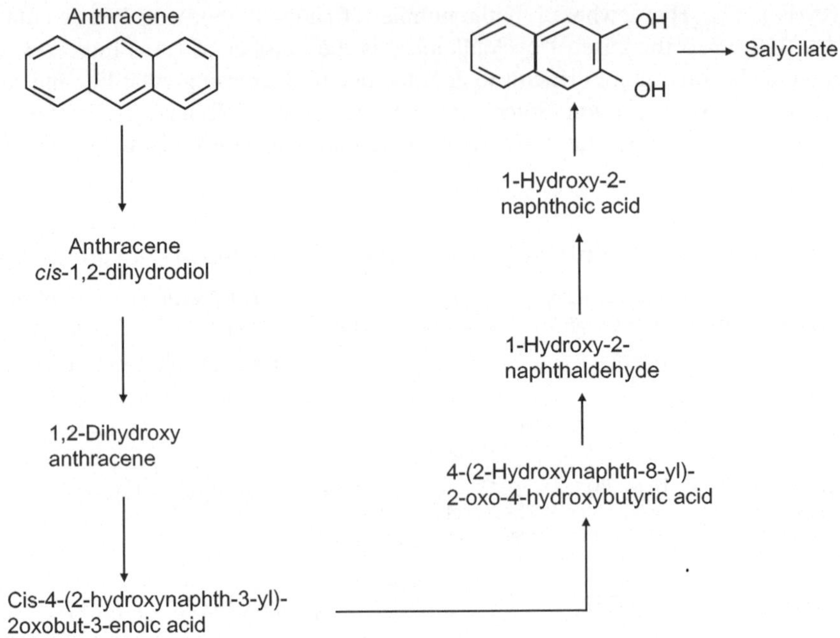

FIGURE 6.8 Biodegradation of anthracene to CO_2. (Adapted from Verschueran, K., *Handbook of Environmental Data on Organic Chemicals*, 2nd ed., Van Nostrand Reinhold, New York, 1983.)

ring at a time. The pathway for biodegradation of anthracene is shown in Figure 6.8. Carboxylates are formed by cleavage of the rings under aerobic conditions. However, simple aromatics must be present to induce the enzymes necessary for PAH degradation (Atlas, 1991). If they are not, limited biodegradation of the PAHs up to five rings may take place. Fungi, due to their oxidase production, can significantly contribute to biodegradation. Nitrate-reducing (McNally et al., 1998) and sulfate-reducing conditions (Zhang and Young, 1997) can lead to microbial degradation of low molecular weights PAHs. Understanding dissolution and sorption are necessary for an understanding of the natural attenuation of PAHs (Luthy et al., 1994; Ghoshal et al., 1996).

6.5.7 MTBE

Methyl tert-butyl ether (MTBE) is an additive to gasoline in concentrations as high as 15% by volume. It reaches the groundwater from inadvertent spills on the surface and through leaking underground gasoline storage tanks. Contaminated groundwater is undrinkable and smells bad. It is highly resistant to biodegradation since it is reactive with microbial membranes. Insertion of an oxygen molecule into MTBE can lead to further breakdown (Steffan et al., 1997). The mechanism for natural attenuation is currently conflicting. Some believe that it is slowly biodegraded (Borden et al., 1997) while others believe that it partially degrades to tert-butyl alcohol, a health hazard (Landmeyer et al., 1998).

6.5.8 Halogenated Aliphatic Compounds

Examples of halogenated aliphatic compounds are pesticides such as ethylene dibromide (DBR) or $CHCl_3$, $CHCl_2Br$, and industrial solvents including methylene chloride and trichloroethylene. Aerobic degradation can be more difficult to achieve because of the lower energy and the higher oxidation state of the compound which is due to the presence of the halogen. Accordingly, anaerobic biodegradation is more beneficial inasmuch as the number of halogens in the compound increases. Aerobic degradation becomes more difficult. Methylene chloride, chlorophenol, and chlorobenzoate are the most aerobically biodegradable. Since methylene chloride can be treated aerobically and anaerobically (Magli et al., 1998), it is highly likely to be removed through natural attenuation.

Because of the decreasing electronegativity, i.e., the ability to accept electrons, the ease of replacement of the halogens in decreasing order can be specified as follows: iodine, bromine, chlorine, and fluorine. Removal of the halogen and replacement by a hydroxide group is often the first step of the degradation process, particularly when the carbon chain length is short.

$$R\text{-}X + H_2O \rightarrow R\text{-}OH + HX \tag{6.7}$$

Methylene chloride is such an example—where formaldehyde, 2-chloroethanol and 1,2-ethanediol are intermediates and carbon dioxide is the final product (Pitter and Chudoba, 1990). If the chain is long, the methyl group at the end of the molecule is oxidized to a halogenated alcohol. This is because the halogen has less influence on the reaction. Biodegradation of chlorinated ethenes involves formation of an epoxide, and hydrolysis to carbon dioxide and hydrochloric acid.

Reductive dehalogenation can occur anaerobically and involves replacement of the halogen with hydrogen or formation of a double bond when two adjacent halogens are removed (dihalo-elimination). As illustrated in Figure 6.9, perchloroethylene (PCE) and TCE can be reduced to form vinylidene and vinyl chloride (VC) that are more toxic and volatile than the original compounds (Freeman and Gossett, 1989). However, rapid oxidation of vinyl chloride to carbon dioxide and water occurs under aerobic conditions. Vinyl chloride has also been found to be more biodegradable than DCE. The results of a pilot test conducted by Scheutz et al. (2010) in a thin clayey till contaminated with cDCE and VC indicated that the injection of an organic substrate such as soybean oil and *Dehalococcoides-vcrA* bacteria can promote the development of a reductive dechlorination bioactive zone. The zone was established in a sand fracture within the clayey till. Complete cDCE dechlorination to ethane occurred within 148 days.

The degradation of the contaminant, 1,1,1-trichloroethane (TCA), was studied in a shallow groundwater in Shanghai (Lu et al., 2015). Natural attenuation processes occurred at the site. The mechanisms included adsorption and biotic dehydrochlorination of TCA. The rates were 0.068 m^3/kg and 0.0045 d^{-1}. The TCA degradation bacterial community were identified as *Arthrobacter, Geobacteraceae, Geobacter, Mucilaginibacter*, and *Trichococcus*.

Another mechanism, induction of mono-oxygenase or dioxygenase enzymes, can lead to the co-metabolism of TCE by methanotrophs (Alvarez-Cohen and McCarty, 1991). Molecular oxygen, and a primary substrate (methane, ethene, phenol, toluene, or other compounds) must be available for natural attenuation by this mechanism.

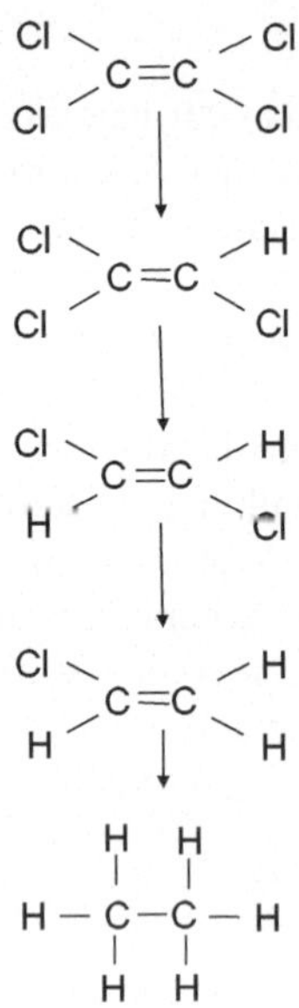

Perchloroethylene C_2Cl_4	ws = 275 ppm; log k_{oc} = 2.42; log k_{ow} = 2.60
Trichloroethylene C_2HCl_3	ws = 1235 ppm; log k_{oc} = 1.81; log k_{ow} = 2.53
1,2,-Dichloroethylene $C_2H_2Cl_2$	ws = 600 ppm; log k_{oc} = 1.77; log k_{ow} = 2.09
Vinyl Chloride C_2H_3Cl	ws = 2763 ppm; log k_{oc} = 0.39; log k_{ow} = 0.60
Ethane CH_3CH_3	ws = NA; log k_{oc} = NA; log k_{ow} = NA

NA = not available

All property values are average values. ws = water solubility at 20 deg. C.

FIGURE 6.9 Schematic showing pathway of degradation of perchloroethylene (PCE) to vinyl chloride (VC), with corresponding values for water solubility and partition coefficients.

6.5.9 Halogenated Aromatic Compounds

The pesticides, DDT, 2,4-D and 2,4,5-T, plasticizers, pentachlorophenol, polychlorinated biphenyls, among others, are examples of halogenated aromatic compounds. Their stability and toxicity are causes of great concern for the environment and public health. The halogenated aliphatic compounds, position, and number of halogens are important in determining the rate and mechanism of biodegradation. In addition, the mechanisms of conversions include hydrolysis (replacement of halogen with hydroxyl group), reductive dehalogenation (replacement of halogen with hydrogen), and oxidation (introduction of oxygen into the ring causing removal of halogen). Unlike the other two, the latter mechanism can occur only in the presence of oxygen. Reductive dehalogenation will more likely take place as the number of halogens increases. Ring cleavage could occur before oxidation, reduction, or substitution of the halogen.

Chlorophenoxy herbicides and chlorobenzene can be oxidized to halocatechols, followed by ring cleavage (Pitter and Chudoba, 1990). Bacterial strains of *Pseudomonas* sp., *Acinetobacter calcoaceticus* and *Alkaligenes eutrophus* have been able to degrade aromatic halogenated compounds in this manner (Reineke and Knackmuss, 1988). The cleavage for chlorobenzene can occur either at the ortho position to form chloromuconic acid or at the meta position to form chlorohydroxymuconic semialdehyde. Subsequent dehalogenation can be spontaneous (Reineke and Knackmuss, 1988).

Chlorinated benzoates (Suflita et al., 1983), 2,4,5-T pesticides, PCBs (Thayer, 1991), 1,2,4-trichlorobenzenes (Reineke and Knackmuss, 1988) are known to undergo reductive dehalogenation under anaerobic conditions. *Rhodococcus chlorophenolicus*

(Apajalahti and Salkinoja-Salonen, 1987) and *Flavobacterium* sp. (Steiert and Crawford, 1986) can aerobically biodegrade pentachlorophenol while anaerobic degradation of 3-chlorobenzoate and PCBs has been identified by methanogenic consortia (Shelton and Tiedje, 1984; Nies and Vogel, 1990). Proteobacteria, Spirochaetes, Firmicutes, and Bacteriodetes have been found in sediment with PCB and TPH contamination with an abundance of chloroflexi-related bacteria (Aldhafiri et al., 2018). Pentachlorophenol (PCP) is an example (shown in Figure 6.10) of this type of reaction, which forms 3,4,5-trichlorophenol, 3,5-dichlorophenol and 3-chlorophenol (Reineke and Knackmuss, 1988). These products can be oxidized aerobically only. Monochlorobenzoate, another product of this reaction, however, degrades aerobically to carbon dioxide. The position of the halogen with respect to other functional groups on the molecule can influence the ease of reductive dehalogenation. In contrast, PCBs can be dechlorinated at the *meta*- and *para*-positions first (Nies and Vogel, 1990).

Pseudomonas sp. and *Arthrobacter* sp. (Reineke and Knackmuss, 1988), are able to hydrolytically dehalogenate 4-chlorobenzoate while *R. chlorophenolicus* (Apajalahti and Salkinoja-Salonen, 1987) and *Flavobacterium* sp. (Steiert and Crawford, 1986) could degrade PCP in this manner. Displacement of the halogen is from *ortho*- and/or *para*-positions. Figure 6.10 is an example of hydrolysis of PCP (Reineke and Knackmuss, 1988). In this reaction, the product, tetrachloro-*p*-hydroquinone, formed by substitution of the chlorine with a hydroxyl group, can only be further degraded anaerobically. In the field, PCP degradation is slow, most likely because the appropriate organisms are not present.

Although PCBs have been banned since the 1970s, they are still found in aqueous and sediment systems (Luthy et al., 1997). If the conditions are appropriate, degradation can occur. Congeners containing fewer chlorines are degraded more quickly than those with more than four chlorine atoms (Harkness et al., 1993). Soluble forms are much more likely to biodegrade through natural attenuation than those sorbed to solids or entrapped in non-aqueous phase liquids (NAPLs).

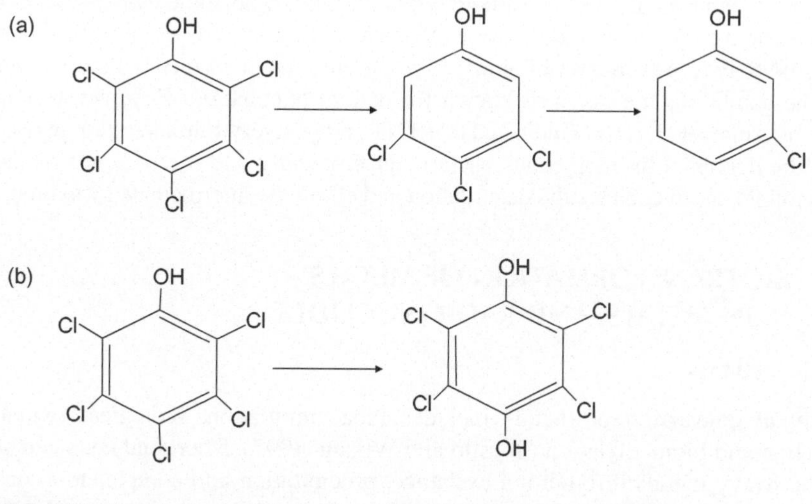

FIGURE 6.10 Reductive dehalogenation (a) and hydrolysis (b) of pentachlorophenol (PCP).

Dioxins such as tetrachlorodibenzo-p-dioxin (TCDD) are by-products of industrial processes and incineration. Since these compounds sorb strongly and are highly complex, there has been limited evidence of their biodegradability (Barkovski and Adriaens, 1996).

6.5.10 Pharmaceuticals

Another group of chemicals that has gained attention is pharmaceuticals due to their incomplete removal from wastewater treatment plants and their endocrine-disrupting potential. Lin et al. (2006) determined that pharmaceutical infiltration in soil was attenuated mainly by sorption and biodegradation. The removal of selected pharmaceuticals (caffeine, acetaminophen, sulfamethoxalone, naproxen and carbamazepine) was studied by Martinez-Hernandez et al. (2017). Acetaminophen and caffeine were easily degradable while carbarmazepine was the least degradable. Concentration levels affected the biodegradation rates. Transformation products were detected for sulfamethoxalone and carbarmazepine.

6.5.11 Nitroaromatics

Nitroaromatics in the form of 2,4,6-trinitrotoluene (TNT), 2,6- and 2,4- isomers of dinitrotoluene, nitrobenzene, 1,3-dinitrobenzene, nitrophenol isomers and nitroaniline can be found in contaminated soils, groundwaters, wastewater and leachates from the manufacture of many products such as explosives (Wujcik et al., 1992). The nitrogen group on an aromatic ring can make biodegradation very difficult and can also be toxic to the bacteria. There is evidence, however, of degradation by two different pathways: nitrogroup reduction, where the nitro (–NO group) is converted to an amine ($-NH_2$) (Preuss et al., 1993), or oxidation of the nitrophenol such that the nitro is replaced by a hydroxyl group (Spain and Gibson, 1991). In the first reaction, production of the amino aromatics may not be beneficial since the products can be toxic and can sorb strongly to the soil (Funk et al., 1995). Although there seems to be some evidence that TNT and Research Department Explosive (RDX) can be used as nitrogen sources (Coleman et al., 1998; Esteve-Nunez and Ramos, 1998), accumulation of metabolites is possible, and the conditions are not well known for natural attenuation. However, recently a study by Amon et al. (2017) indicated that TNT and pentaerythritol tetranitrate (PETN) could be degraded by indigenous bacteria anaerobically and the presence of mono-rhamnolipid could enhance the degradation and eliminate intermediate formation.

6.6 BIOTRANSFORMATION OF METALS, NON-METALS, AND RADIONUCLIDES

6.6.1 Metals

Microbial conversion of metals includes bioaccumulation, biological oxidation/reduction and biomethylation (Soesilo and Wilson, 1997). Microbial cells can accumulate heavy metals through ion exchange, precipitation and complexation on and within the cell surface containing hydroxyl, carboxyl and phosphate groups. Bacterial oxidation/reduction could be used to alter the mobility of the metals. For example, some bacteria can reduce Cr(IV) in the form of chromate ($CrO_4{}^{2-}$) and dichromate ($Cr_2O_7{}^{2-}$) to Cr(III) which is less toxic and mobile due to precipitation above pH 5 (Bader et al., 1996). Microorganisms of the species *Geobacter sulfurreducens* are able to reduce Ag+ and form silver nanoparticles which could reduce the toxicity of these ions (Law et al., 2008). Several reasons for this reaction have been postulated including detoxification, metabolism with other components and use as an electron acceptor. Rhamnolipids have also been found to be able to reduce Cr(VI) to Cr(III) (Ara and Mulligan, 2015). Conversion can also be indirect by microbial production of Fe(II), sulfide, and other components that reduce chromium. Environmental conditions must be favorable for chromium reduction to occur.

Mercury can be found as Hg(II), volatile elemental mercury (Hg(0)), methyl- and dimethyl forms. Metabolism occurs through aerobic and anaerobic mechanisms through uptake, conversion of Hg(II) to Hg(0), methyl and dimethylmercury or to insoluble Hg(II) sulfide precipitates. Although volatilization or reduction during natural attenuation would still render mercury mobile, Hg(II) sulfides are immobile if sufficient levels of sulfate and electron donors are available.

Arsenic can be found as the valence states As(0), As(II), As(III) and As(V). Forms in the environment include As_2S_3, elemental As, arsenate ($AsO_4{}^{3-}$), arsenite (AsO^{2-}) and other organic forms such as trimethyl arsine and methylated arsenates. All forms are toxic and in particular, the anionic forms are mobile and highly toxic. Microbial transformation under aerobic conditions produces energy through oxidation of arsenite. Arsenate (As(V)) can be used as an electron acceptor anaerobically. Other mechanisms include methylation, oxidation, and reduction under anaerobic or aerobic conditions.

Biomethylation involves the addition of a methyl ($-CH_3$) group to a metal such as arsenic, mercury, cadmium or lead (Smith et al., 1994). The methylated forms are more mobile and can migrate into the groundwater. Although, methylation increases volatility, it is not likely that methylation of metals such as arsenic will be performed for remediation since the by-products are more toxic. None of these are really suitable for treatment by natural attenuation since the arsenic remains toxic and mobile.

At low concentrations, selenium is a micronutrient for animals, humans, plants, and some microorganisms. At high concentrations, however, it is toxic. It can be found naturally in four major species, selenite ($SeO_3{}^{2-}$, IV), selenate ($SeO_4{}^{2-}$, VI), elemental selenium (Se (0)), and selenide (–II) (Frankenberger and Losi, 1995; Ehrlich, 2002). Oxidation of selenium can occur under aerobic conditions while selenate can be transformed anaerobically to selenide or elemental selenium. Methylation of selenium detoxifies selenium for the bacteria by removing the selenium from the bacteria. Volatilization of selenium from contaminated agricultural soils has shown some promise (Thompson-Eagle and Frankenberger, 1992). This form is very mobile and toxic for animals and humans. Immobilization of selenate and selenite is accomplished via conversion to insoluble selenium. Due to the many forms of selenium, selenium decontamination by micro-organisms is not promising.

Oxygenated ions called oxyanions include nitrate ($NO_3{}^-$), chlorate ($ClO_3{}^-$), and perchlorate ($ClO_4{}^-$). Nitrates can cause respiratory stress disease methemoglobinemia in infants or lead to the formation of nitrosamines. They, however, can be removed by natural attenuation due to denitrifying bacteria that convert nitrate to nitrogen gas. The oxygenated chlorine compounds can also be transformed microbially since they act as electron acceptors. However, very little is known about them. As long as sufficient amounts of electron donors are present, such as carboxylic acids, hydrogen, amino acids, and carbohydrates, formation of the non-toxic chlorine ion is possible (Malmqvist et al., 1991).

There is the potential for biomethylation of iodine-129 to methyl-iodide (EPA, 2010) in wetlands. Low redox conditions and high concentrations of organic matter in this environment favor this mechanism.

Radionuclides are common where preparation and production of nuclear fuel are involved, and where nuclear weapons are produced. They include uranium and plutonium and neptunium. Most common forms of uranium are the insoluble $U(IV)O_2$ and soluble $U(VI)O_2{}^{2+}$ (Lovley, 1995). While metabolizing organic compounds, micro-organisms can anaerobically convert the soluble uranium form to the insoluble. Although this process may be effective for natural attenuation, the introduction of oxygen will redissolve uranium. In contrast, plutonium can be reduced by sulfate-reducing bacteria from the insoluble Pu(IV) to the soluble Pu(III), thereby making this contaminant more mobile. On the other hand, neptunium can be immobilized by micro-organisms by reduction to Tc(IV) and Tc(V) oxides from Tc(VII)—indicating that adequate electron donors and acceptors must be present for immobilization of the contaminants.

6.6.2 Bacterial Metabolism of Nitrogen

In nitrate respiration, nitrate is reduced to NO_2^-, then N_2O, NO, and finally, nitrogen gas. This reaction, known as *denitrification*, is a naturally occurring process that removes nitrogen from the soil and releases it into the atmosphere. This process is also called dissimulative nitrate reduction and can be performed by bacteria only. When ammonia and then amino acids and proteins are produced for growth, this process is called *assimilative nitrate reduction.* Bacteria, fungi, and plants are able to perform this process. This latter process can occur when oxygen is present, whereas the former process cannot. Micro-organisms that can use oxygen for respiration if it is available, or other compounds if it is not, are called facultative anaerobes. Denitrifying bacteria are always facultative. These are commonly found in activated sludge processes. Obligate anaerobes do not require oxygen that can also be toxic to some species. When both ammonium and nitrite/nitrate ions are present (Caschetto et al., 2018), then anaerobic oxidation of ammonium ions may occur in a process known as *annamox* as shown in Equation (6.8):

$$NH_4^- + NO_2^- \rightarrow N_2 + 2H_2O \tag{6.8}$$

6.6.3 Bacterial Metabolism of Sulfur

Sulfate-reducing bacteria can convert sulfate to hydrogen sulfide, H_2S, or sulfur, S. Sulfur can be used for cell synthesis in assimilatory reactions or as an electron acceptor for energy generation in dissimilatory reactions. In the latter case, hydrogen sulfide is the final gaseous product. Fewer than 12 species of bacteria are capable of performing this process. Less energy is obtained from sulfate than from oxygen or nitrate. Acetate, lactate, or pyruvate can be used as organic substrates by some sulfate-reducing bacteria, while hydrogen gas can be used by some lithotrophic bacteria.

Oxidation of hydrogen can be performed by hydrogen lithotrophic bacteria. If hydrogen is used to form cell material, the following reaction takes place:

$$6H_2 + 2O_2 + CO_2 \rightarrow CH_2O + 5H_2O \tag{6.9}$$

As in most cases, oxygen is required and carbon dioxide is used as the carbon source. CH_2O is the general formula of cell material. Some bacteria are facultative and can use organic material for energy if hydrogen is not available. Some can also use organic compounds as the carbon source but use hydrogen for energy, such as the bacteria *Escherichia coli* which are used to indicate contamination in water.

Sulphur bacteria are facultative autotrophs which use sulfur, hydrogen sulfide, thiosulphate and organic sulfides or organic materials for energy. Typical reactions are:

$$\begin{gathered} H_2S + 2O_2 \rightarrow SO_4^{2-} + 2H^+ \\ S^0 + H_2O + \frac{3}{2}O_2 \rightarrow SO_4^{2-} + 2H^+ \\ S_2O_3^{2-} + H_2O + 2O_2 \rightarrow 2SO_4^{2-} + 2H^+ \end{gathered} \tag{6.10}$$

The conversion of hydrogen sulfide to sulfate is a multistep process in which sulfur is produced. As hydrogen sulfide is depleted in some species, the sulfur inside the cell can then be used as the energy source. Those micro-organisms that use the very insoluble sulfur must grow on the sulfur solids, slowly oxidizing the gradually soluble sulfur. In all of these reactions, sulfuric acid is produced, lowering the pH to 1 in high concentrations. *Beggiatoa* are frequently found in sediments containing high levels of hydrogen sulfide and require organic materials for carbon (Eweis et al., 1998).

Sulfate-reducing bacteria (SRB) form metal (Me) sulfides that are insoluble as shown in the following reactions:

$$\begin{gathered} CH_3COOH + SO_4^{2-} \rightarrow 2HCO_3 + HS^- + H^+ \\ H_2S + Me^{2+} \rightarrow MeS + 2H^+ \end{gathered} \tag{6.11}$$

As can be seen from the equations, sulfate, low redox conditions and an electron donor such as methanol are required. Oxygen should not be there and nutrients must be added. Stimulating sulfate reduction can increase also pH and form metal hydroxides and oxides that precipitate and do not migrate in soils and groundwater. A project was proposed by the Flemish Institute for Technological Research for NATO/CCMS (1999). It involved bioprecipitation of the heavy metals (zinc, cadmium, arsenic, lead, chromium, nickel, and copper) within biological reactive zones or biowalls. Results from batch systems using groundwater and aquifer samples were presented in NATO/CCMS (2003). The first site contained zinc (0–150 mg/L), cadmium (0.4–4 mg/L), and arsenic (20–270 μg/L) with high concentrations of sulfate (400–700 mg/L), ideal for SRB activity. No zinc removal could be obtained unless acetate was added at low concentrations. Zinc could be reduced to 213 μg/L. Arsenic and cadmium were also removed through precipitations. If high concentrations of acetate were added, methanogenic bacteria dominated over the SRB bacteria, resulting in low metal removal efficiencies. Some tests with aquifer materials and addition of *Desulfovibrio desulficans* Dd8301 indicated that metals were removed only the redox potential was below −220 mV. Groundwater away from the site with lower levels of contamination could be treated only if SRBs were added.

The second site had cadmium (up to 17 mg/L), copper (up to 92 mg/L), chromium (up to 8.3 mg/L), zinc (up to 78 mg/L), and nickel (up to 3.5 mg/L), with high levels of sulfate (up to 3000 mg/L) with a pH of 2–4. Sulfate concentrations greater than 2000 mg/L or addition of zero valent iron were necessary to obtain a redox potential of lower than −200 mV for total removal of nickel and removal of lead, zinc, chromium, and cadmium.

The third site was an industrial landfill site that had arsenic (up to 1.2 mg/L), zinc (up to 880 μg/L) and sulfate (up to 3800 mg/L) in the leachate. pH was in the range of 2.6 and 6.0. High concentrations of arsenic caused toxicity problems, but when concentrations were lower the addition of acetate and a redox-manipulating compound were able to enhance arsenic immobilization.

6.7 CONCLUDING REMARKS

The basics of microbiology are essential if one is to be able to understand one of the most important processes of natural attenuation. Growth constants whether based on first order or Monod kinetics are important inputs for models to assess the potential and predict the efficiency of natural attenuation as a remediation technology. In addition, the essential components such as oxygen and other electron acceptors (i.e., nitrate, sulfate and iron (III) etc.) and nutrients must be available for micro-organisms in sufficient quantities for them to function in an optimal manner. Other factors include the presence of water, and mineral nutrients. Toxicity and bioavailability of the contaminant may also play an important factor in biodegradation. In addition, knowledge of the pathways of degradation enables us to identify by-products that can be used as evidence of biodegradation of the contaminant in the subsurface.

A wide variety of hydrocarbons can be degraded by micro-organisms through electron transfer by various mechanisms. Most of the knowledge related to natural attenuation is related to the degradation of BTEX compounds. Although the rate of aerobic biodegradation is higher than anaerobic, the latter type may be more dominant. Products of aerobic degradation are carbon dioxide and water, while the products of anaerobic degradation include carbon dioxide, water, methane, hydrogen, nitrogen, and others.

Aerobic degradation of alkanes or aromatic compounds can lead to the formation of a fatty acid or catechol that is less volatile, more soluble, and more degradable. Biodegradation of PAHs may lead to the formation of more volatile, lower molecular weight products, particularly through fungal metabolism. Conversion of chlorinated compounds such as TCE by reductive dehalogenation through anaerobic means can form ethene, which can be more persistent in the environment. Alternating aerobic and anaerobic conditions may be required to eliminate these types of compounds.

Although substantial information exists concerning the biodegradation of organic contaminants, there is considerably less data available on biotransformation of metals and other inorganics. Table 6.2 indicates the likelihood and understanding of the natural attenuation of various contaminants. As can be seen, the level of understanding with the exception of nitrate (which is high) is at best moderate. This means that the biotransformation process is not well understood scientifically but that it can be a major attenuation process. Therefore, substantial research is still required in this area. Natural attenuation of metals such as chromium, non-metals such as selenium and various radionuclides has potential once the right conditions are understood and are available. The reversibility of these microbial reactions must also be further investigated.

TABLE 6.2
Potential Microbial-Inorganic Contaminant Interactions

Contaminant	Microbial Contaminant Interaction Mechanism	Transformation
Cu, Ni, Zn	Sorption by extracellular polymers and biomass	Immobilization by sorption
Cd, Pb	Sorption by extracellular polymers and biomass	Immobilization, possible methylation
Fe, Mn	Anaerobic electron acceptor, oxidation to hydroxides, sorption by extracellular polymers and biomass	Increase in mobility by reduction, immobilization by precipitation and sorption
Cr	Oxidation or reduction, cometabolism, sorption by extracellular polymers and biomass	Immobilization by precipitation
Hg	Oxidation or reduction, sorption by extracellular polymers and biomass	Volatilization, immobilization by precipitation and sorption
As	Anaerobic electron acceptor, cometabolism, oxidation to hydroxides, sorption by extracellular polymers and biomass	Volatilization or immobilization by precipitation or sorption
Se	Anaerobic electron acceptor, oxidation to hydroxides, sorption by extracellular polymers and biomass	Volatilization or immobilization by precipitation or sorption
Nitrate	Anaerobic electron acceptor	Conversion to nitrogen
Perchlorate	Anaerobic electron acceptor	Reduction to chloride
U	Anaerobic electron acceptor, sorption by extracellular polymers and biomass	Immobilization by precipitation
Pu	Cometabolism, sorption by extracellular polymers and biomass	Increased mobility by reduction, immobilization by precipitation and sorption
Tc	Anaerobic electron acceptor, cometabolism sorption by extracellular polymers and biomass	Immobilization by precipitation

Source: NRC (National Research Council) Natural Attenuation for Groundwater Remediation, Committee on Intrinsic Remediation, Water Science and Technology Board and Board on Radioactive Waste Management, Commission on Geoscience, Environment and Resources, National Academy Press, Washington, DC, pp. 274, 2000.

REFERENCES

Alexander, M. (1994), *Biodegradation and Bioremediation*, Academic Press, San Diego, CA.

Alvarez-Cohen, L., and McCarty, P.L. (1991), Effects of toxicity, aeration and reductant supply on trichloroethene transformation by a mixed methanotropic culture, *Appl. Environ. Microbiol.*, 57(1):228–235.

Aldhafiri, S., Mahmoud, H., Al-Sarawi, M., and Ismail, W.A. (2018), Natural attenuation potential of polychlorinated biphenyl-polluted marine sediments, *Pol. J. Microbiol.*, 67:37–48.

Amon, M.M., Khanahmad, H., Teimouri, F., Sadani, M., Karami, M.A., and Hatamzadeh, M. (2017), Effect of monorhamnolipid contribution on anaerobic-natural attenuation of explosives in contaminated soils, *J. Environ. Eng.*, 143(8):04017035.

Anderson, T.T., and Lovley, D.R. (1997), Ecology and biogeochemistry of in situ groundwater Bioremediation, *Adv. Microb. Ecol.*, 15:289–350.

Apajalahti, J.H.A., and Salkinoja-Salonen, M.S. (1987), Complete dechlorination of tetrahydroquinone by cell extracts of pentachloro-induced *Rhodococcus chlorophenolicus*, *J. Bacteriol.*, 169:5125–5130.

Ara, I., and Mulligan, C.N. (2015), Reduction of chromium in water and soil using a rhamnolipid biosurfactant, geotechnical engineering, *Journal of the SEAGS & AGSSEA*, 46(4):25–31.

Arab, F., and Mulligan, C.N. (2018), An eco-friendly method for heavy metal removal from mine tailings, *Environ. Sci. Pollut. Res.*, 25(16):16202–16216.

Aronstein, B.N., Calvillo, Y.M., and Alexander, M. (1991), Effects of surfactants at low concentrations on the desorption and biodegradation of sorbed aromatic compounds in soil. *Environ. Sci. Technol.*, 25(10):1728–1731.

Atlas, R.M. (1991), Microbial hydrocarbon degradation-bioremediation of oil spills, *J. Chem. Technol. Biotechnol.*, 52:149–152.

Aüllo, T., Berlendis, S., Lascourreges, J.-F., Dessort, D., Duclerc, D., Saint-Laurent, S., Schraawers, B. et al. (2016), New bio-indicators for long term natural attenuation of monoaromatic compounds in deep terrestrial aquifers, *Front. Microbiol.*, 7:1221.

Aust, S., Fernando, T., Brock, B., Tuisel, H., and Bumpus, J. (1988), Biological treatment of hazardous wastes by *Phanaerochaete chrysosporium*, in G. Lewandowski, B. Blatzis, and P. Armenante (Eds.), *Proceedings, Conference on Biotechnology Applications in Hazardous Waste Treatment*, Longboat Key, FL. Engineering Foundation, New York.

Bader, J.L., Gonzales, G., Goodell, P.C., Pilliand, S.D., and Ali, A.S. (1996), Bioreduction of hexavalent chromium in batch cultures using indigenous soil microorganisms, *HSRC/WERC Joint Conference on the Environment*, Albuquerque, NM.

Barkovski, A.L., and Adriaens, P. (1996), Microbial dechlorination of historically present and freshly spiked chlorinated dioxins and diversity of dioxin-dechlorinating populations, *Appl. Environ. Microbiol.*, 62:4556–4562.

Bartha, R. (1986), Biotechnology of petroleum pollutant biodegradation, *Microb. Ecol.*, 12:155–172.

Borden, R.C., Daniel, R.A., LeBrun, L.E., and Davis, C.W. (1997), Intrinsic biodegradation of MTBE and BTEX in a gasoline-contaminated aquifer, *Water Resour. Res.*, 33:1105–1115.

Brock, T.D., Smith, D.W., and Madigan, M.T. (1984), *Biology of Microorganisms*, 4th ed., Prentice-Hall, Englewood Cliffs, NJ.

Brown, R.A., Hicks, P.M., Hicks, R.J., and Leahy, M.C. (1995), Postremediation bioremediation, in R.E. Hinchee, J.T. Wilson, and D.C. Downey (Eds.), *Intrinsic Bioremediation*, Batelle Press, Columbus, OH.

Caschetto, M., Robertson, W., Petitta, M., and Aravena, R. (2018), Partial nitrification enhances natural attenuation of nitrogen in a septic system plume, *Sci. Total Environ.*, 625:801–808.

Cerniglia, C.E. (1984), Microbial metabolism of polycyclic aromatic hydrocarbons, *Adv. Appl. Microbiol.*, 30:31–39.

Chappell, J. (1997), Phytoremediation of TCE using *Populus*, Report prepared for U.S. EPA Technology Innovation Office.

Coleman, N.V., Nelson, D.R., and Duxbury, T. (1998), Aerobic biodegradation of hexachloro-1,3,5-trinitro-1,3,5-trazine (RDX) as a nitrogen source by a *Rhodococcus* sp., strain DN 22, *Soil Biol. Biochem.*, 30:1159–1167.

Dupont, R.R., Sims, R.C., Sims, J.L., and Sorensen, D. (1988), In situ biological treatment of hazardous waste-contaminated soils, in D.L. Wise (Ed.), *Biotreatment Systems*, vol. II, CRC Press, Boca Raton, FL.

Ehrlich, H.L. (2002), *Geomicrobiology*, 3rd ed., Marcel Dekker, New York.

EPA (2010), *Monitored Natural Attenuation of Inorganic Contaminants in Ground Water, Volume III—Assessment for Radionuclides Including Americium, Cesium, Iodine, Plutonium, Radium, Radon, Strontium, Technetium, Thorium, Tritium, and Uranium*, EPA 600-R-10-093, Office of Research and Development, Washington, DC.

Esteve-Nunez, A., and Ramos, J.L. (1998), Metabolism of 2,4,6-trinitrotoluene *by Pseudomonas sp.* JLR11, *Environ. Sci. Technol.*, 32:3802–3808.

Evans, W.C., and Fuchs, G. (1988), Anaerobic degradation of aromatic compounds, *Annu. Rev. Microbiol.*, 42:289–317.

Eweis, J.B., Ergas, S.J., Chang, D.P.Y., and Schroeder, E.D. (1998), *Bioremediation Principles*, WCB McGraw-Hill, Boston, MA.

Frankenberger, W.T. Jr., and Losi, M.E. (1995), Applications of bioremediation in the cleanup of heavy metals and metalloids, in H.D Skipper and R.F. Turco (Eds.), *Bioremediation Science and Applications*, SSSA, Special Publication Number 43, SSSA, Madison, WI, pp. 173–210.

Frobisher, M., Hinsdill, R.D., Crabtree, K.T., and Goodheart, C.R. (1974), *Fundamentals of Microbiology*, Saunders, Philadelphia, PA.

Freeman, D.L., and Gossett, J.M. (1989), Biological reductive dechlorination of tetrachloroethylene and trichloroethene to ethylene under methanogenic conditions, *Appl. Environ. Microbiol.*, 55(9):2144–2151.

Fulton, D.E. (1996), Selection and application of effective LNAPL recovery techniques, hazardous and industrial waste, *Proceedings of the 28th Mid-Atlantic Industrial Hazardous Waste Conference*, p. 619.

Funk, S.B., Crawford, D.L., Crawford, R.L., Mead, G., and Davis-Hoover, W. (1995), Full-scale anaerobic bioremediation of trinitrotoluene (TNT) contaminated soil, *Appl. Biochem. Biotechnol.*, 51–52:625–633.

Ghoshal, S., Ramaswami, A., and Luthy, R.G. (1996), Biodegradation of naphthalene from coal tar and heptamethylnonane in mixed batch systems, *Environ. Sci. Technol.*, 30:1282–1291.

Gibson, D.T. (1988), Microbial metabolism of aromatic hydrocarbons and the carbon cycle, in S.R. Hagedorn, R.S. Handson, and D.A. Kunz (Eds.), *Microbial Metabolism and the Carbon Cycle*, Harwood Academic Publishers, New York.

Grbic-Galic, D. (1990), Methanogenic transformation of aromatic hydrocarbons and phenols in groundwater aquifers, *Geomicrobiol. J.*, 8:167–200.

Harkness, M.R., McDermott, J.B., Abramawicz, D.A., Salvo, J.J., Flanagan, W.P., Stephens, M.L., Mondella, F.J. et al. (1993), In situ stimulation of aerobic PCB biodegradation in Hudson River sediments, *Science*, 259:503–507.

Hazen, T.C. (1991), Test plant for in situ bioremediation demonstration of the Savannah River integrated demonstration project, DOE/OTP TTP NO.SR 0566-01, WSRC-RD-91-23.

Higgins, I.J., and Gilbert, P.D. (1978), *The Biodegradation of Hydrocarbons*, The Oil Industry and Microbial Ecosystems, Heyden & Son, London, UK, pp. 80–115.

Hollinger, C., and Schumacher, W. (1994), Reductive dehalogenation as a respiratory process, *Antoine Leeunhoek*, 66:239–246.

Khovrychev, M.P., Ivanova, I.I., and Taptykova, S.D. (1974), Chemical composition and physiological properties of *Candida utilis* in the presence of inhibition of growth by copper ions, *Microbiology*, 43:405–409.

Landmeyer, J.E., Chapelle, F.H., Bradley, P.M. Pankow, J.F., Church, C.D., and Tratnyek, P.G. (1998), Fate of MTBE relative to benzene in a gasoline-contaminated aquifer (1993–1995), *Ground Water Monit. Rev.*, 18(4):93–102.

LaRossa, R.A., and Schloss, J.V. (1984), The sulfonyl berbicie sulfometuron methyl is an extremely potent and selective inhibitor of acetolactate synthase in *Salmonella typhimurium*, *J. Biol. Chem.*, 259:8753–8757

Law, N., Ansari, S., Livens, F.R., Renshaw, J.C., and Lloyd, J.R. (2008), Formation of nanoscale elemental silver particles via enzymatic reduction by *Geobacter sulfurreducens*, *Appl. Environ. Microbiol.*, 74(22):7090–7093.

Lin, A. Y-C., Plumlee, M.H. and Reinhard, M. (2006), Natural attenuation of pharmaceuticals and alkylphenol polyethoxylate metabolites during river transport: photochemical and biological transformation. *Environmental Toxicology & Chemistry*. 25(6), 1458–1464.

Lovley, D.R. (1995), Bioremediation of organic and metal contaminants with dissimilatory metal Reduction, *J. Ind. Microbiol.*, 14:85–93.

Lu, Q., Zhu, R.-L., Yang, J., Li, H., Liu, Y.-D., Lu, S.-G. Luo, Q.-S., and Lin, K.-F. (2015), Natural attenuation model and biodegradation for 1,1,1-ttrichloroethane contaminant in shallow groundwater, *Front. Microbiol.*, 6:839.

Luthy, R.G., Dzombak, D.A., Peters, C.A., Roy, S.B., Ramaswami, A., Nakles, D.V., and Nott, B.R. (1994), Remediating tar-contaminated soil at manufactured gas plant sites, *Environ. Sci. Technol.*, 28:266A–276A.

Luthy, R.G., Dzombak, D.A., Shannon, M.J.R., Unterman, R., and Smith, J.R. (1997), Aqueous solubility of PCB congeners from an Arochlor and an Arochlor/hydraulic oil mixture, *Water Res.*, 31(3):561–573.

Magli, A., Messmer, M., and Leisinger, T. (1998), Metabolism of dichloromethane by the strict anaerobe *Dehalobacterium formicoaceticum*, *Appl. Environ. Microbiol.*, 64:646–650.

Malmqvist, A., Welander, T., and Gunnarson, L. (1991), Anaerobic growth of microorganisms with chlorate as an electron acceptor, *Appl. Environ. Microbiol.*, 57:2229–2232.

Martinez-Hernandez, V., Meffe, R., Kohfahl, C., and de Bustamante, I. (2017), Investigating natural attenuation of pharmaceuticals through unsaturated column tests, *Chemosphere*, 177:292–302.

Matsumura, F., and Esaac, E.G. (1979), Degradation of pesticides by algae and microorganisms, in N.A.Q. Kahn (Ed.), *Pesticides and Xenobiotic Metabolism in Aquatic Organisms*, American Chemical Society, Washington, DC, ACS Symposium Series 99.

McCarty, P. (1988), Bioengineering issues related to in situ remediation of contaminated soils and groundwater, in G.S. Owen (Ed.), *Environmental Biotechnology*, Plenum Publishing Company, New York, pp. 143–162.

McCarty, P.L. (1965), Thermodynamics of biological synthesis and growth, in *Proceedings, Second International Conference on Water Pollution Research*, Pergamon Press, New York, p. 169.

McCarty, P.L., Semprini, L., and Dolan, M.E. (1991), In-situ methanotrophic bioremediation for contaminated groundwater at St. Joseph, Michigan, in R.E. Hinchee and R.F. Offenbuttel (Eds.), *On-Site Bioremediation Processes for Xenobiotic and Hydrocarbon Treatment*, Butterworth-Heinemann, Stoneham, MA, pp. 16–40.

McNally, D.L., Mihelcic, J.R., and Lueking, D.R. (1998), Biodegradation of three and four-ring polycyclic aromatic hydrocarbons under aerobic and denitrifying conditions, *Environ. Sci. Technol.*, 32:2633–2639.

Monod, J. (1949), The bacterial cultures, *Annu. Rev. Microbiol.*, 3:371.

Morrison, R.T., and Boyd, R.N. (1973), *Organic Chemistry*, Allyn & Bacon, Boston, MA.
Mulligan, C.N., Yong, R.N., and. Gibbs, B.F. (1999), On the use of biosurfactants for the removal of heavy metals from oil-contaminated soil, *Environ. Progr.*, 18(1):50–54.
NATO/CCMS (1999), Pilot study evaluation of demonstrated and emerging technologies for the treatment and clean up of contaminated land and groundwater (phase III), 1998 Annual Report Number 228, North Atlantic Treaty Organization, EPA-542-R-98-002.
NATO/CCMS. (2003), Pilot study evaluation of demonstrated and emerging technologies for the treatment and clean up of contaminated land and groundwater (Phase III), 2002 Annual Report, Number 255, North Atlantic Treaty Organization, EPA-542-R-02-010.
Nies, L., and Vogel, T.M. (1990), Effects of organic substrates on dechlorination of Arochlor 1242 in anaerobic sediments, *Appl. Environ. Microbiol.*, 56:2612–2617.
NRC (National Research Council) Natural Attenuation for Groundwater Remediation, Committee on Intrinsic Remediation, Water Science and Technology Board and Board on Radioactive Waste Management, Commission on Geoscience, Environment and Resources, National Academy Press, Washington, DC, 2000, 274 pp.
Park, K.S., Sims, R.C., and Dupont, R.R. (1990), Transformation of PAHs in soil systems, *J. Environ. Eng.*, 116(3):632–640.
Perry, J.J. (1984), Microbial metabolism of cyclic alkanes, in R.M. Atlas (Ed.), *Petroleum Microbiology*, MacMillan Publishing, New York, pp. 61–97.
Pirnik, M.P. (1977), Microbial oxidation of methyl branched alkanes, *CRC Crit. Rev. Microbiol.*, 5:413–422.
Pitter, P., and Chodoba, J. (1990), *Biodegradability of Organic Substances in the Aquatic Environment*, CRC Press, Boca Raton, FL.
Porgues, N., Jaiswicz, L., and Hoover, S.R. (1953), Biological oxidation of dairy waste VII, *Proceedings, 24th Industrial Waste Conference*, Purdue University, West Lafayette, IN.
PRC Environmental Management (1997), Recent developments for in situ treatment of metal contaminated soils, *Prepared for U.S. Environmental Protection Agency, Office of Solid Waste and Emergency Response*, Technology Innovation Office, Washington, DC.
Preuss, A., Fimpel, J., and Diekart, G. (1993), Anaerobic transformation of 2,4,6-trinitrotoluene (TNT), *Arch. Microbiol.*, 159:345–353.
Reineke, W., and Knackmuss, H.J. (1988), *Annual Review of Microbiology*, L.N. Ornston, A. Balows, and P. Baumann (Eds.), Annual Reviews, Palo Alto, CA.
Reynolds, T.D., and Richards, P.A. (1996), *Unit Operations and Processes in Environmental Engineering*, PWS Publishing Company, Boston, MA.
Salinitro, J.P. (1993), The role of bioattenuation in the management of aromatic hydrocarbon plumes in aquifers, *Ground Water Monit. Remediat.*, 13:150–161.
Scheutz, C., Broholm, M.M., Durant, N.D., Weeth, W.B., Jorgensen, T.H., Dennis, P., Jacobsen, C.S., Cox, E.E., Chambon, J.C., and Bjerg, P.L. (2010), Field evaluation of biological enhanced reductive dechlorination of chlorethenes in clayey till, *Environ. Sci. Technol.*, 44:5134–5141.
Schiewer, S., and Volesky, B. (1998), Biosorption processes for heavy metal removal, in D. Lovely (Ed.), *Environmental Metal-Microbes Interactions*, Chapman Hill, New York.
Semprini, L., Roberts, P.V., Hopkins, G.D., and McCarty, P.L. (1990), A field evaluation of in-situ biodegradation of chlorinated ethenes: Part 2, Results of biostimulation and biotransformation experiments, *Groundwater*, 28:715–727.
Shelton, D.R., and Tiedje, J.M. (1984), Isolation and partial characterization of bacteria in an anaerobic consortium that mineralizes 3-chlorobenzoic acid, *Appl. Environ. Microbiol.*, 48:840–848.
Sims, J.L., Sims, R.C., and Mathews, J.E. (1990), Approach to bioremediation of contaminated soil, *Hazard. Waste Hazard. Mater.*, 4:117–149.

Sinclair, J.L., Kampbell, D.H., Cook, L., and Wilson, J.T. (1993), Protozoa in subsurface sediments from sites contaminated with aviation gasoline or jet fuel, *Appl. Environ. Microbiol.*, 59(2):46–47.

Smith, L.A., Alleman, B.C., and Copley-Graves, L. (1994), Biological treatment options, in J.L. Means and R.E. Hinchee (Eds.), *Emerging Technology for Bioremediation of Metals*, Lewis Publishers, Boca Raton, FL, pp. 1–12.

Soesilo, J.A., and Wilson, S.R. (1997), *Site Remediation Planning and Management*, Lewis Publishers, New York.

Spain, J.C., and Gibson, D.T. (1991), Pathway for biodegradation of p-nitrophenol in a *Moraxella* sp., *Appl. Environ. Microbiol.*, 57:812–819.

Stanier, R.Y., Ingraham, J.L., Wheelis, M.L., and Painter, P.R. (1986), *The Microbial World*, Prentice-Hall, Englewood Cliffs, NJ.

Steffan, R., McClay, K., Vainberg, S., Condee, W., and Zhang, D. (1997), Biodegradation of the gasoline oxygenates methyl tert-butyl ether, ethyl tert-butyl ether, and tert-amyl methyl ether by propane oxidizing bacteria, *Appl. Environ. Microbiol.*, 63:4216–4222.

Steiert, J.G., and Crawford, R.L. (1986), Catabolism of pentachlorophenol by a *Flavobacterium* sp, *Biochem. Biophys. Res. Commun.*, 141:825–830.

Suflita, J.M., Robinson, J.A., and Tiedje, J.M. (1983), Kinetics of microbial dehalogenation of haloaromatic substrates in methanogenic environments, *Appl. Environ. Microbiol.*, 45:1466–1473.

Thayer, A.M. (1991), Bioremediation: Innovative technology for cleaning up hazardous waste, *Chem. Eng. News*, 69(34):23–44.

Thomas, J.M., and Ward, C.H (1989), In situ biorestoration of organic contaminants in the subsurface, *Environ. Sci. Technol.*, 23:760–766.

Thompson-Eagle, E.C., and Frankenberger, W.T. Jr. (1992), Bioremediation of soils contaminated with selenium, *Adv. Soil Sci.*, 17:261–310.

U.S. Environmental Protection Agency (1983), *Guidance Manual for POTW Pretreatment Program Development*, Washington, DC.

U.S. Environmental Protection Agency (1985), *EPA Guide for Identifying Cleanup Alternatives at Hazardous Waste Sites and Spills: Biological Treatment*, EPA 600/3-83/063, Washington, DC.

Vergani, L., Mapelli, F., Marasco, R., Crotti, E., Fusi, M., Di Guardo, A., Armiraglio, S., Dafonchio, S., and Borin, S. (2017), Bacteria associated to plants naturally selected in a historical PCB polluted soil show potential to sustain natural attenuation, *Front. Microbiol.*, 8:1385.

Verschueran, K. (1983), *Handbook of Environmental Data on Organic Chemicals*, 2nd ed., Van Nostrand Reinhold, New York.

Weidemeier, T.H., Rifai, H.S., Newell, C.J., and Wilson, J.T. (1999), *Natural Attenuation of Fuels and Chlorinated Solvents in the Subsurface*, John Wiley & Sons, New York.

Whittaker, R.H. (1969), New concepts of kingdoms or organisms. Evolutionary relations are better represented by new classifications than by the traditional two kingdoms, *Science*, 163(863):150–160.

Widdel, F. (1988), Microbiology and ecology of sulphate- and sulphur-reducing bacteria, in J.B. Zehner (Ed.), *Biology of Anaerobic Microorganisms*, John Wiley & Sons, New York, pp. 469–585.

Wujcik, W.J., Lowe, W.L., and Marks, P.J. (1992), Granular activated carbon pilot treatment studies for explosives removal from contaminated groundwater, *Environ. Progr.*, 11(3):178–196.

Zhang, X., and Young, L.Y. (1997), Carboxylation as an initial reaction in the anaerobic metabolism of naphthalene and phenanthrene by sulfidogenic bacteria, *Appl. Environ. Microbiol.*, 63:4759–4654.

7 Field Performance and Assessments

7.1 INTRODUCTION

Before the 1990s, the only options available for remediation of contaminated sites were through engineered bioremediation. The more common ones involved addition of oxygen and nutrients or other practices such as pump and treat or excavation for treatment offsite. Prior to the 1970s, the subsurface was actually considered to be devoid of biological activity. We now know better. There is microbial activity almost throughout the entire subsurface. Since the 1990s, it has been determined that there are numerous subsurface processes that lead to the natural attenuation of contaminants. The processes (described in the previous chapters) include biodegradation, dilution, assimilation processes and partitioning, volatilization, and biological or chemical reactions. It is important at this stage to distinguish between the use of the term *natural attenuation* (NA) as a process, and the term *natural attenuation capacity* as a property of a soil. When we refer to the property of *natural attenuation*, this term will be italicized in this chapter.

Contaminants can be sorbed to the soil, in the pore spaces in the vapor phase or dissolved in the groundwater. This is where most of the processes for NA of contaminants will occur. In the U.S., remediation by NA increased at Environmental Protection Agency (EPA) Superfund sites from 6% in 1990 to 23% in 1997 (Renner, 2000). It is used extensively at sites where there are leaking underground storage tanks (ASTM, 1995; USEPA, 1996) and in 1997 was used at 15,000 such sites (Renner, 2000). As there are, however, 358,000 releases or leaking underground storage tanks, there is the potential for extensive use of NA at these sites—attributable to the extensive cost of engineered remediation.

However, as the use of natural attenuation processes for application to remediation of contaminated sites has been frequently used, we need to ask whether the protocols, procedures and capabilities for scientific evaluation of the effectiveness of natural attenuation processes have been appropriate and satisfactory? This concern is being driven by awareness of the fact that the use of monitored natural attenuation (MNA) as a remediation tool is a knowledge-based technique. In this chapter, we will be looking at reported performances of MNA at various sites, with a view to determining whether the stakeholders consider the technique to be successful and/or effective. Assessment procedures, monitoring protocols, modelling capabilities used for the reported field application cases, and field data in support of effectiveness (or otherwise) for natural attenuation as a process will be examined. The various protocols and requirements for MNA application are discussed in the next chapter.

7.2 ASSESSMENT OF POTENTIAL FOR NATURAL ATTENUATION

The capacity for natural attenuation of contaminants should be determined to evaluate the rate of natural attenuation and to determine the level of contaminant removal to be achieved and the potential risks to receptors. Figure 7.1 shows some of the principal mechanisms involved in attenuating contaminants associated with the pollutant plume originally shown in Figure 1.2. Reliance on the many assimilative and transformation processes associated with the natural attenuation of contaminants to protect the biotic receptors from the threats posed by the contaminants requires one to be very confident that the soil-water system will function as planned. Chapter 8 provides the protocols and requirements for determining whether natural attenuation processes can be effectively used to remediate a contaminated site. The "lines of evidence" and "evidence of success" protocols will be discussed in detail in that chapter. For this chapter, we will turn our attention to lessons we can learn from indirect and direct field applications of NA. It is well accepted that for successful application of natural attenuation as a tool for remediation of contaminated sites, the geological, biogeochemistry, and hydrological conditions for each site which must be evaluated on a case by case basis. A detailed and comprehensive understanding of the site-specific conditions, including availability of substrates and nutrients, will lead to higher success rates.

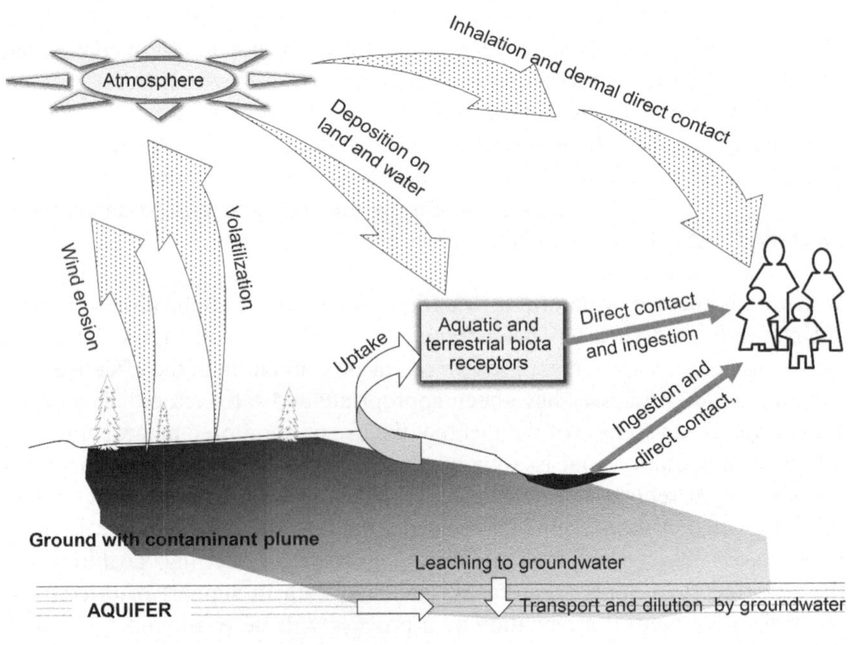

FIGURE 7.1 Schematic showing contaminated ground with a pollutant plume as the potential source of health and environmental threats to environmental and human receptors. Also shown are the processes associated with the transport of the contaminant plume.

Fate and transport models are used to provide predictions of distribution and concentrations of contaminants at various times and over specified distances from the source. These predictions provide guidance for initial evaluation and also throughout the natural attenuation process. They are particularly useful in instances where biodegradation is a major mechanism contributing to the total attenuation process. That being said, decay rates which need to be determined for input in to the models are problematic issues. These have been reviewed by Suarez and Rifai (1999) for petroleum and chlorinated hydrocarbons. Methods for improving our understanding of the processes involved, and for assessing the potential for application of natural attenuation processes in remediation of contaminated sites are still under development.

7.2.1 Processes and Mechanisms Involved in Natural Attenuation

Future land use will also be a major consideration regarding the potential for application of NA as a remediation technology in the area being considered for remediation treatment. If contaminants are continually released at an existing facility, these will have the potential to affect human health and the environment. Thus, application of NA as a remediation strategy could be a viable option if control of pollutant source can be obtained. Various factors at a site need to be evaluated. For natural attention of metals and radionuclides to be realized, the primary attenuating mechanisms of sorption and dilutions must be available. In the case of radioactive materials, daughter nuclides will be produced and it must be determined whether they are more hazardous than the parent. The primary mechanisms contributing directly to the natural attenuation of contaminants for the three types of contaminants are shown in Figure 7.2. Various factors and issues need to be considered in some fair detail, to determine the potential for NA. Some of these include the following considerations:

- At high pH values, net negative charges in the clay minerals are expected to be high and the potential for increased cation sorption will be correspondingly increased. The likelihood of precipitation of metals is also expected to be high.
- Anion sorption increases as the pH decreases as shown previously in Chapter 4. Low pH conditions also affect the solubility of iron hydroxides and carbonates. The result of these, in turn, will affect sorption.
- Calcium is known to provide an indication of calcium carbonate stability.
- Low redox values dissolve iron hydroxides by leading to sulfide formation.
- The presence of chelating agents such as ethylenediaminetetraacetic acid (EDTA), nitrilotriacetic acid (NTA), diethylenetriamine penta-acetic acid (DTPA), citrate and oxalic acids can significantly influence the rate of transport of metals.
- The biodegradation rates of the chelating agents mentioned above will affect partitioning of the metals during transport through soils. Waters et al. (1998) have shown that the chelating agents degrade in the following order: citrate~oxalate>>NTA>EDTA>DTPA. These agents influence sorption by decreasing sorption or increasing complexes with the contaminant and the soil surface.

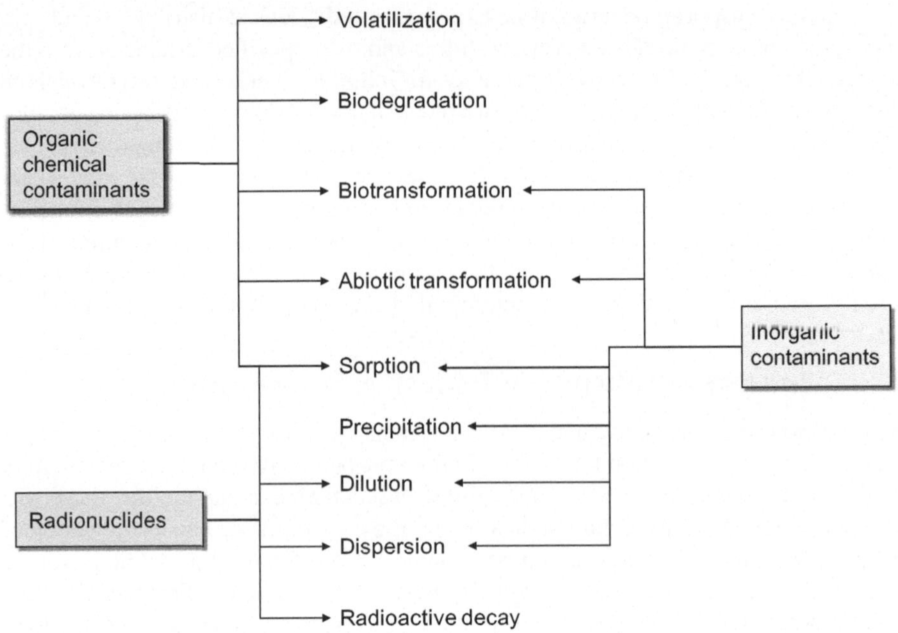

FIGURE 7.2 Primary mechanisms and processes contributing directly to natural attenuation of contaminants.

- Excessive carbon dioxide produced can also increase porosity from calcite and dolomite dissolution under acidic conditions (Bennett et al., 2000). Changes in porosity can occur as a result of dissolution processes.
- Other reactions under anoxic conditions such as carbonate and bicarbonate saturation with calcite can plug pore spaces and decrease permeability.

Since the pollutant source responsible for contaminating the soil is usually removed from the contaminated site, the sorbed metals that are left need to evaluated for potential desorption or detachment, since the desorbed metals will be mobile and will pose a threat to the well-being of the immediate environment. Performance assessment calculations usually assume reversible desorption; although these rates can be very slow, they can be the determining factors in the mobility of these types of contaminants.

7.2.2 Bioattenuation and Bioavailability

Determination of the capacity for bioattenuation has not received a great deal of attention in assessment of the natural attenuation of organic chemical pollutants. Substrates can become less bioavailable via interaction with negatively charged clay particles and organic material (Alexander, 1994). Sorption and sequestration can be influenced by pH, organic matter content, temperature, and pollutant characteristics. The biodegradation of polycyclic aromatic hydrocarbons (PAHs) is particularly affected by sorption.

Bioavailability of electron acceptors can influence microbial activity. Solid Fe(III) must be available and must be in direct contact with the micro-organisms or obtained from humic acids that can chelate iron. Their presence can significantly increase iron bioavailability (Lovley, 2001). Nutrients such as nitrogen and phosphorus can also be limiting. Micro-organisms attached to the soil minerals are capable of degrading soils that release phosphorus. There have been good indications that the rate of degradation or weathering can correlate well with microbial activity (Bennett et al., 2000). This leads to the suggestion that determination of sediment or soil contents could be interpreted in terms of biodegradation potential. Whilst this thesis is tenable, much work needs to be done to support this line of reasoning—given that many other chemical and physico-chemical factors contribute significantly to the weathering of rocks and soils.

7.2.3 Other Factors

The presence of other predatory organisms such as protozoa can also be detrimental. As the contaminant levels increase, so do the levels of microorganisms. This also increases the likelihood of grazing by protozoa. The rate of predation is influenced by the soil and pollutant characteristics, groundwater velocity, and redox conditions (Kota et al., 1999).

According to the Office of Solid Waste and Emergency Response (OSWER) directive (USEPA, 1999), the contaminated site will need to be characterized for the nature and concentrations of contaminants and potential impact to receptors (Figure 7.2). Contributions of sorption, dilution, and dispersion of the contaminants should be evaluated in the groundwater regime in addition to the hydraulic regime. This includes recharge, discharge areas, and volumes. For biodegradation, the presence of nutrients, electron donors and acceptors, metabolites and by-products, and presence of available microbial populations should be evaluated. These pieces of information need to be incorporated into a site-specific fate and transport model. Technological limitations for application of NA may include the presence of solution channels, fractures, joints or foliations in the rock aquifers. Materials with anisotropic properties can make it difficult to provide accurate predictions of groundwater flow and direction. The difficulties and errors can be minimized with recognition and proper accounting of the anisotropic properties.

Evaluation of the potential and capability of NA at a site will be enhanced as information of bacterial species and their activities is catalogued and evaluated. It is necessary to bear in mind that the tools presently available for assessment of the rates of potential biodegradation are still not reliable. Determination of biodegradation rates in the laboratory can be one to two orders greater than those in the field (Kota et al., 1998). Surface heterogeneity is one of the complicating factors. No accepted methods exist for incorporating rate data into mathematical models. In the Kao and Wang (2001) mass flux method for assessment of natural attenuation of BTEX, corrections were made for the biodegradation rate using dispersion, dilution, volatilization and sorption. The basis for the corrections was obtained by using the recalcitrant tracer, 1,2,4-trimethylbenzene (TMB). Biodegradation was estimated using a first order decay rate. Evidence for BTEX removal included the following:

1. Exhaustion of dissolved oxygen concentration,
2. Production of Fe(II), carbon dioxide and methane,
3. An increase in pH,
4. Decreases in BTEX carbon to TOC ratio as it is transported,
5. Decreases in BTEX mass in comparison to the tracer, and
6. Faster removal of some components in the BTEX.

Toluene degraded the fastest, followed by o-xylene, then m- and p-xylene, then xylene and finally ethylbenzene. The technique of using mass flux and an in situ tracer seems to be a more direct way for determination of the natural attenuation of the BTEX instead of relying on microcosm tests and fate and transport models. Approximately 87% of BTEX was removed by biodegradation.

A later study by Höhener et al. (2006) compared the first order biodegradation rates of 13 volatile petroleum hydrocarbon vapors in the lab and field in Denmark. For the compounds with low reactivity (n-hexane, 3-methylpentane, the cycloalkanes, and isooctane) the rate coefficients were between 0.005 and 0.35 day^{-1}. For these compounds, there was a lack of sensitivity for the inverse model approaches and the microcosm experiments, in particular. The highly reactive compounds (toluene, m-xylene, 1,2,4-trimethyl benzene, n-octane and n-decane) showed biodegradation rates higher than 0.35 day^{-1}. The column tests showed a better ability to predict the field rates than the microcosm approach. However, there was a two- to three-fold uncertainty factor. Benzene showed higher rates in the lab than the field studies. However, slowly degrading compounds underestimated rate constants by a factor of up to 5. Benzene was particularly problematic. The lab systems were not able to predict the degradation rates shown in the field data, which ranged from 0.45 to 1.95 day^{-1} depending on the model used. Although microcosm data did not show degradation, column tests resulted in rates of 0.21 ± 0.12 day^{-1}. Thus, the rate was underestimated by a factor 3–10. The three models showed field rates differing by up to a factor of 4. This shows the problem of predicting biodegradation rates of a common contaminant at sites.

7.3 ASSESSMENT OF SUSTAINABILITY

An approach that can be used to evaluate NA is to determine its sustainability (NRC, 2000). The mechanisms for natural attenuation of contaminants need to be sustainable, particularly if the rate of contaminant release is high. This can occur in the case of a pool of non-aqueous phase liquid (NAPL) or a tailings pond. The various contaminant attenuating processes and mechanisms need to be evaluated in respect to long-term capabilities since groundwater and soil qualities need to be assessed in view of the fate and transport of these contaminants. The availability of electron acceptors from the groundwater or soil as well as electron donors should be evaluated on a long-term basis. In the case of soil mineralogy, sites for sorption, electron

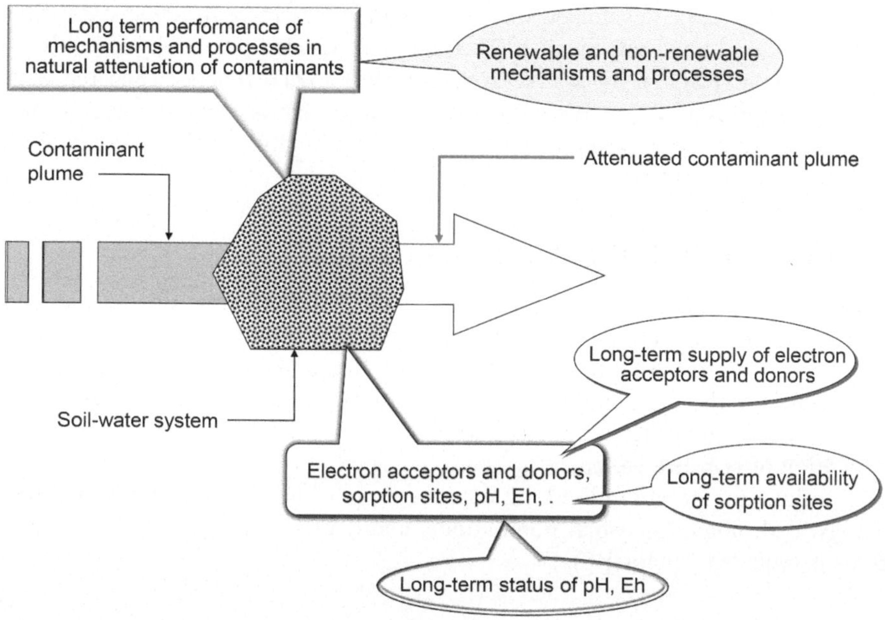

FIGURE 7.3 Schematic showing transport of pollutant plume through a soil-water system. The concerns for long-term performance of the mechanisms and processes contributing to the natural attenuation of the pollutants are central to the problem at hand. Also shown are the various issues regarding long-term supply and availability of electron acceptors and donors, sorption sites, and long-term status of pH and Eh.

acceptors and alkalinity are not in infinite supply. Renewable and non-renewable mechanisms and processes will need to be determined as shown in Figure 7.3. Mass budgeting can be used as a tool for evaluating sustainability as it determines the contaminant destruction or sequestration and compares it to production of components. This compares the importance of natural attenuation processes against each other.

The advantage of determination of sustainability of natural attenuation processes on a long-term basis can be seen in the example of a non-sustainable NA application at a former copper mining site where acid drainage (pH 2–3) was released from an unlined pond. High concentrations of sulfate, calcium, iron, manganese, copper, aluminum, and zinc were present in the groundwater plume 12 km from the source. Initial studies (Stollenwerk, 1994) showed that the soil carbonate raised the pH from 5 to 6 and led to precipitation or sorption of iron, copper, zinc, and other contaminants. As the carbonate became depleted, the pH dropped and the metals were remobilized. This tells us that we need to evaluate the sensitivity of natural attenuation to changing environmental conditions. In addition, we need to structure or develop methods to assess the effects of changing environments and their effects on the long-term sustainability of NA and its long-term effectiveness. Changes in the immediate environment could be the result of interactions in the soil-water system with other contaminants, formation of toxic by-products, climate change, and adverse effects from other remediation technologies.

Lebron et al. (2013) evaluated methods for the assessment of the sustainability of MNA at chlorinated solvent sites. The methods included potentially bioavailable organic carbon (PBOC), the source zone depletion (SZD) function, and the long-term sustainability (LTS) of PBOC and dissolved oxygen concentrations. All three components were successfully used, but depend on availability of data, particularly the source zone mass flux.

7.4 PROCEDURES FOR MONITORING

If NA is effective, the mass and toxicity of the contaminants will be reduced. Monitoring of the quality of both the groundwater (porewater) and soil material permits one to verify the effectiveness of NA, and to deduce the processes responsible for the attenuated results. As will be seen in Chapter 8, there are strict protocols and criteria that have to be met. If we are to be comfortable with application of NA as a treatment process, we need to be assured that the process will continue to operate in the long term. A proper knowledge of the subsurface geology, hydrology, microbial populations, and degradation and conversion processes is required (Chapter 8).

7.4.1 Importance of Monitoring

The USEPA directive (USEPA, 1999) says that when NA is used for remediation of soil and groundwater, it cannot be a *no action* approach. The application of NA is dependent on extensive monitoring. The designation of monitored natural attenuation (MNA) is used to reflect this. Lines of evidence have been established and include:

- Historical groundwater and soil chemistry data showing decreases in contaminant mass or concentration of time at monitoring or sampling points.
- Indirect evidence from hydrogeological and geochemical data indicating that natural attenuation is occurring such as by sorption, dilution, volatilization or biodegradation and the required levels will be achieved.
- Use of field or microcosm studies for direct demonstration of natural attenuation or degradation of the contaminant. The studies are conducted in situ or with actual contaminated site media.

If the first line of evidence is not conclusive, then the second will be required. If the second cannot be proven, then the third will be necessary. Site-specific factors such as size and nature of contaminants, availability of and risk to receptors, and environmental factors such as climatic conditions, hydrogeology, hydrology, subsurface geology, and others can influence the amount of information required. The EPA states that MNA is most appropriate if a reasonable time frame for remediation can be demonstrated. To accomplish this, it can be combined with source control or groundwater extraction or other remediation processes. Therefore, the plume should be no longer increasing or should be decreasing in size.

Monitoring is thus a substantial part of MNA. According to the EPA (1999), the main objectives include:

- Showing that natural attenuation is performing as expected,
- Detection of hydrologic, geochemical, microbiological or other changes that could affect the natural attenuation process,
- Identify by-products that are toxic and/or mobile,
- Determining if the plume is expanding in a downgradient, lateral or vertical direction,
- Ensuring that there is no impact on receptors and that controls are sufficient to protect them,
- Determining if new releases of contaminants are occurring that could affect the natural attenuation process,
- Verifying that the remediation objectives are achieved.

Figure 7.4 shows the basic elements involved in a monitoring scheme. Note that the number of monitoring and sampling wells shown in the diagram is not representative of what will generally be required in the field. Experience from field studies have shown that the number of wells can be well over 100 wells and sampling points—depending on the size of the management zone shown in Figure 7.4. Monitoring and sampling must continue until one is assured that there are no longer any threats to the health of humans or to the environment. This is typically from one to three years according to EPA (1999) but this period can be substantially longer. As with the "lines of evidence," monitoring will be dealt with in greater detail in Chapter 8.

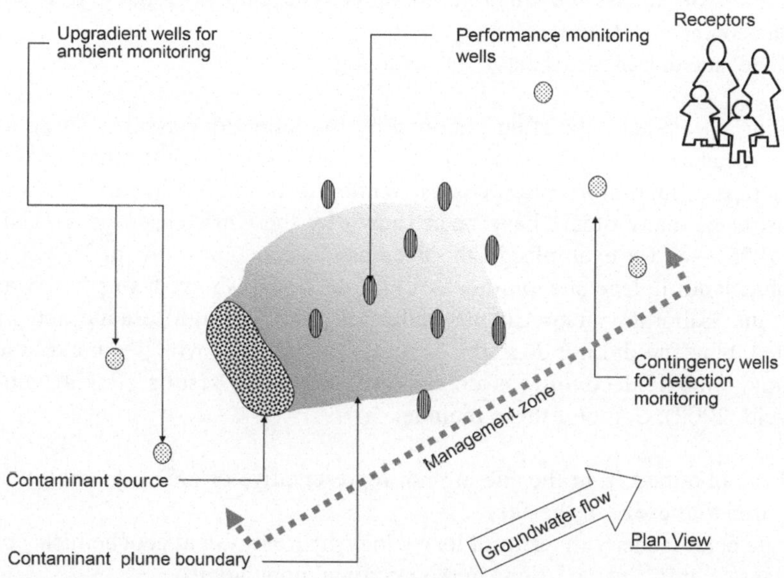

FIGURE 7.4 Monitoring network for performance, detection, and ambient monitoring as required in application of NA as a remediation tool.

7.4.2 Components of Monitoring

The term monitoring can have different meanings, depending on the context in which it is used. In NA, *monitoring* involves determination of the status of the pollutant plume and the effectiveness of the NA treatment. In that sense, there are several components which comprise the monitoring scheme and requirement. Monitoring for the NA of heavy metals in the subsurface soil is a good case in point. The various components that constitute the monitoring (sampling and testing) requirements include the following:

- Distribution and concentration of target heavy metal contaminants; Precipitated metals as distinct species;
- Speciations;
- pH and Eh.

Because heavy metals are more likely to be sorbed onto soil solids, the monitoring program must include sampling of the subsurface soils for determination of sorbed metal species and solutes in the pore water. The other components such as speciation, pH and Eh, etc. are necessary as monitoring components because they provide guidance into the status of the metal pollution plume.

In respect to organic chemical pollutants, according to the National Research Council (NRC) protocol (NRC, 2000), metabolism by microorganisms is documented by:

- Loss of the contaminant,
- Laboratory assays showing microorganisms have the potential to transform the contaminant, and
- Evidence that biodegradation is occurring.

Modelling parameters are often obtained by the laboratory assays. These assays involve measuring the decrease in the contaminant concentration caused by introduction of laboratory microcosms. While all these criteria cannot always be demonstrated, many others have been shown by the Christensen group (Nielsen et al., 1995)—as for example, in the decrease in concentrations of BTEX in the anaerobic landfill leachate plume. A chloride tracer showed sorption was not significant. Laboratory experiments indicated that although toluene and xylene degraded, benzene did not degrade. Several reasons exist which can explain why laboratory and field conditions do not give the same results (Roling and van Verseveld, 2002). Some of these include:

- Data obtained from the site are not representative or are very small compared to the size of the site;
- Site heterogeneity in terms of its geology, hydrology, and geochemistry can cause significant variations in the microbial populations;
- Microbial assays can be misleading or can be inaccurate, resulting in erroneous determination of the microbial activities and population;

- Changes in the chemical or physical properties of the sample can alter microbial diversity. While tests should be performed as soon as possible, we need to recognize that incubation tests take a long period of time. Caution should be exercised when interpreting lab tests concerning the microbial activities and community from in situ samples.

The US Department of Energy (DOE) as provided technical guidance for natural attenuation at their sites (DOE, 1999). As shown in Figure 7.4, three types of monitoring are utilized:

a. Performance monitoring within and adjacent to the plume,
b. Detection monitoring at the boundary of the management zone, and
c. Ambient monitoring (up gradient of the plume).

The USEPA (2007a) has also indicated that there are three basic categories for monitoring including: (1) monitoring to determine background contaminant levels and relevant ambient geochemical indicators such as redox potential [Eh] and pH, (2) monitoring to evaluative the potential efficiency of attenuation processes, and (3) monitoring to determine expansion of the plume. Performance monitoring is used to determine the progress or the effectiveness of the natural attenuation process. At the same time, checks on the various parameters that can impact directly on the process can be made. The protocols, future directions, and detailed requirements are discussed in detail in Chapter 8. In this section, we look at present concerns in monitoring procedures. For proper capture of subsurface information, monitoring wells need to be located not only in the contaminant plume, but also along the sides and ahead of the plume. Figure 7.4 shows the major details of such a scheme. The number of sampling locations will depend on several factors. These include:

a. The nature of the pollutants and source location,
b. The dimensions and features of the existent pollutant plume,
c. The hydrogeology and the subsurface features of the site, and
d. Model predictions of plume advance.

The intervals between monitoring stations should be sufficient to show the progress of the natural attenuation process, frequent at the beginning and then decreasing in frequency as the progress slows. Data should be able to demonstrate lines of evidence including:

- Reduction in mass concentrations of the contaminant over time
- Presence of conditions to allow geochemical or biological attenuation processes to occur
- Indicators that geochemical or biological processes are occurring such as degradation processes
- Analysis from the field of soil samples indicating precipitation or adsorption onto aquifer materials.

It is important to anticipate (to plan beforehand) what will be measured or detected in the monitoring system. In particular, we need to know what types of contaminants and concentrations can be expected at the various locations, and at various time periods. Taking the example of the six groups of contaminants found at DOE sites—fuel hydrocarbons, chlorinated organics, high explosives, metals, inorganic anions, and tritium—procedures and techniques required to detect each group will be different. Each group requires different and specific considerations in implementation of monitoring and testing schemes. For metals, since sorption can be either reversible or irreversible, it is necessary to incorporate measures that would permit examination of surface processes in the monitoring process. Sequential soil extractions can be used during the development of the conceptual model stage to determine the bioavailability of the metals. Mulligan et al. (2001) have shown that remedial techniques can be monitored by application of selective sequential extraction techniques. Sequential leaching tests can also be used during long-term monitoring of cationic metals. Measurements of pHs are required since: (a) cationic metals sorb strongly to the soil solids above pH 5, and (b) precipitation and co-precipitation of these same metals will occur at high pH levels. Redox potential should also be determined because of its potential effect on, for example, iron hydroxide stability for cations. Redox conditions will also impact directly on the reduction of chromium and technetium anions by soil organic matter. Regarding analytical methods, EPA and American Society for Testing and Materials (ASTM) standards can be used.

7.4.3 Established Monitoring Techniques

Chemical methods are used to determine the changes in concentrations of contaminants, and the production of by-products and/or co-reactants. Tracers are used to determine transport or sorption. Decreases in concentration of organic chemical pollutants, however, do not mean that complete degradation is taking place. Other pieces of evidence are required. These include: (a) carbon dioxide and methane production, and/or (b) changes in the concentration of electron acceptors (oxygen, nitrate, sulfate, and ferric ion). A good review of many of the problems associated with evaluation of electron acceptors can be found in Christensen et al. (2000). Sorption, precipitation and volatilization can complicate verification of subsurface bioattenuation. The pH and temperature are significant factors in determining the influence of the factors on biodegradation.

Determination of intermediate metabolites can provide significant information regarding the biodegradation of the parent compound of interest. For metal contamination, potential by-products such as sulfide should also be monitored. The compound should also be stable chemically and biochemically. Knowledge of the various microbial pathways is required. Energy requirements can also have a significant influence on determining the probably of existence of a compound. For example, if a compound requires significant energy for production, it is unlikely to be produced (Zwolinski et al., 2000). Table 7.1 provides some of the more significant parameters that need to be determined in the soil-water system for a better appreciation of the effectiveness of the NA treatment option. In addition to those parameters, we could add other measurements such as methane, to indicate anaerobic production, alkalinity for carbon dioxide production, chloride for verification of dechlorination and

TABLE 7.1
Parameters for Monitoring in Soil

Parameter	Reason for Monitoring
Volatile organic compounds	Degree of soil contamination
PAHs	Degree of fuel components
Chlorinated solvents	Degree of soil contamination
Heavy metals	Degree of soil contamination
Total organic carbon	Evaluation of bioactivity in vadose zone
Dissolved oxygen	Evaluation of bioactivity in vadose zone
Carbon dioxide	Evaluation of bioactivity in vadose zone
Nitrate	Potential electron acceptor for organic compounds
Iron (III)	Potential for iron reduction
Iron (II)	Electron donor
Sulfate	Electron acceptor for organic compounds
Chloride	Monitoring of chlorinated solvent reduction
Oxidation-reduction potential	Indication of the nature of degradation
Methane	Indicator of anaerobic degradation
Alkalinity	Buffering capacity
Temperature	Influences rate of biodegradation
pH	Influences rate of biodegradation
Hydroxide and carbonate content	Potential for metal sorption or precipitation

oxidation-reduction potential (ORP) for electron transfer potential. The frequency of sample procurement will depend on the rate of groundwater movement and spacing of monitoring wells. The fundamental requirement is to be able to determine the rates of contaminant attenuation. One should be prepared to alter the frequency of monitoring and sampling dependent on the results forthcoming from the ongoing monitoring and testing program. The sampling process should continue until it is proven that there is no risk to the health of humans or the environment.

7.4.4 Development of Monitoring Techniques

For the determination of metals and radionuclides in soil-water system and groundwater, the techniques used in the field must provide information that would permit one to determine mobility and bioavailability. Some of these, according to the Sandia National Laboratory (Waters et al., 1998) include scanning electron microscopy (SEM), isotope exchange techniques and soil digestion.

Newer techniques for detecting and evaluating populations of microorganisms that are culture-independent are being developed and utilized to provide one with a better understanding of the natural attenuation process. Polymerase chain reaction (PCR) is one of these techniques. This is based on small quantities of gene fragments. The potential for treatment of specific contaminants can be determined by evaluating the presence of various genes responsible for degradation or transformation (Brockman, 1995; Stapleton et al., 1998). Probes can be developed to evaluate the presence of various microbial strains. Most of the current information available is related to the species responsible for the aerobic biodegradation of BTEX and PAHs. These are the *Pseudomonas* species. The database called Genbank (https://www.ncbi.nlm.nih.gov/genbank/) contains the available sequence information. This method is complicated by the fact that a specific pollutant can be biodegraded via various pathways. For this method to be viable, a large amount of information on gene fingerprints and environmental data will need to be collected from many sites. Davis et al. (2002) examined 16S RNA gene sequences from a chlorinated ethane contaminated aquifer to characterize the microorganisms in the sediment. The reductive and oxidative mechanisms occurring in the aquifer were supported by the types of organisms present. Lu et al. (2015) identified *Trichococcus*, *Geobacteraceae*, *Geobacterlovleyi*, *Mucilaginibacter*, and *Arthrobacter* using the 16S rRNA gene clone library.

Hunkeler et al. (2011) also examined chlorinated ethane degradation using quantitative PCR. The presence of *Dehalococcoides* sp. genes were detected which provided support for reductive dechlorination of DCE. Němeček et al. (2017) used real time PCR to examine relative amounts of dechlorinating bacteria which resulted in the identification of *Dehalococcoides mccartyi* and *Dehalobacter* sp. Next generational sequencing analysis determined the presence of species for oxidation of the chlorinated ethenes in specific areas of the aquifer. Bælum et al. (2014) indicated that multiple functional genes should be used as biomarkers for reductive dechlorination.

The results of studies on natural attenuation of uranium in a tailings disposal site reported by Abdelouas and his coworkers (Abdelouas et al., 1998, 2000, 2002) indicate that uranium does not significantly adsorb on the surfaces of the sandstone minerals due to the negative charges of the carbonate minerals. Biological reactions to precipitate U(VI) by conversion to U(IV) are also possible by sulfate-reducing bacteria. Other contaminants such as technetium can also potentially be removed from subsurface soils and groundwater via reduction of Te(VII) and precipitation of TeO_2 and/or TeS_2 by sulfate-reducing bacteria.

Sulfate-reducing bacteria (SRB) are found in many types of environments including metal contaminated media including soil, mining residues, and wastewaters, among others. They can function in both anaerobic and aerobic environments. The results of the study of the distribution of sulfate-reducing bacteria at a uranium mill tailings disposal site (Chang et al., 2001) showed the presence of dissimilatory sulfite reductase (DSR) gene sequences, suggesting the dominance of the genus *Desulfotomaculum*, in up to 1500 ppb of uranium in the groundwater. The role of these SRB in reducing U(VI) to U(IV) and Te(VI) to Te(IV) need to be studied.

Other techniques under development include:

a. Measurements of the isotopes $^{13}C/^{12}C$ to determine if hydrocarbon degradation occurs,
b. Ribonucleic acid (RNA) probes from specialized bacteria for chlorinated solvent degradation, and
c. Determination of oxygenase activity.

Oxygenases are required for the co-metabolic degradation of chlorinated solvents (Sinke, 2001). An evaluation of the natural attenuation of explosives in soil was performed by Pennington et al. (1999). They determined that the stable isotope of soil organic matter was too small for detection and could not be related to changes in TNT concentration. However, a correlation of stable isotope ratios of nitrogen with changes in TNT in groundwater was found.

Compound-specific isotope analysis (CSIA) has been used to characterize hydrolysis of parathion in an aquifer at a pesticides landfill (Wu et al., 2018). Enrichment of the carbon isotope ratio compared to parathion indicated evidence for in situ hydrolysis. Another study by Alberti et al. (2017) showed that site characterization by CSIA provided evidence of the natural attenuation of chlorinated benzenes by anaerobic biodegradation. Damgaard et al. (2013) employed a combination of techniques (chemical, microbial, and CSIA) in an integrated manner to identify the biotic and abiotic degradation processes of chlorinated ethenes and ethanes in a clay. Lv et al. (2018) also used a combination of geochemical, microbial, isotopes (^{13}C, ^{14}C and ^{34}S) to show that sulfate reduction and methanogenesis were the main natural attenuation mechanisms at a petroleum contaminated groundwater site.

Most of the studies to evaluate the effectiveness of NA monitor groundwater to determine the presence of electron acceptors. Since, however, Fe(III) and SO_4^{2-} result in the mineral form of Fe and S, the mineral forms of iron sulfide and solid Fe(II) and Fe(III) need to be determined (Kennedy et al., 1998). This is because aqueous forms may not show noticeable changes during the natural attenuation process. A mild acid (0.05 N HCl) can be used to extract Fe(II) followed by spectrometric analysis. Fe(III) and sulfides can be analyzed after strong acid extraction (6N HCl) and further chromium and zinc precipitation for the sulfides (acid volatile sulfides, AVS) (Kennedy et al., 1998). Background conditions should also be examined in non-contaminated sections of the aquifer. This analysis can lead to a much better estimate of whether sulfate reduction is a factor in the development of the various processes contributing to the natural attenuation of contaminants.

The presence of chromium(III) solids was determined to evaluate reduction of Cr(VI) reduction in a bedrock (Zhao et al., 2017). Cr(VI) was extracted with NaOH. Subsequent analysis for Cr was performed by an ICP-MS. The method involved extraction with hydrogen peroxide. Complementary mineralogical analysis indicated that the presence of Fe(II), chlorite and biotite could enable reduction of the Cr(VI).

There is consensus that better sensor technology needs to be developed if we are to meet the requirements for proper monitoring of NA. Some of the newer sensor tools include:

- Direct push smart probes (Table 7.2) and cone penetrometer (CPT). The CPT can be coupled with various detectors such as membrane interface probe system (MIPS) for determining VOCs in the subsurface.
- Laser-induced fluorescence (LIF) or discrete multi-layer samples (DMLS) for vertical contaminant determination. Laser-induced fluorescence can be used to measure hydrocarbons real-time in undisturbed soil and groundwater.
- CPT laser-induced fluorescence delineation. This technique was used in a qualitatively manner to show the soil areas affected by contamination from Underground storage tank (UST) and piping by BTEX at Keesler Air Force Base in Biloxi, Mississippi (Weidemeier et al., 1999). Information of qualitative and semiquantitative nature can be obtained.

TABLE 7.2
Direct Push Instruments Being Developed for Monitoring Natural Attenuation Processes, Particularly Related to Soil Properties

Instrument	Application
Cone pressure and sleeve friction	For identification of flow pathways and confining layers
Laser Induced Fluorescence (LIF)	Defining hydrocarbon plumes and possible degradation (currently experimental) through emission spectra
Modeling Software with Global Positioning System (GPS)	Conceptual models for geological and chemical mapping, chemical fate and transport
Moisture probe (time domain reflectometry) and conductivity	Determination of unsaturated hydraulic conductivity and location of ionic fronts from salt water intrusion, river recharge and infiltration
Laser Induced Breakdown Spectroscopy (LIBS)	Determination of metal contaminants in soil
Soil gas sampler	Detection of VOCs from contaminants or biological activity, detection of respiration rates
Video microscope	Grain size determination, permeability and hydraulic conductivity estimation, cone pressure confirmation and free-phase NAPL (under development)

Source: Meuzelaar, H.L.C. ed., *Field Analytical Chemistry and Technology*, Vol. 2, No. 2, John Wiley & Sons, New York, 1998.

Various types of sensors can be integrated with cone penetrometer systems (CPT). Hydrocarbons such as fuel oil, motor oil, grease, coal tar, gasoline, jet fuel and diesel fuels can be screened by this method. The boundaries of a plume can be detected before and during treatment to monitor the progress of the remediation. Figure 7.5 shows a fiber-optic LIF sensor coupled to a CPT. A specific wavelength of light generated by a laser is conducted downward to a fiber-optic cable towards a sapphire window at the cone tip which advances in the subsurface. Two- and three-ring aromatic compounds and PAHs are detected as the laser causes them to fluoresce. The induced fluorescence returns over the second fiber to the surface detector system. The intensity of the signal is a measure of the concentration of the contaminant. In general, sandy soils offer higher fluorescence responses than clay soils. Other components in the subsurface can also give fluorescent response signals including de-icing agents, antifreeze and other detergents. The availability of a response signal database is essential if proper differentiation between the various receptors is to be obtained. Detection limits are generally 50–1000 mg/kg and responses are linear up to about 10,000 mg/kg. Calibration is performed by spiking known quantities into representative soil samples. The main advantage of this type of system is that real time data can be obtained in the field. The CPT pushes (penetration into the subsurface) of 65–100 m can be done in a 10-h period, and can only be undertaken in unconsolidated materials such as sediments and soft-to-medium clays. Limitations in the optical fiber constrain the CPT push to a depth of 50 m. This type of analysis

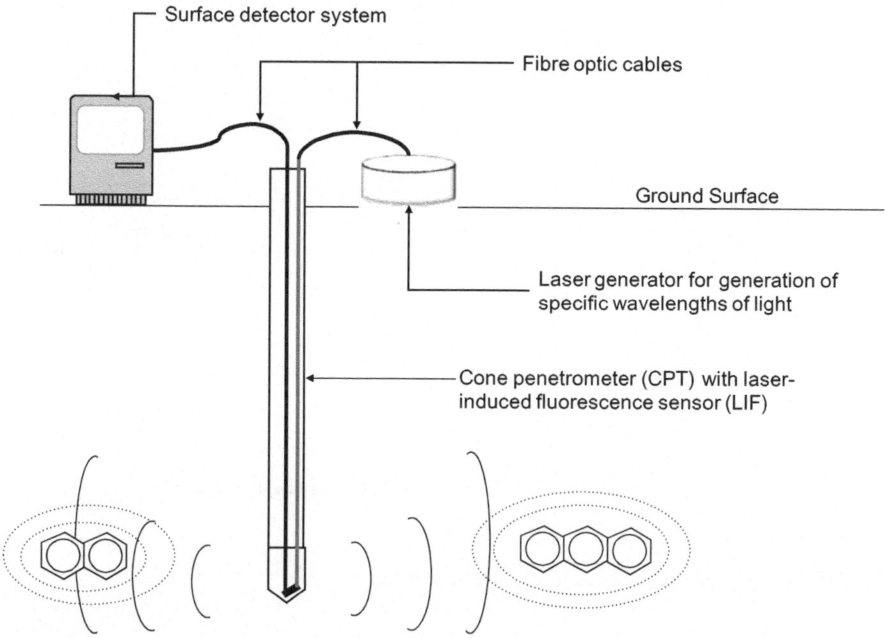

FIGURE 7.5 Laser-induced fluorescence sensor attached to a cone penetrometer for detection of PAHs.

is most cost-effective for large sites. A direct-push based BACTRAP® (in situ microcosms) probe based on Geoprobe® was evaluated in a field trial for a BTEX contaminated unsaturated soil zone (Schurig et al., 2014). The probe can be used where there are insufficient monitoring wells or there is a lack of accessibility.

Geophysical methods traditionally used in the petroleum and mining industries have recently been adapted for hazardous waste sites and indirect detection of aqueous and nonaqueous phase contaminants such as light non-aqueous phase liquid (LNAPLs) and dense non-aqueous phase liquid (DNAPLs). These methods are useful for detection of buried drums and other structures at hazardous waste sites that are sources of contamination. Geophysical methods which can be used for monitoring natural attenuation phenomena include electromagnetometry, magnetometry, and ground penetrating radar (Figure 7.6). The main types of contaminants that can be characterized by these methods include petroleum compounds, chlorinated solvents, polyaromatic hydrocarbons, and inorganic compounds. Electrical conductivity and resistivity (Figure 7.7), electromagnetometry, and to some extent ground penetrating radar can directly detect contaminants by determining changes in soil conductivities by the chemicals. Other techniques include magnetometry, seismic reflection and refraction, and gamma logging. These are able to identify subsurface characteristics that provide pathways for contaminant movement. Other techniques that are used increasingly for monitoring natural attenuation are summarized in Table 7.3.

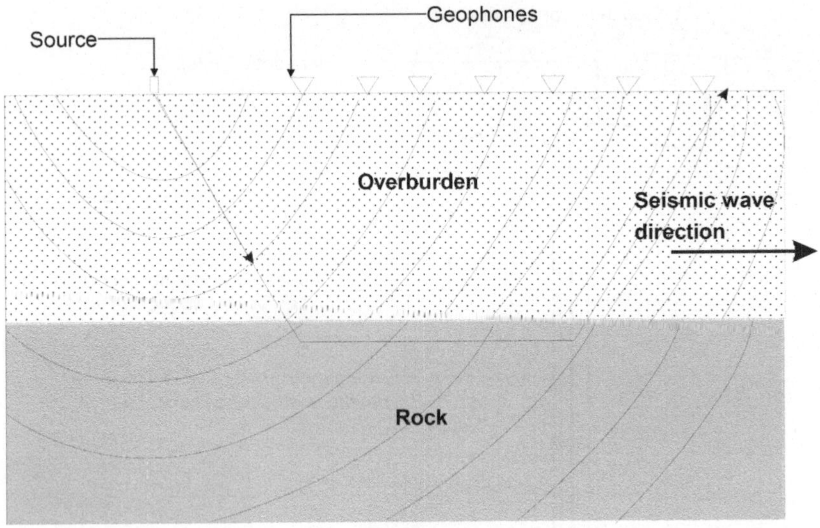

FIGURE 7.6 Schematic of ground penetrating radar.

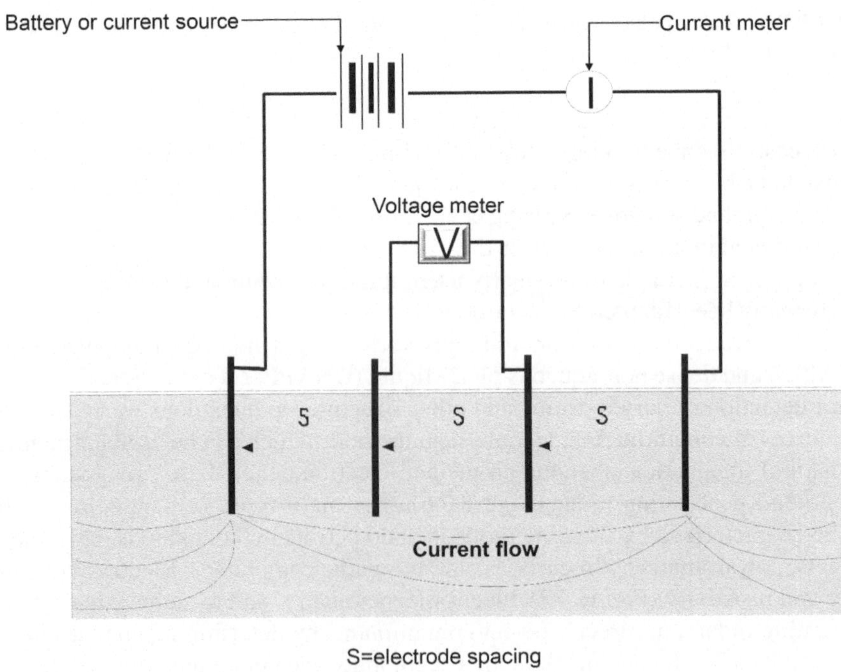

FIGURE 7.7 Schematic of resistivity survey.

TABLE 7.3
Summary of Available Natural Attenuation Models Soils and Groundwater

Model name	Applicability	Developer or reference
BIO1D	1 D model for biodegradation and sorption of hydrocarbons	GeoTrans Inc.
BIOREDOX	3D model for chlorinated solvents and petroleum hydrocarbons. Couples biodegradation and reduction of oxygen, nitrate, sulphate and carbon dioxide	Carey et al. (1998)
BIOPLUME II	2D model for transport of a single dissolved hydrocarbon species with oxygen limited biodegradation, first order decay, linear sorption, advection and dispersion	Rifai et al. (1988)
BIOPLUME III	2 D model for multiple hydrocarbons	Development commissioned by AFCEE
BioTracker	1 D Natural attenuation screening model with visualization tools for groundwater. Multispecies transport and transformation. Used with Sequence and BioTrends. Based on Bioredox	Scisoftware
CHEMFLO-2000	To simulate water movement and chemical transport in unsaturated soils and the convection-dispersion equation (chemicals). Public domain.	https://www.epa.gov/water-research/chemflo-2000-interactive-software-simulating-water-and-chemical-movement-unsaturated
3DFATMIC	3D model to simulate subsurface flow, transport, and fate of contaminants which are undergoing chemical and/or biological transformations for both saturated and unsaturated zones.	https://cfpub.epa.gov/si/si_public_record_report.cfm?Lab=NRMRL&dirEntryId=124866
Mass flux toolkit	The Mass Flux Toolkit, to calculate mass flux from transect data, and apply mass flux values to determine the effect of natural attenuation processes;	https://www.gsi-net.com/en/software/free-software/mass-flux-toolkit.html
MODFLOW	Model for estimating the vertical migration of dissolved organic solutes through the vadose zone to ground water and is a closed-form analytical solution of the advective-dispersive-reactive transport equation.	McDonald and Harbaugh (1988)

(*Continued*)

TABLE 7.3 (*Continued*)
Summary of Available Natural Attenuation Models Soils and Groundwater

Model name	Applicability	Developer or reference
MT3D	For simulation of unsaturated zone flow and transport of oily wastes. Partitioning of pollutant between the liquid, soil, vapor, and oil phases by linear equilibrium isotherms. Degradation of pollutant and oil is described as first-order process.	S.S. Papadopoulos & Associates Inc.
Natural attenuation software (NAS)	NAS is for the estimation of the time frame for MNA in groundwater and includes advection, dispersion, sorption, NAPL dissolution, and biodegradation	https://www.nas.cee.vt.edu/index.php
REMChlor	REMChlor, or Remediation Evaluation Model for Chlorinated Solvents, is an analytical solution for simulating source and plume remediation. Decay rates and parent/daughter yield coefficients are variable functions of time and distance.	https://www.epa.gov/water-research/remediation-evaluation-model-chlorinated-solvents-remchlor
RT3D	Modification of MT3D. For multispecies transport of chlorinated compounds, by-products and solid-phase species. Instantaneous aerobic degradation, BTEX degradation with multiple electron acceptors, sequential anaerobic degradation of PCE/TCE, and combined aerobic/anaerobic degradation of PCE/TCE.	https://bioprocess.pnnl.gov/
SEVIEW (SESOIL/AT123D)	SESOIL-vadose zone transport and fate model linked with AT123D groundwater model. Can be combined with BIOSCREEN and MODFLOW	http://www.groundwatersoftware.com/seview.htm
SWIFT-3D	3 D finite difference model to simulate contaminant, fluid and heat transport in porous and fractured media. Linear and nonlinear desorption, dispersion, diffusion, dissolution, leaching and decay. Public domain	http://www.mpassociates.gr/software/environment/swift98.html

(*Continued*)

TABLE 7.3 (*Continued*)
Summary of Available Natural Attenuation Models Soils and Groundwater

Model name	Applicability	Developer or reference
SWMS_2D	SWMS_2D is a computer program for simulating water and solute movement in two-dimensional saturated media, includes Richards equation for saturated-unsaturated water flow and the convection-dispersion equation for solute transport. provisions for linear equilibrium adsorption, zero-order production, and first-order degradation. Applicable for water and solute movement in unsaturated, partially saturated, or fully saturated porous media.	https://www.ars.usda.gov/pacific-west-area/riverside-ca/us-salinity-laboratory/docs/swms-2d-model/
VLEACH	1 D finite difference model for evaluating effects on ground water from the leaching of volatile, sorbed contaminants through the vadose zone. Includes liquid-phase advection, solid-phase sorption, vapor-phase diffusion, and three-phase equilibration in terms of soil properties, recharge rates, depth of water, or initial conditions. Public domain.	http://www.groundwatersoftware.com/vleach.html

A technique for sampling soil vapors or gases which has been described by Tartre (2001) involves purging the soil with nitrogen or other inert gas to determine the volatile phase in equilibrium with the dissolved, sorbed or free hydrocarbon phase. The purging technique can be applied to both the vadose and saturated zones. Mass transfer and mass production from a small volume of soil can be determined.

Guidance documents have been prepared regarding the newer chacterization tools. The Interstate Technology and Regulatory Council (ITRC) (2013) prepared descriptions of chemical technologies such as CSIA and molecular biological techniques. A guidance protocol also exists for application of nucleic acid-based tools for chlorinated solvent sites (ESTCP, 2011).

7.5 MODELS TO SIMULATE NATURAL ATTENUATION

7.5.1 Background on Modelling

Models are used to predict the fate and transport of contaminants for the vadose and saturated zones. Some of the more complex ones reviewed by Weidemeier et al. (1999) include SESOIL, VADSAT, Jury's model and SAM. These models require from 19 to 40 site and chemical parameters for input, depending on the model. SESOIL is the only one that considers chemical processes such as cation exchange capacity, hydrolysis and metal complexation. However, groundwater dilution of the leachate is not taken into account. A comparison of these models by Connor et al. (1994) shows that the main difference is: (a) volatilization is neglected by the SAM model, and (b) better correlation of results between models are obtained for transport in clay than in sandy soil.

Models have been developed for evaluation of NA and prediction of impacts on receptors. They can also be used to determine the effects of source removal or reduction, pump and treat, and the impact of barriers on NA. Models can either predict plume length without calibration or for fitting with observed data as a calibration procedure for the model. Risk-based corrective action programs used models without calibration whereas the Air Force Natural Attenuation Initiative is based on the second approach and the modeling results are used as lines of evidence.

Both analytical and numerical models can be used to simulate the conditions at a contaminated site. Analytical models provide exact solutions consistent with their mathematical formulations. This by no means should be construed to mean that they provide exact solutions to the real field problem. This will only happen when the analytical model accurately represents all the processes and boundary conditions in the field problem. Numerical models provide approximate solutions and are generally favored because of its capability to handle a wide range on hydrogeological conditions. Many processes must be approximated for analytical models. The heterogeneity of the subsurface can lead to over or underestimation of contaminant movement. Decisions concerning what model to use should be based on available data, the complexity of the site and the accuracy of the information required. Assumptions within each of the models must be known to understand their limitations.

Analytical models are used to simulate advection, biodegradation, dispersion and sorption in one, two or three dimensions. They are more useful if limited data is available. Numerical models are more applicable for cases where there is heterogeneity in the aquifer. Therefore, distribution of the hydraulic and contaminant properties within the aquifer is required.

7.5.2 Available Models for Natural Attenuation

Various packages are available for predictive modeling of natural attenuation (NA). The analytical model BIOSCREEN developed for the Air Forces Center for Environmental Excellence by Groundwater Services Inc. (Houston, TX) is used to simulate the natural attenuation from petroleum fuel releases. The Domenico analytical transport model is the basis for the model which is in a Microsoft Excel format and includes the assumption that the source is infinite and concentrations from the source do not change. The BIOSCREEN model assumes a declining source concentration. Attenuation processes for the soluble hydrocarbons include advection, dispersion, adsorption, in addition to aerobic and anaerobic degradation. Since only vertical zones are considered from the source, the program is more applicable for LNAPL than DNAPL sites (Wiedemeier et al., 1999). Regarding biodegradation, decreases in electron acceptors such as dissolved oxygen, nitrate, and sulfate and metabolic by-product formation (iron(II) and methane) are also incorporated. These parameters are accepted as evidence that biodegradation is occurring. It is used to determine if NA is feasible or whether other lines of evidence are needed. The more sophisticated BIOPLUME model has been at Superfund and Resource Conservation and Recovery Act (RCRA) sites. Outputs from the model include graphs of the plume centerline, plume concentrations as a 3D color plot, and mass balances showing removal by each electron acceptor. A mass flux calculator has been added to Version 1.4 to show the mass fluxes at various points in the plume.

BIOCHLOR is a model that simulates the natural attenuation of chlorinated solvents. Some of the features included in the program include: (a) determination of reductive dechlorination of chlorinated ethenes to ethenes and chlorinated ethanes to ethanes in sequential steps, and (b) separation of higher dechlorination rates into a different zone from lower dechlorination rates, (c) development of a database to assist in the prediction of first-order decay rates for input in to BIOCHLOR, and (d) incorporation of advection, dispersion and adsorption in the program scheme. Both this software and BIOSCREEN are available from the USEPA (https://www.epa.gov/water-research/bioscreen-natural-attenuation-decision-support-system.).

The program BIOSCREEN was used to estimate the time for remediation of BTEX by natural attenuation processes at a facility contaminated with fuel hydrocarbons (Suarez and Rifai, 2002). Field data for monitoring along a flow line was used to calibrate the model, and input information included field data such as hydraulic conductivity and gradient, porosity, retardation factor, and biodegradation decay rate. Information on biodegradation rate was obtained from a well at the source, and benzene was used as the key compound to obtain the attenuation rate. With this calibration, the data agreed fairly closely with the measured values, and an estimate of a 250-year time period was made for effective biodegradation of the benzene in the contaminated site, with an initial concentration of approximately 900 mg/L at the source.

One of the major drawbacks of using models such as BIOSCREEN is that a first order decay coefficient is used to calibrate the models (Odencrantz et al., 2002). The Domenico (1987) solution for advection-dispersion-biodegradation is given by:

$$C(x,t) = \frac{C_0}{2} \exp\left[\frac{x}{2\alpha_x}\left(1 - \sqrt{\frac{1 + 4\lambda\alpha_x}{v}}\right)\right] erfc\left[\frac{x - vt\sqrt{1 + \frac{4\lambda\alpha_x}{v}}}{2\sqrt{\alpha_x vt}}\right] erf\left[\frac{Y}{2\sqrt{\alpha_y x}}\right] \quad (7.1)$$

where:

C_0 = Initial concentration
α_x = Longitudinal dispersivity
α_y = Transverse dispersivity
λ = Rate of biodegradation
t = Time
x = Distance from source
v = Retarded velocity due ($v = v_s/R$, R is the retardation factor and v_s is seepage velocity)
Y = Width of the source.

The effect of the various factors on the calculations is shown in Figures 7.8 and 7.9.

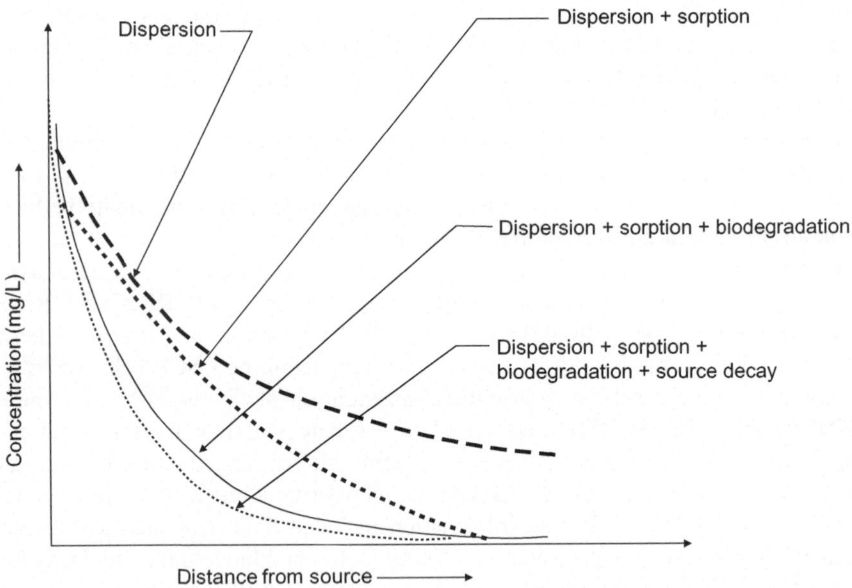

FIGURE 7.8 Impact of attenuation processes at a site. The more the attenuation processes, the higher the overall attenuation rate will be. (From Newell, C.J. et al., *Calculation and Use of First-Order Rate Constants for Monitored Natural Attenuation*, Office of Research and Development, National Risk Management Research Laboratory, EPA/540/S-02/500, Cincinnati, OH, 2002.)

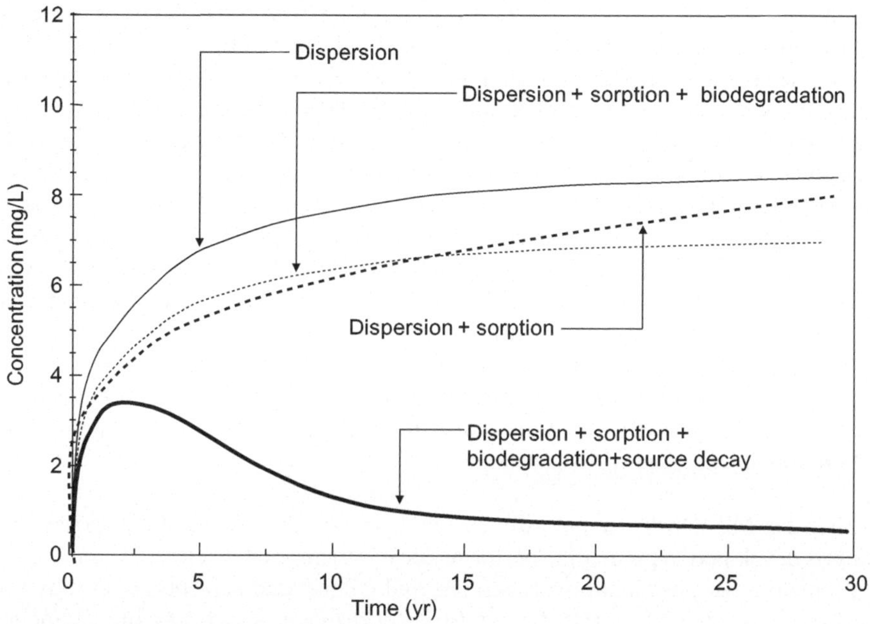

FIGURE 7.9 Effect of the different mechanism on contaminant concentration versus time near the source. (From Newell, C.J. et al., *Calculation and Use of First-Order Rate Constants for Monitored Natural Attenuation*, Office of Research and Development, National Risk Management Research Laboratory, EPA/540/S-02/500, Cincinnati, OH, 2002.)

The first order decay coefficient, i.e., the rate of biodegradation λ, is adjusted until it coincides with field analyses. All information regarding dispersion, sorption and biodegradation are represented as one parameter. This can lead to significant inaccuracies. More details can be found at https://www.epa.gov/water-research/bioscreen-natural-attenuation-decision-support-system.

The BIOPLUME III two-dimensional model (Rifai et al., 1997) simulates natural attenuation by taking into account advection, dispersion, ion exchange and biodegradation. Both aerobic and anaerobic electron acceptors are included (https://www.epa.gov/water-research/bioplume-iii). The U.S Geological Survey (USGS) formed the basis of the model in July 1989 (Konikow and Bredehoeft, 1978). It is mainly used for evaluation of groundwater flow and transport. Biodegradation reactions can be simulated by instantaneous degradation, first-order decay, and Monod kinetics. Electron acceptors include oxygen, nitrate, iron, sulfate, and carbon dioxide. One of the main limitations is that components such as BTEX are treated as one component not as individual ones. This is particularly problematic since benzene and toluene are more biodegradable than xylene.

Other models are available such as BIOPLUME II (Rifai et al., 1988), BIOPLUME III (Rifai et al., 1997) and Bio1D (Srivinsan and Mercer, 1988). They are applied mainly for evaluation of the transport and biodegradation of petroleum compounds. BIOPLUME III can simulate the aerobic and anaerobic biodegradation of organic contaminants. The electron acceptors, oxygen, nitrate, iron (III), sulfate, and carbon dioxide, are depleted sequentially. Transport and production of Fe(II) can also be simulated if it is used as an electron acceptor. First order decay, instantaneous reactions, and Monod kinetics can be simulated. It can be used to determine the extent of the plume and its persistence with or without engineered controls. As an example, this model was applied to examine the feasibility of natural attenuation at Patrick Air Force Base, Florida (Rifai et al., 2000). It was determined that it would take 30–40 years to stabilize the plume at a distance of less than 170 m from the source area. The breakdown of aerobic and anaerobic biodegradation was 30% and 65% of the total losses, respectively.

The models MODFLOW and RT3D (Reactive Transport in Three Dimensions, Sun et al., 1996) are also used extensively. Both sorbed (sorption and desorption processes) and aqueous species can be simulated, in addition to dissolution of NAPL. Contaminants include heavy metals, explosives, petroleum, and/or chlorinated hydrocarbons. They are three dimensional models for simulation of solution solute and microbial transport with applications for natural attenuation and enhanced bioremediation processes.

Bioredox-MT3DMS is a three-dimensional multi-species fate and transport model that was developed to predict the performance of intrinsic or accelerated bioremediation remedies for petroleum hydrocarbons and chlorinated solvents. BioRedox was developed from the MT3DMS fate and transport model and works in conjunction with MODFLOW. An oxidation-reduction reaction database is incorporated into the model. Iron hydroxide depletion in the soils can be modeled. Biological reactions can be first order or instantaneous.

Natural attenuation software (NAS) (Kram et al., 2007) is designed as a screening tool for estimation of the time for remediation by MNA. It is applicable for relatively homogeneous aquifers and consists of a of analytical and numerical solute transport models for advection, dispersion, sorption, NAPL dissolution, and biodegradation.

Geochemical modeling is applicable for providing secondary lines of evidence for MNA. Speciation programs can provide estimates of species concentrations as function of pH and redox potential and dissolved groundwater concentrations. PHREEQC (Version 2) and MINTEQA2 are the most commonly used: https://wwwbrr.cr.usgs.gov/projects/GWC_coupled/phreeqc/index.html.

https://www.epa.gov/ceam/minteqa2-equilibrium-speciation-model

Other models are indicated in Table 7.3.

7.5.3 Application, Calibration, and Verification of Models

Utilization of analytical and computer models to a particular site requires contaminant transport data. In addition to usage of models for prediction and analysis of attenuation events at a site, models can also be used to determine if more or extra data are needed. These can be location specific or general. In the final analysis, models provide us with an appreciation of the effectiveness of NA at the site under investigation. An approach for a model methodology can be seen in Figure 7.10. A well thought out conceptual model (Figure 7.11) is the key to a successful natural attenuation model. It aids in the understanding of the site, assessment of available data and

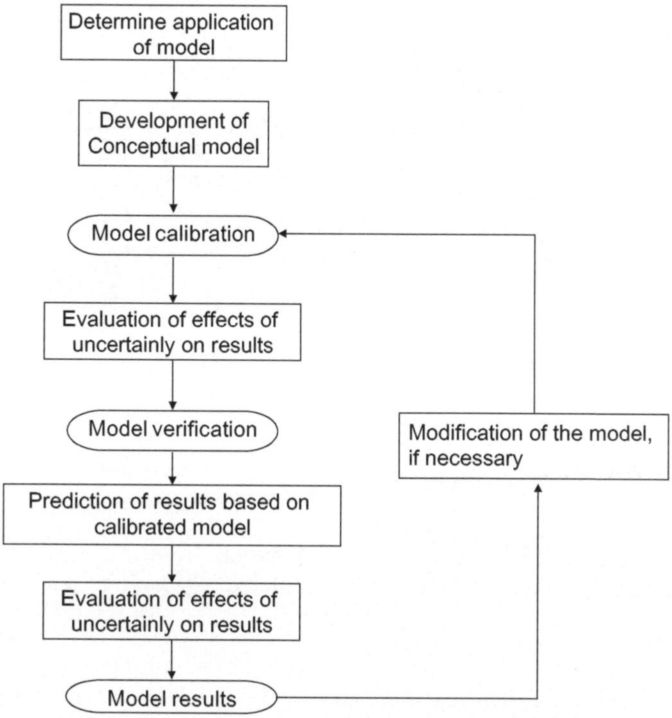

FIGURE 7.10 Simple procedure for model development.

determination of other required inputs and where further sampling is required. The choice of model and the suitability of the particular model will depend on the limitations, assumptions, and site characteristics such as the hydrogeology (Corapcioglu and Baehr, 1987; Carey et al., 1998). The scope, availability, and quality of input data will be the key to a successful model, particularly if the site is complex. Calibration, verification, and prediction are all essential to improve the accuracy of the model for site conditions. Calibration is used to adjust the results so that they can accord better with measured data. Biodegradation rate constants are often modified when the results obtained are not close to field data. Use of laboratory data can lead to serious errors when these are extrapolated to field scale. To a very large extent, this is because we lack the proper scaling laws. The use of data from other sites for application to the site under investigation can lead to significant errors in predictions because of the differences between the native conditions in the two sites. At all times, site specific field data should be inputted into the model. The contribution of biotic factors should be determined so that the models can be used to predict accurately the movement and concentration of contaminants and effectiveness of NA (Rifai et al., 1995; Mercer, 1998).

Models such as BIOSCREEN (Newell et al., 1996) and BIOCHLOR (Aziz et al., 2000) use first order rates for attenuating contaminants as they leave the source zone.

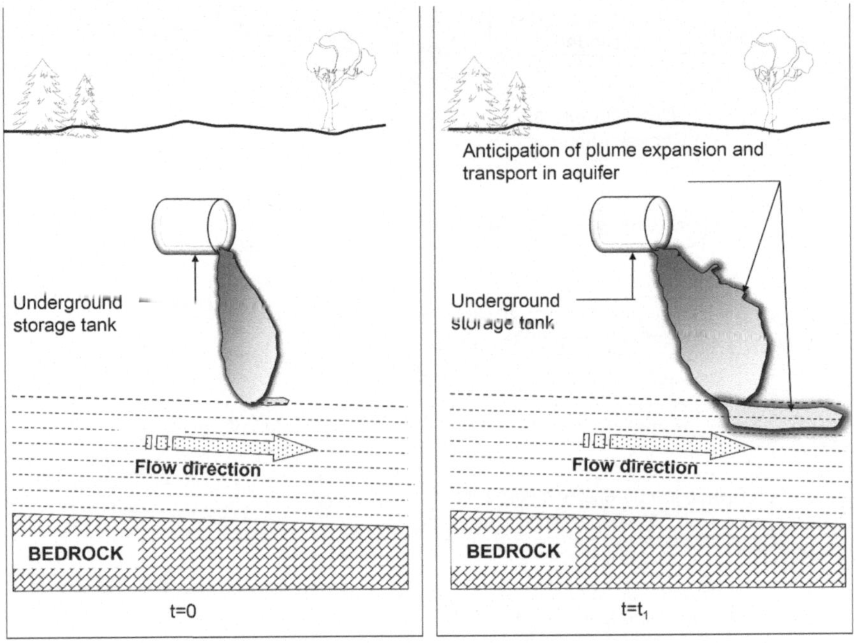

FIGURE 7.11 Schematic showing conceptual model for NAPL release from a leaking underground storage tank.

The rates and their applications are not clearly defined, and natural attenuation processes must be clearly understood for proper application of the rate constants. Three types of rate constants are used (Newell et al., 2002).

- Pollutant concentration versus time constant (inverse time) to show the time for achievement of remediation goals. It is determined by plotting the natural log concentration at a specific location versus time and determining the slope.
- k-concentration versus distance constant (inverse time) which is a bulk attenuation rate to show if the plume is changing in size due to dispersion, biodegradation, or other attenuation processes. It is determined by plotting the natural log concentration at a specific location versus distance and determining the slope and multiplying by the groundwater seepage velocity.
- λ – Biodegradation rate constant (inverse time) for soluble contaminants to determine the effect on solute transport. Determined various ways such as comparison of contaminant and tracer transport or by calibrating a model (BIOSCREEN, BIOCHLOR, BIOPLUME, MT3D, etc.) with field data. Field studies can utilize a tracer or the steady-state contaminant plume method of Buscheck and Alcantar (1995). For the latter method, the number of wells must be sufficient and the plume must be at steady state to give appropriate results.

Source degradation and reduction due to the various attenuation processes (Figures 7.8 and 7.9) can have significant influences on contaminant concentrations. In addition, uncertainties in the calculations for the rate of biodegradation can originate from sources including the placement of monitoring wells, variations throughout the year, sampling and analytical procedures as well as the heterogeneity at the site (Newell et al., 2002). High molecular weight compounds were not determined, and it is not known how the presence of higher molecular weight compounds influences the degradation of the lower molecular weight ones. The models typically incorporate linear and or reversible performances. However, for PAHs, sorption is often nonlinear and hysteric and highly dependent on the type of media. In the future, models will need to determine the attenuation of PAHs by dilution, sorption and biodegradation in a more specific manner and less of a "black box" depletion approach (Rogers et al., 2002).

Regarding the use of models, the National Resource Council (NRC) (2000) discussed several common problems with their application. We cite three categories as follows:

- The first category is related to the model framework which includes conceptual models and data used as the basis for these models. The present and future site conditions must agree with the assumed boundary conditions, equations and parameters used. Data should not be taken from another site, and observations must be weighted appropriately. The most appropriate model for the site should be used.
- The second category essentially states that a closed mind should not be maintained when evaluating the model and the field data. Models should not be forced if they do fit the data. Other models or an alternate conceptual model should be considered if this problem arises.
- The final category involves a model's end results. The model should not be extrapolated beyond its ability, nor should its accuracy be overestimated. Uncertainties in the data, data assumptions, methods of verification and validation, parameters inputted, and the complexity of the site can also influence the model's results.

7.6 PROTOCOLS DEVELOPED FOR NATURAL ATTENUATION

The various protocols for use of NA as a remediation tool, together with the lines of evidence and evidence of success of NA plus Enhanced natural attenuation (ENA) and EngNA are developed in Chapter 8 in terms of "where we should go from here." These recognize the need or usefulness in obtaining as much supporting information as can be obtained, to support the "knowledge-based" requirements for application of natural attenuation processes in the treatment of contaminated ground. In this section, we will pay more attention to present usage of protocols and requirements as evidenced from field applications of NA.

7.6.1 Various Technical Protocols

Various technical protocols have been established such as the *Designing Monitoring Programmes to Effectively Evaluate the Performance of Natural Attenuation* by the Air Force Center for Environmental Excellence (Weidemeier and Haas, 1999). Sampling type, frequency and location, and analyses required for NAPL contaminants are described. Concentrations of dissolved oxygen, nitrate, Fe(II), sulfate and methane can be determined with contaminant data obtained downgradient of the plume. Monitoring wells should be located to obtain this data and sampling frequency should be determined on the basis of distribution and location of monitoring stations, evidence of attenuation, and hydrogeology. Computational models and statistical methods can be used to determine the time of clean up and plume behavior, to aid in the planning of monitoring frequency and monitoring intensity. A summary of well locations and analytical parameters to be monitored is shown in Table 7.4 for a LNAPL plume. Background groundwater quality information can be obtained from up- and side-gradient wells (see Figure 7.4). Those monitoring stations within the plume should provide information that indicates the progress of the natural attenuation process. Downgradient wells can be employed to monitor the plume progress and contaminant concentrations. A full set of geochemical data for hydrocarbon contaminants includes dissolved oxygen, nitrate, Fe(II), sulfate, and methane concentrations, pH, conductivity, alkalinity and redox potential. Additional parameters are required for chlorinated hydrocarbons including chloride, total organic carbon, and hydrogen.

TABLE 7.4
Well Location and Sampling Requirements for Monitoring of a NAPL Plume

Well Location	Purpose	Long-Term Sampling
Up or side-gradient	Background water quality monitoring	No analysis or a limited set of parameters including dissolved oxygen, redox potential, temperature, pH unless changes are indicated or indicator contaminants
Source area	Determination of composition or source concentration	Analysis of indicator contaminants in the NAPL and groundwater below NAPL and limited set of geochemical parameters
Downgradient of source along plume center-line	Plume monitoring over time	Analysis of indicator contaminants unless change in plume behaviour when full set required
Downgradient of plume	Detection of plume migration	Analysis of indicator contaminants unless change in plume behaviour when full set required
Contingency well	Determination of migration towards receptors	Analysis of indicator contaminants unless change in plume behaviour when full set required
Surface water	Determination of impact on surface water	Analysis of indicator contaminants unless change in plume behaviour when full set required

Source: Weidemeier, T.H. et al., *Natural Attenuation of Fuels and Chlorinated Solvents in the Subsurfaces*, John Wiley & sons, New York, 1999.

The EPA 1998 *Technical Protocol for Evaluation of Natural Attenuation of Chlorinated Solvents in Ground Water* (Weidemeier et al., 1998) was established to demonstrate mechanisms of chlorinated solvent natural attenuation. The EPA (2001) points out, however, that field information can be substantially different from laboratory research, particularly regarding dechlorination rates and product concentrations. For chloromethane, chloroethanes, chlorinated benzenes and chlorinated ethers, the reductive dechlorination rates need to be compared in the laboratory and field. Field information is not available for many processes and their reaction rates. To complicate matters, there are other uncertainties such as those related to interactions with other contaminants, high concentrations at the source area, and mechanisms for degradation other than reductive dechlorination.

The protocol by the ASTM applies to petroleum contamination from underground storage tank releases into the groundwater (ASTM, 2015). Included in the protocol are requirements and considerations such as:

- Site characterization, evaluation of potential risks and the ability to meet remediation goals;
- Evaluation of the size of the pollutant plume;
- Monitoring requirements and implementation should be based on site-specific conditions and should be maintained until objectives are met;
- Comparison of remediation by natural attenuation performance to remedial goals;
- Comparison of remediation to natural attenuation with other options;
- Structuring of contingency plans and implementation, if the objectives cannot be met by natural attenuation.

To prove that natural attenuation is occurring, lines of evidence are established to indicate decreases in contaminant concentrations (NRC, 1993). They include:

- Decreases in contaminant concentration and/or plume size over time. This is used to indicate that biodegradation is occurring faster than plume size increases.
- Chemical indicators of microbiological activity in the groundwater chemistry such as consumption of oxygen, nitrate, and sulfate and production of Fe(II), Mn(II) and methane.
- Laboratory microcosm studies are used to simulate aquifer conditions to determine if bacteria at the site can biodegrade the contaminants and at what rate. This step is mainly used if neither of the first two clearly indicates significant trends.

Of the three lines of evidence indicated by the NRC (1993), it is suggested that primary lines are usually required. Secondary or tertiary are required only if primary lines are insufficient. Site specific attenuation rates can be used as a secondary line of evidence showing that attenuation is occurring and the rate at which is occurring. The time frame establishment of goals and the requirement for source control are determined by the regulatory agency. The remediation goals must be established earlier which differs from the US Air Force Protocol (Weidermeier et al., 1995). Monitoring frequency is according to the potential impact on receptors, ability to meet remedial goals, plume behavior and institutional controls used. Contingency plans are necessary only if the goals cannot be met. The details of protocols for lines of evidence, evidence of success, and monitoring provided in Chapter 8 differ somewhat from these. The requirements and protocols discussed in Chapter 8 pay attention to the same issues and factors discussed in this chapter, but the soil-contaminant interaction components are added to the assessment procedures.

The EPA guidance document covers all contaminants in soil and groundwater at RCRA and Superfund sites, and for underground storage tanks (UST) (EPA, 1999). For the OSWER policy on natural attenuation (*Use of Monitored Natural Attenuation at Superfund, RCRA Corrective Action and Underground Storage Tank Sites*), if natural attenuation is chosen for remediation, the following must be demonstrated:

- That natural attenuation is occurring as expected,
- Toxic by-products are identified (TCE, DCE, vinyl chloride, methyl mercury, etc.),
- Extent of plume expansion,
- Identification of new releases or other environmental conditions that could affect the progress of natural attenuation,
- That clean up objectives can be obtained, and
- Contingency plans must be made in the event that natural attenuation is not effective.

A subsequent guidance was issued in 2015 (USEPA, 2015) and focuses on inorganic contaminants. It also indicates that multiple lines of evidence should be used to evaluate the feasibility of MNA as the previous guidance. However, this guidance recommends a phased analytical approach for inorganic contaminants that builds on the pre-tiered analysis approach involving a detailed analysis of site characteristics for controlling and sustaining attenuation.

Other protocols have been reviewed by the NRC (2000). It is useful to note, from the document, that the number of samples, parameters to be monitored and contaminant concentrations to be obtained can vary substantially among the protocols. For example, in addition to those mentioned above, 14 federal, state, profession, and industry protocols were evaluated and reported in the document, together with community concerns, scientific and technical issues, and implementation issues. Only seven have been peer-reviewed. In general, the NRC (2000) noted that only the Department of Energy (DOE) guidance entitled *Site Screening and Technical Guidance for Monitored Natural Attenuation at DOE Sites* (Brady et al., 1998) and the EPA protocol address inorganic contaminants, but in a very limited way. The DOE protocol gives some guidance regarding the sorption and sequestration of inorganics by giving some default values for sorption coefficients and suggesting that sequential extraction or isotopic pulsing could be used to evaluate irreversible uptake. The US Air Force *Technical Protocol for Implementing Intrinsic Remediation with Long-Term Monitoring for Natural Attenuation of Fuel Contaminants* (Weidemeier et al., 1995) was evaluated as one of the most scientifically sound, as it describes ways to estimate biological activity, dilution, sorption and dispersion. Methyl tert-butyl ether (MTBE) is not addressed, however. A more recent document by the EPA (2001) has made various recommendations concerning MTBE including:

- Evaluating the biodegradability of MTBE and other oxygenates under field conditions
- Determining with more certainty the rate of dissolution of MTBE and other fuel components
- Establish a database on the natural attenuation of MTBE.

Wilson et al. (2005) reported the state of knowledge of MTBE at leaking underground storage sites. Recommendations were provided regarding data requirements for risk assessment of MNA. It was particularly aimed at regulators to allow decision making regarding the suitability of MNA at a site. More recently, a technical protocol by the American Petroleum Institute (API, 2007) was published. It builds on previous reports of the USEPA and ASTM. Information on data collection, evaluation, and interpretation is included. A tiered approach is utilized for interpreting lines of evidence to evaluate natural attenuation on a site-specific basis.

The USEPA (2001) has also tried to establish procedures for evaluating the natural attenuation of inorganics. For example, it is suggested that: (a) the natural attenuation mechanisms for arsenic and other inorganics should be determined, (b) the effect of geochemical conditions on remobilization need to be established, (c) the effect of organics on inorganic contaminant behavior and vice versa should be studied, (d) guidelines need to be developed to understand immobilization processes based on laboratory and field data and the use of models, and (e) uncertainty analysis needs to be incorporated into decision-making.

There are further documents on inorganics. Three reports have been subsequently issued by the USEPA (2007a, 2007b, 2010) to address inorganics including MNA of metals, nitrate, perchlorate, and radionuclides. Following these, a 2015 guidance report was issued to specifically address inorganic contaminants (USEPA, 2015). A tiered approach for site characterization is shown to reduce data uncertainty as shown in Figure 7.12. Truex et al. (2011) and NJDEP (2012) have also has issued guidance for inorganic and radionuclide contaminants, and organic compounds, particularly for petroleum hydrocarbons and chlorinated solvents, inorganics, and radionuclides.

The majority of the available protocols for organic contaminants address only fuel hydrocarbons or chlorinated solvents and infrequently PAHs, PCBs, explosives, and pesticides. More recently however, Rectanus et al. (2015) has summarized the mechanisms of MNA in soil and groundwater for munitions.

There are major shortcomings in these protocols. The following are some of the recommendations made by the NRC (2000):

- Agreement is required on the use of protocols for natural attenuation and how they are used to obtain regulatory approval for natural attenuation.
- All protocols should be peer-reviewed by independent experts.
- Conceptual models are preferred over scoring systems as decision guidelines.

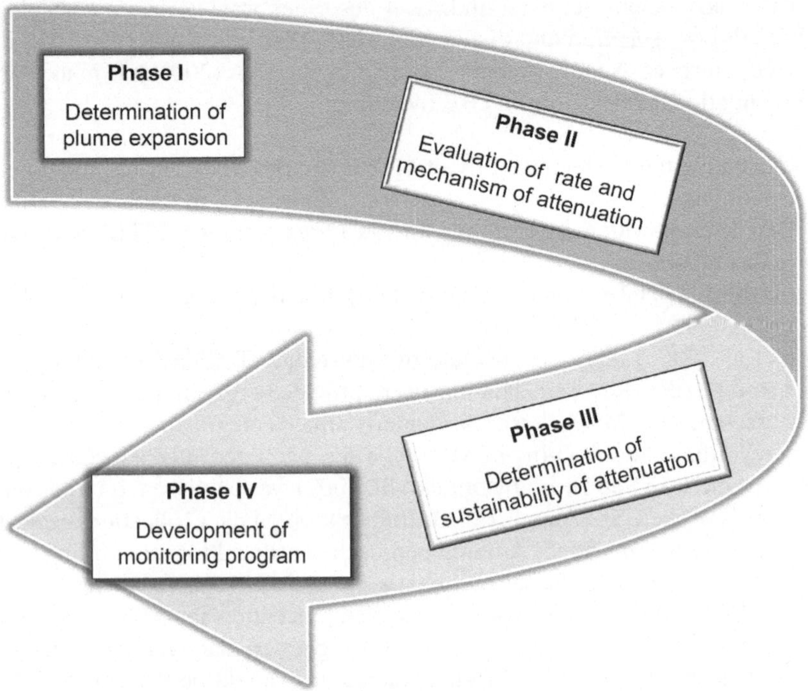

FIGURE 7.12 Tiered analysis approach for inorganic contaminants. (From USEPA, *Use of Monitored Natural Attenuation for Inorganic Contaminants in Groudwater at Superfund Site,* U.S. Environmental Protection Agency, Office of Solid Waste and Emergency Response, Directive 9283.1–36, 2015.)

- Documents to support the protocols should be easy to use.
- Training by neutral organizations should be provided for proper use of the protocol.

7.6.2 Inclusion of Soils and Sediments in Protocols

One aspect not considered by the NRC, since their focus was on groundwater quality, was that most protocols are designed for groundwater natural attenuation and not for natural attenuation of contaminants in soils or sediments. Few protocols exist for soil with the exception of the USEPA (1999) and those by the DOE. More recently some efforts have been put forward. A framework for inorganic non-volatile contaminant was prepared for evaluation of MNA in the vadose zone (Truex and Carroll, 2013), Magar et al. (2009) developed a technical guidance for natural recovery of sediments. This was followed by USEPA (2014) describing methodologies for assessment and prediction of natural processes in sediments.

Petroleum contaminants that are of low solubility will tend to remain in the soil. It is imperative that these contaminants do not pose threats in the soil through direct contact or leaching to the groundwater. Leaching and environmental mobility of sorbed contaminants are significant agents for release of these contaminants for transport to biotic receptors. As another example, chlorinated solvents have the potential to slowly leach into the groundwater over long periods of time. For inorganic contaminants in soils, it is particularly important to identify the mechanisms of interaction such as the type of sorption or redox reaction for retention. Some precipitation and absorption reactions are quite stable such as precipitation of cesium on to clay minerals, whereas others such as surface adsorption of uranium on iron-oxide minerals and organic complexation are less stable or more reversible. The presence of chelating agents such as EDTA can also increase the mobility of trivalent chromium. Other factors such as pH changes, redox, chemical speciation or concentration changes may also influence mobility.

Radionuclides can also decay to other daughter products such as Pu-241 to Am-241 and Np-237. These can be more mobile and toxic than the parent. Radionuclides are also a concern since direct contact is not necessary for harm to come to humans. Gamma radiation or x-rays can increase the risk to humans at the near surface. Natural attenuation of inorganic contaminants can be effectively applied if the process or mechanism for attenuation is demonstrated to be irreversible.

7.7 CASE STUDIES OF NATURAL ATTENUATION

7.7.1 Natural Attenuation of Organic Contaminants

Most of the reported case studies of NA have focused on the groundwater aspects of the application. This is not surprising, since the initial thrust of NA has been directed towards management of groundwater quality. Reported case studies on NA in soils and sediments are relatively scarce. However, the evidence shows that there will be considerably more application of NA to contaminated sites—as knowledge and understanding of the abiotic and biotic processes contributing to the natural attenuation of contaminants in soils and sediments increases. Some of the reported case studies of NA in soils and sediments will be described in this section.

Fuel components such as benzene, toluene, ethylbenzene and xylene, known as BTEX, are readily biodegraded in the subsurface. Other contaminants such as chlorinated hydrocarbons are not believed to degrade easily and are difficult to understand. Weatherill et al. (2018) have reviewed field experience and monitoring technologies with regards to chlorinated ethenes. They have indicated that future research should focus on identifying limiting reactants, residence times and microbial ecophysiology in relation to biotic and abiotic factors.

Compounds such as tetrachloroethylene (PCE) are converted by reductive dechlorination to dichloroethylene (DC), and vinyl chloride (VC).

PCE → TCE → DCE → VC → ethane or ethane
Increasing water solubility →

An analysis of 61 BTEX-contaminated sites showed no reductive chlorination occurred at 23 of them, accumulation of DCE was detected at 18 and vinyl chlorine production was at 20 sites (McNab et al., 2000). It had previously been expected that vinyl chloride would accumulate (Vogel et al., 1987). However, no accumulation of vinyl chloride, ethylene or ethane were detected since it was likely that bacteria in the subsurface could have converted these components to carbon dioxide with various electron acceptors including iron (III), manganese (IV) or oxygen. Anaerobic oxidation of VC or *cis*-DCE could have occurred with the electron acceptors oxygen, iron (III), manganese (IV) or natural organic matter.

Schiefler et al. (2018) studied the attenuation of chlorinated ethenes at a site in Denmark. An integrated approach was used including chemical, isotopic, and microbial analyses. The degree of dechlorination varied according to geological layer and location. Near the surface, aerobic co-metabolism and/or abiotic geochemical processes were natural attenuation processes.

The natural attenuation of chlorinated solvents was studied in a difficult to remediate fractured bedrock (Figure 7.13) (Şimşir et al., 2017). *In situ* measurements were performed and showed that reductive dechlorination occurred in the sediment. Microcosms in the sediment or by using incubated Bio-Sep beads showed that C1 to C3 chlorinated solvents were dechlorinated completely or partially. Quantitative PCR and 16S rNA gene amplicon sequencing showed the presence of *Dehalococcoides* and *Dehalabacter* genes, particularly in the deep sediment. The zone of fractured bedrock and sediment was particularly important for attenuation of the contaminants before reaching the water column.

Lv et al. (2018) studied the mechanism of natural at a petroleum contaminated site. An integrated approach using petroleum, geochemical, microbial, and isotope analyses was performed in a spatial and temporal manner. Hydrocarbon contamination decreased by 60% over a 4-year period. Biodegradation was dominant and was

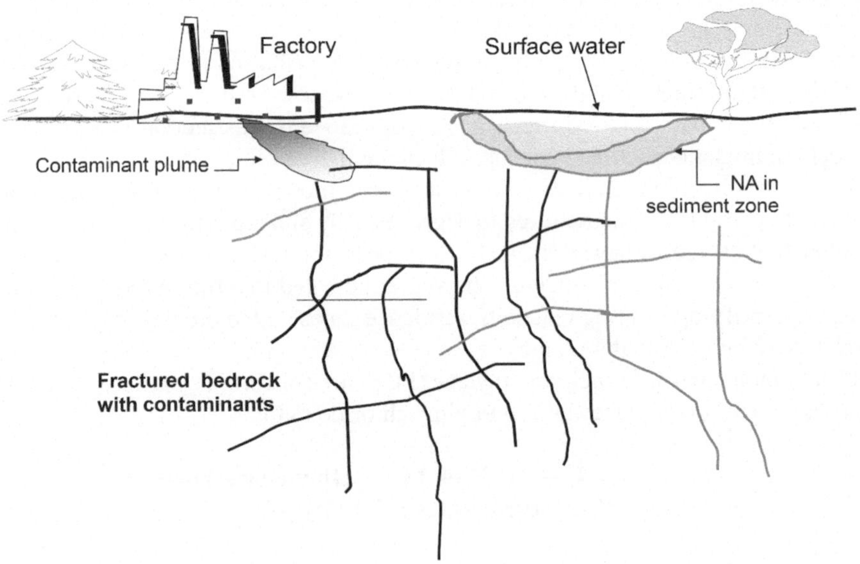

FIGURE 7.13 Natural attenuation in sediment zone protecting surface water from contaminated fractured bedrock. (From Şimşir, B. et al., *Environ. Sci. Technol.*, 51, 4821–4830, 2017.)

at a rate of 0.0015 d^{-1}. Sulfate reduction and methanogenesis are major mechanisms of natural attenuation of the petroleum.

As high volumes of hydraulic fracturing fluids are injected to enhance oil and natural gas extraction, the fate of these fluids is important to know. The fluids contain a mixture of polyglycol ether surfactants, nonylphenol ethoxylates, and polypropylene glycols. Anaerobic conditions led to complete removal of the surfactants within three weeks. 16S rNA gene analysis showed the presence of the diol dehydratase gene cluster (pduCDE) in *Firmicutes* metagenomes that coincided with surfactant chain shortening. An accumulation of acetate was seen due to enzyme cleavage of the ethoxylate chain into aldehydes and then alcohols and carboxylic acids such as acetate. Polypropylene glycol concentrations decreased the slowest. Acetone was produced as a by-product. Therefore, the overall mechanism of the fracturing fluid components was due to biotransformation and chain cleavage.

The natural attenuation of MTBE, particularly in the groundwater, has been a matter of much discussion. Emphasis has been placed on its potential for biodegradation. In the Odencrantz (1998) evaluation of field studies across North America on MTBE mobility, the plumes were found to be very large despite the distance from the source. These cases showed no real evidence of biodegradation. Plumes were up to 1082 ft (330 m) long in the sandy aquifers with aquitards 20–55 ft (6–17 m) below water table. This is in contrast to hydrocarbon fuel sites in California (Rice et al., 1995) and 75% of benzene plumes in Texas (Mace et al., 1997) where they were less than 262 ft (80 m) long. Jansen et al. (2002) have used NA as part of, or as a sole procedure, for treatment of MTBE contaminated sites with concentrations in soil or groundwater less than 5 ppb. This treatment procedure has been used at 15 gas station sites—with NA being the lone remedial technique at six of them.

The review by Seagren and Becker (2002) on the natural attenuation of BTEX and MTBE concluded that MTBE was biodegradable in shallow aquifers but at rates slower than BTEX. Dispersion and dilution are important mechanisms for MTBE natural attenuation (Landmeyer et al., 1998). The NRC (2000) rates the level of understanding for natural attenuation of MTBE as moderate with a low chance of success. Field studies have generated highly variable results. Present information shows that MTBE is much more mobile than BTEX. This could lead to restrictions in the number of sites where natural attenuation can be used. However, there are indications that the MTBE plumes stabilize eventually.

The review of the status of natural attenuation of MTBE in groundwater at several sites (Fiorenza et al., 2002)—Base Borden in Ontario, Canada; Sampson County site and the Elizabeth City site in North Carolina; 74 Amoco service stations in the U.S.; and the Laurel Bay Exchange site in South Carolina—indicated that:

- Mechanisms for natural attenuation seem to be highly uncertain.
- Only 3% of the MTBE remained but there did not seem to be evidence of biodegradation at the Borden site.
- Degradation of MTBE occurred only close to the source at the Sampson County site. A first order decay coefficient of 0.0010 per day for MTBE was obtained. Further positions downgradient from the source gave a decay coefficient of zero for MTBE, indicating a potential risk distant from the source,

- A plume of methane (7 mg/L) was concurrent with that of MTBE at the Elizabeth City site.
- MTBE biodegradation rates decreased according to the groundwater seepage velocity, 5.0 per year for the highest velocity to 2.2 per year for the lowest velocity. At that rate, it was estimated that MTBE concentrations could decrease to 30 μg/L in 60 years, from an initial source of 1200 μg/L.
- For the 74 gasoline sites, biodegradation rates were determined for six of them and shown to range from 0.3 to 10.9 per year. There also seemed to be some association with methanogenic sites although there is little support for degradation by methanogenic bacteria.
- At the Laurel Bay exchange site, although there could have been biodegradation, dilution and dispersion seemed to be the major attenuation mechanisms.

7.7.2 Natural Attenuation of Inorganic Contaminants

The reduction mechanisms in soil following release of radiocesium at Fukushima has been studied in catchment soil (Konoplev et al., 2018). Vertical and lateral migration of the contaminant was occurring. This decreased the particulate and dissolved concentrations in surface runoff and river water. The soil and sediment in the Niida River floodplain vertical radiocesium concentrations were determined. The rate constants of natural attenuation in these areas were 0.2–0.4 year^{-1}. These rates are higher than Chernobyl as precipitation and temperatures were higher, slopes were steeper and biological activity was higher in the soil.

Metals and arsenic attenuation was studied following acid mine drainage from copper mines into the Danube River (Dordievski et al., 2018). Calcite dissolution from limestone bedrock and bicarbonate concentrations of 250 mg/L neutralized the river water and precipitated metals (Al, Fe, Cu, and Pb) and arsenic. Dilution of Mn, Ni, Zn, and Cd. was also a mechanism. Therefore, overall concentrations of metals and arsenic in river water were reduced by the processes of neutralization, dilution, precipitation, and transport of precipitates downstream. The presence of limestone bedrock can play an important role in reducing metal concentrations results from Acid mine drainage (AMD) discharges.

Caschetto et al. (2018) examined the natural attenuation of nitrogen in a plume from a camp ground in Ontario, Canada. Chemical and isotopic characterization of ammonia and nitrate was performed. The results indicated that 60%–80% of N removal was occurring along the path of the septic tank effluent flow. Both ammonium and nitrate ions were present in the plume because the nitrification in the vadose zone was incomplete due to the shallow water table at levels of about 0.2–0.7 m below the infiltration pipes. Both annamox and denitrification processes were evident in the reduction of total inorganic nitrogen. It was therefore suggested that controlling the water table depth near the septic tank infiltration pipes would promote nitrogen attenuation. Water table depths required in local regulations may need to be revised to promote natural attenuation.

7.7.3 Enhanced Natural Attenuation

Natural attenuation by itself may not be sufficient to achieve effective remediation of contaminated sites in a reasonable time frame. When this is judged to be the case, NA can be combined with engineered treatment processes to reduce risk to biotic receptors, and to enhance remediation rates and efficiency. Source reduction or oxygen addition is a method of enhancing NA. Source reduction technologies include free product removal, soil vapor extraction, bioventing, and bioslurping as described in Mulligan (2002). Oxygen can be achieved by biosparging, and the addition of oxygen releasing compounds.

An example of an attempt to enhance NA is the field tests presently being conducted using sodium lactate (an electron donor in many biochemical reactions) injected into the groundwater to enhance biodegradation of trichloroethene. Initially, 1135 L of lactate were injected weekly. After 5 weeks, concentrations of 3800 μg/L of TCE dropped to 10 μg/L. The injection program began in 1998 and currently 5000 L of lactate are injected every two months (Strzelecki, 2002).

A similar approach can be adopted by adding hydrogen release compound (HRC, Regenesis, San Clemente, CA). After injection of the compound into the groundwater, lactic acid is slowly released and hydrogen gas is produced by the bacteria. The hydrogen serves as an electron donor for chlorinated hydrocarbon and nitroaromatic compounds. An application of HRC as aid to NA can be found in the treatment of the Army Chemical Depot in Texas. Heavy metal and volatile organic compounds (VOCs) contamination of the groundwater and soil in the facility occurred from the 1940s to 1974 (Koenigsberg and Vigue, 2002). The explosives that were discharged contaminated the upper aquifer with 2,4-6-trinitrotoluene (TNT), 2,4-dinitrotoluene (2,4-DNT), hexahydro-1,3,5-triazine (RDX), 1,3,5 trinitrobenzene (1,3,5-TNB) and nitrate. Whilst the TNT naturally degraded, the other products did not. A hydrogen release compound (HRC) was injected into the soil at 30 locations to enhance biodegradation. Monitoring was undertaken at 15 well locations. The monitored results showed presence of TNT by-products 2,4-DNT and 1,3-5 TNB. 105 days after injection of the HRC, significant reduction of all by-products was observed as follows: 72%–98% for 2,4-DNT, 45%–99% for RDX, 86%–100% for 1,3,5-TNB and 52%–94% for nitrate.

A combination of NA and phytoremediation has been used at a site contaminated from underground storage tanks (Nzengung and Ramaley, 2001). Groundwater samples taken at the site showed the presence of PCE, TCE and cis-1-2DCE, and surface water lake samples showed PCE, cis-1,2 DCE, 1,1 DCE, and VC while sediment sample indicated PCE and TCE. This indicated that many different processes contributing to natural attenuation were ongoing. Field and laboratory analyses showed that halorespiration mechanisms were occurring in one part of the sediments in the source area (group I in Figure 7.14), while co-metabolic processes occurred in the sediments in the non-source area (group II). Sorption was also an important removal mechanism since the f_{oc} in group II sediments was greater than 2.5%. Group III sediments were deep and test results indicated limited amounts of biodegradation (only PCE to DCE occurred) and sorption. As natural attenuation processes would not completely dehalogenate PCE in the non-source area and deep sediments,

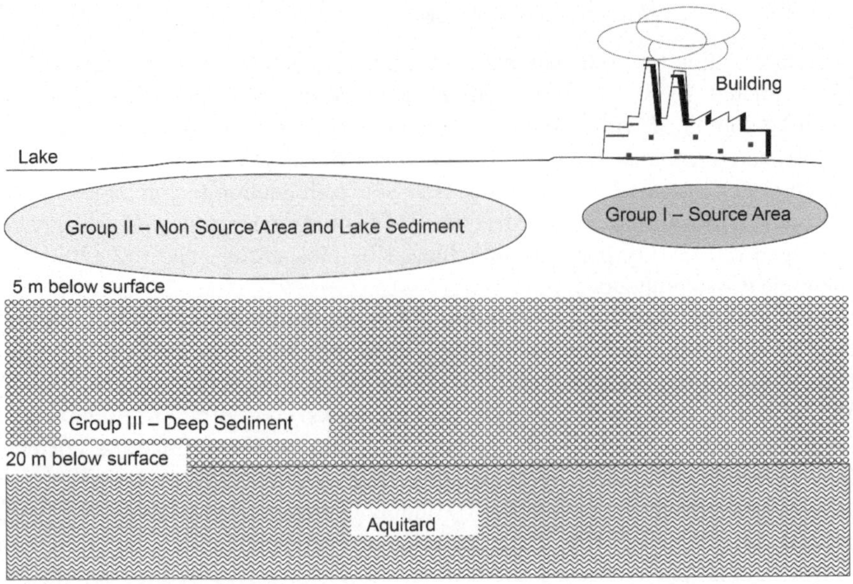

FIGURE 7.14 Zones of perchloroethylene (PCE) attenuation. (From Nzengung, V.A. and Ramalay, S., Coupling natural attenuation and phytoremediation to clean up a shallow chlorinated solvent plume at the former Naval Training Center in Orlando, Florida, *Proceedings of 2001 International Containment & Remediation Technology Conference and Exhibition*, Orlando, FL, 2001.)

it could not be used as the only technology for remediation. Therefore, amendments were used to accelerate reductive dechlorination. Laboratory experiments showed this could be accomplished through addition of carbon sources, particularly for the deep sediments. Phytoremediation was utilized since reductive dechlorination and mineralization of PCE and TCE occurs with willows and cottonwood trees. The trees would enable: (a) removal of residual chlorinated solvents after source removal, (b) reduction of the high discharge from the aquifer into the lakes, and (c) enhance natural attenuation by increasing the dissolved organic carbon in the aquifer. Therefore, willow trees were planted near the source area and a wetland was constructed near the lake.

While most engineered remedial technologies are designed to have a positive effect on attenuation of contaminants, there can be negative effects (Weidemeier and Haas, 2002). This factor should be considered when making contingency plans. For example, if air is introduced by chemical oxidants, oxygen releasing materials, air sparging, bioslurping, or pump and treat an aquifer contaminated with chlorinated solvents, it can disrupt the natural reductive dechlorination processes. Other processes may lead to the spread of contaminants into previously uncontaminated areas such as pump and treat, and in-well circulation for petroleum hydrocarbons and soil vapor extraction for non-degradable VOCs. These factors should always be considered if natural attenuation is to be used with an engineering remedial technology.

Scheutz et al. (2010) injected *Dehalococcoides* bacteria in a soybean oil diluted in groundwater into a sand-filled fracture. A dechlorinating bioactive zone was established a month after injection. Complete cis, 2 dichloroethene (c-DCE) dechlorination to ethane was obtained within 148 days. The zone extended into the nearby clay till as indicated by microbial, electron donor and fermentation product analysis, thus showing promise for enhanced natural attenuation.

Further studies were performed by injection of molasses with dechlorinating bacteria (Dangaard et al., 2012). The long-term study took place over four years. The presence of the *Dehalococcoides* spp., degradation products, and isotopic fractionation of trichloroethene, cis-dichloroethene, and vinyl chloride indicated natural attenuation of chlorinated ethenes in clay till, sand lenses, and fractures. Mobility of soluble products such as vinyl chloride was shown. A bioactive zone of 1.8 m was established in the clay till. The mass of the ethenes was reduced by 24%. Modelling studies indicated that 85% removal could be obtained in 50 years.

Bioaugmentation was then investigated at three sites in Denmark. Lactate (an electron donor) was added with microcosms of two cultures of *Dehalococcoides* (Scheutz et al., 2014). Complete dechlorination of PCE, TCE, 1,1DCE, and c-DCE to ethane occurred at two sites despite the presence of TCA. However, TCA and 1,2 dichloroethane dechlorinated to chloroethane only. Bioaugmentation enhanced dechlorination rates but did not allow treatment of TCA.

Agnello et al. (2015) compared four bioremediation strategies for heavy metal and petroleum contaminated soil in a pot experiment: (a) natural attenuation, (b) phytoremediation with alfalfa (*Medicago sativa* L.), (c) bioaugmentation with *Pseudomonas aeruginosa*, (d) bioaugmentation-assisted phytoremediation. The soil contained moderate levels of heavy metals (Cu, Pb, and Zn at 87, 100 and 110 mg/kg, respectively) and petroleum hydrocarbons (3800 mg/kg). Heavy metals concentrated mainly in the alfalfa plant roots. Total petroleum hydrocarbon removal was highest for bioaugmentation-assisted phytoremediation treatment (68%). Bioaugmentation removed 59%, phytoremediation (47%), and natural attenuation (37%) over a period of 90 days. The results of this study demonstrated that the use of plants and bioaugmentation is a potentially useful approach for remediation of mixed contaminated soil. The longer periods of time needed for natural attenuation could be shortened by using such techniques for enhancement.

7.8 CONCLUDING REMARKS

In this chapter, we have examined the various aspects of field application of NA. We have looked at assessing the potential for natural attenuation, the methods for monitoring, the models and protocols used, and so me reported case studies.

We have seen that there are various advantages to natural attenuation including:

- Possibility of destroying the contaminant completely;
- Remediation method that is potentially more acceptable to the public than other remediation technologies;

- Can be used with other remediation technologies as pre- or post-treatment at a site;
- Can potentially reduce remediation costs significantly.

Potential disadvantages are:

- Longer remediation times compared to other technologies;
- Lack of knowledge concerning mechanisms for remediation particularly with regards to inorganic contaminants;
- Substantial and complex requirements for monitoring;
- By-products can be more environmental hazardous or mobile than the parent compounds;
- Desorption or resolubilization or transfer to other media of contaminants can occur;
- Changes in the hydrological or geochemical conditions can influence stability and mobility of contaminants, such as naturally occurring metals;
- Public acceptance is slow, therefore more education is required;
- Modelling data have to be evaluated with caution and examined from where the data originated, and the assumptions that were made.

Many similarities exist between existing protocols and guidelines for NA, particularly concerning lines of evidence and data requirements. Very few consider the soil and soil gas in their protocols and most have been adapted for hydrocarbon and chlorinated solvent contamination and some for inorganic contaminants. The natural attenuation of many other contaminants has not been investigated extensively—such as PAHs, PCBs, and pesticides. Monitoring techniques will need to be substantially improved, particularly in the soil. The factors influencing the rate of NA of compounds that are not readily attenuated, including metals, chlorinated solvents and high molecular weight organics will need to be determined. For monitoring, direct in situ indicators of bioremediation, particularly for NA (e.g., genetic markers, metabolic byproducts and techniques for characterizing the fate of contaminants, including degradation products) during natural attenuation are being developed.

Available computational models are mainly concerned with groundwater plumes. Integration of soil and groundwater models is complicated. Care should also be exercised regarding source and quality of input data and evaluation of uncertainty. Interaction of organic and inorganic contaminants with the soil components are important factors in the processes that contribute to the natural attenuation of contaminants. Inorganic compounds such as manganese and iron oxides can actively promote catalytic activity for the remediation of soils containing organic chemical pollutants.

REFERENCES

Abdelouas, A., Lutze, W. and Nuttall, E. (1998), Chemical reactions of uranium in groundwater at a mill tailings site, *J. Contamin. Hydrol.*, 34:343–361.

Abdelouas, A., Lutze, W., Gong, W., Nuttall, E.H., Strietelmeier, B.A. and Travis, B.J. (2000), Biological reduction of uranium in groundwater and subsurface soil, *Sci. Total Environ.*, 250:21–35.

Abdelouas, A., Fattahi, M., Grambow, B., Vichot, L. and Gautier, E. (2002), Precipitation of technetium by subsurface sulphate-reducing bacteria, *Radiochim. Acta.*, 90:773–777.

Agnello, A.C., Bagard, M., Van Hullebusch, E.D., Esposito, G. and Huguenot, D. (2016), Comparative bioremediation of heavy metals and petroleum hydrocarbons co-contaminated by natural attenuation, phytoremediation, bioaugmentation and bioaugmentation-assisted phytoremediation, *Sci. Total Environ.*, 563–564:693–703.

Alexander, M. (1994), *Biodegradation and Bioremediation*, Academic Press, London, UK.

Alberti, L., Marchesi, M., Trefiletti, P., and Aravena, R. (2017), Compound-specific isotope analyss (CSIA) application for source apportionment and natural assessment of chlorinated benzenes, *Water*, 9:872.

American Petroleum Institute (API). (2007), Technical Protocol for Evaluating the Natural Attenuation of MtBE, API Publication 4761, 2007.

American Society for Testing and Materials (ASTM). (1995), *Standard Guide for Remediation of Groundwater to Natural on Petroleum Release Sites*. ASTM Designation. E1729–95, ASTM, West Conshohocken, PA.

ASTM E1943-98. (2015), *Standard Guide for Remediation of Ground Water by Natural Attenuation at Petroleum Release Sites*, ASTM International, West Conshohocken, PA, 2015, www.astm.org.

Aziz, C.E., Newell, C.J., Gonzales, J.R., Haas, P.E., Clement, T.P. and Sun, Y. (2000), *BIOCHLOR Natural Attenuation Decision Support System*, User's Manual Version 1.1, US EPA, Office of Research and Development, EPA/600/R-00/008, Washington, DC, https://www.epa.gov/water-research/biochlor-natural-attenuation-decision-support-system.

Bælum, J., Scheutz, C., Chambon, J.C., Jensen, C.M., Brochmann, R.P., Dennis, P., Laier, T., Broholm, M.M., Bjerg, P.L., Binning, P.J. and Jacobsen, C.S. (2014), The impact of bioaugmentation on dechlorination kinetics and on microbial dechlorinating communities in subsurface clay till, *Environ. Pollut.*, 186:149–157.

Bennett, P.C., Hiebert, F.K. and Roger, J.R. (2000), Microbial control of mineral-groundwater equilibria: Macroscale to microscale, *Hydrogeol. J.*, 8:47–92.

Brady, P.V., Spalding, B.P., Krupka, K.M., Waters, R.D., Zhang, P., Borns, D.J. and Brady, W.D. (1998), *Site Screening and Technical Guidance for Monitored Natural Attenuation at DOE Sites*, Sandia National Laboratories.

Brockman, F.J. (1995), Nucleic-acid based methods for monitoring the performance of in situ bioremediation, *Mol. Ecol.*, 4:567–578.

Buscheck, T.E.G. and Alcantar, C.M. (1995), Regression techniques and analytical solutions to demonstrate intrinsic bioremediation. in *Proceedings of the 1995 Battelle International Conference on In-Situ and On-Site Bioreclamation*, Columbus, OH, Battelle Press, pp. 109–116.

Carey, G.R., van Geel, P.J., Murphy, J.R., McBean, E.A. and Rover, F.A. (1998), Full-scale field application of a coupled biodegradation-redox model (BIOREDOX). in Wickramanayake, G.B., and Hinchee, R.H. (Eds.), *Natural Attenuation of Chlorinated Solvents*, Batelle Press, Columbus, OH, pp. 213–218.

Caschetto, M., Robertson, W., Petitta, M. and Aravena, R. (2018), Partial nitrification enhances natural attenusation of nitrogen in a septic system plume, *Sci. Total Environ.*, 625:801–808.

Chang, Y.J., Peacock, A.D., Long, P.E., Stephen, J.R., McKinley, J.P., MacNoughton, S.J., Hussain, A.K.M.A., Saxton, A.M. and White, D.C. (2001), Diversity and characterization of sulphate-reducing bacteria in groundwater at a uranium mill tailings site, *Appl. Environ. Microbiol.*, 67:3149–3160.

Christensen, T.H., Bjerfg, P.L., Banwart, S.A., Jakobsen, R., Heron, G. and Albrechtsen, H.J. (2000), Characterization of redox condition in groundwater contaminant plumes, *J. Contamin. Hydrol.*, 45(3):165–242.

Connor, J.A., Newell, C.J., Nevin, J.P., and Rifai, H.S. (1994), Guidelines for use of groundwater spreadsheet models in risk-based corrective action design, *Proceedings of the National Ground Water Association Petroleum Hydrocarbons and Organic Chemicals in Ground Water Conference*, Houston, TX, pp. 43–55.

Corapcioglu, M.Y., and Baehr, A.L. (1987), A compositional multiphase model for groundwater contamination by petroleum products. 1. Theoretical considerations, *Water Resour. Res.*, 23 (1):191–200.

Damgaard, I., Bjerg, P.L., Bælum, J., Scheutz, C., Hunkerler, D., Jacobsen, C.S., Tuxen, N. and Broholm, M.M. (2013), Identification of chlorinated solvents degradation zones in clay till by high resolution chemical, microbial and compound specific isotope analysis, *J. Contam. Hydrol.*, 146:37–50.

Dangaard, I., Bjerg, P.L., Jacobsen, C.S., Tsitonaki, A., Kerrn-Jespersen, H. and Broholm, M. (2012), Performance of full-scale enhanced reductive dechlorination in clay till. *Groundwater Monit. Remediation*, 33(1):48–61.

Davis, J.W., Odom, J.M., DeWeerd, K.A., Stahl, D.A., Fishbain, S.S., West, R.J., Klecka, G.M. and DeCarolis, J.G. (2002), Natural attenuation of chlorinated solvents at Area 6, Dove Air Force Base: Characterization of microbial community structure, *J. Contam. Hydrol.*, 57:41–59.

DOE, U.S. Department of Energy. (1999), *Technical Guidance for the Long-Term Monitoring of Natural Attenuation Remedies at Department of Energy Sites*, October 8.

Domenico, P.A. (1987), An analytical model for multidimensional transport of a decaying contaminant species, *J. Hydrol.*, 91:49–58.

Dordievski, S., Ishiyama, D., Ogawa, Y. and Stevanovic, Z. (2018), Mobility and natural attenuation of metals and arsenic in acidic waters of the drainage system of Timok River from Bor copper mines (Serbia) to Danube River, *Environ. Sci. Pol. Res.*, 25:25005–25019.

Environmental Security Technology Certification Program (ESTCP). (2011), Guidance Protocol. Application of Nucleic Acid-Based Tools for Monitoring Monitored Natural Attenuation (MNA), Biostimulation, and Bioaugmentation at Chlorinated Solvent Sites, Environmental Restoration Project ER-0518, January 2011.

Fiorenza, S., Suarez, M.P. and Rifai, H.S. (2002), MTBE in groundwater: Status and remediation, *J. Environ. Eng.*, 128(9):773–781.

Höhener, P., Dakhel, N., Christophersen, M., Broholm, M and Kjeldsen, P. (2006) Biodegradation of hydrocarbons vapors: Comparison of laboratory studies and field investigations in the vadose zone at the emplaced fuel source experiments, Airbase Værløse, Denmark, *J. Contam. Hydrol.*, 88:337–358.

Hunkeler, D., Abe, Y., Broholm, M., Jeannottat, S., Westergaard, C., Jacobsen, C.S., Aravena, R. and Bjerg, P.L. (2011), Assessing chlorinated ethane degradation in a large-scale contaminant plume by dual carbon-chloride isotope analysis and quantitative PCR, *J. Contam. Hydrol.*, 119:69–79.

Interstate Technology & Regulatory Council (ITRC). (2013), Environmental Molecular Diagnostics: New Site Characterization and Remediation Enhancement Tools, Environmental Molecular Diagnostics Team. EMD-2, Apr 2013.

Jansen, R., Moyer, E., Woodward, R. and Sloan, R. (2002), *MTBE Remediation Seminar*, Spring.

Kao, C.M. and Wang, Y.S. (2001), Field investigation of the natural attenuation and intrinsic biodegradations rates at an underground storage tank site, *Environ. Geol.*, 4:622–631.

Kennedy, L.G., Everett, J.W., Ware, K.J., Parsons, R. and Green, V. (1998), Iron and sulphur mineral analysis methods for natural attenuation assessments, *Bioremediat. J.*, 2(3):259–271.

Koenigsberg, S.S and Vigue, B.W. (2002), The next big thing, Cost-effective DOD sites using slow-release compounds, *Pollution Engineering*, 34(3):14–17.

Konikow, L.F., and Bredeheoft, J.D. (1978), Computer model of two-dimensional solute transport and dispersion in ground water. in *Techniques of Water Resources Investigation of the United States Geological Survey*, Book 7, U.S. Geological Survey, Reston, VA.

Konoplev, A., Golosov, V., Wakiyama, Y., Takase, T., Yoschenko, V., Yoshihara, T., Parenyuk, O. et al. (2018), Natural attenuation of Fukushima-derived radio-cesium in soils due to its vertical and lateral migration, *J. Environ. Radioact.*, 186:23–33.

Kota, S., Borden, R.C. and Barlaz, M.A. (1998) Influence of protozoa grazing on contaminant biodegradation, *FEMS Microbiol. Ecol.*, 29, 179–189.

Kota, S., Borden R.C. and Barlaz, M.A. (1999), Influence of protozoa grazing on contaminant biodegradation, *FEMS Microbiol. Ecol.*, 29:179–189.

Kram, M.L., Widdowson, M.A., Chapelle, F.H. and Casey. C.C. (2007), *Estimating Cleanup Times Associated with Combining Source-Area Remediation with Monitored Natural Attenuation Using Natural Attenuation Software (NAS).* Environmental Security Technology Certification Program (ESTCP) Project ER-0436. October 2007.

Landmeyer, J., Chapelle, F., Bradley, P., Pankow, J., Church, C. and Tratnyek, P. (1998), Fate of MTBE relative to benzene in a gasoline-contaminated aquifer (1993–1998), *Ground Water Monit. Rev.*, 18(4):93–102.

Lebrón, C.A., Chapelle, F.H., Widdowson, M.A., Novak, J.T., Parker, J.C. and Singletary, M.A. (2013), *Verification of methods for assessing the sustainability of monitored natural attenuation.* CR-NAVFACEXWS-EV1302, Final Report for Environmental Security Technology Certification Program, Alexandria VA, January 2013.

Lovley, D.R. (2001), Reduction of iron and humics in subsurface environments. in Fredickson, J.K. and Fletcher, M. (Eds), *Subsurface Microbiology and Biogeochemistry.* Wiley-Liss, New York.

Lu, W., Zhu, R.L., Yang, J., Li, H., Liu, Y.D., Lu, S.G., Luo, Q.S., and Lin, K.F. (2015), Natural attenuation model and biodegradation for 1,1,1-trichlorethane contaminant in shallow groundwater, *Front. Microbiol.*, 6:839.

Lv, H., Su, X., Wang, Y., Dai, Z. and Liu, M. (2018), Effectiveness and mechanism of natural attenuation at hydrocarbon site. *Chemosphere*, 206:293–301.

Mace, R.E., Fisher, R.S., Welch, D.M. and Parra, S.P. (1997), Extent, mass and duration of hydrocarbon plumes from leaking petroleum storage tank sites in Texas, *Geol. Circ.*, 97–101.

Magar, V.S., Chadwick, D.B., Bridges, T.S., Fuschman, P.C, Conder, J.M., Dekker, T.J., Stevens, J.A., Gustavson, J.E., and Mills, M.A. (2009), *Technical Guide: Monitored Natural Recovery at Contaminated Sediment Sites*, ESTCP project ER-0622, 276.

McDonald, G. and Harbaugh, A.W. (1988), *A Modular Three-Dimensional Finite-Difference Ground Water Flow Model*, U.S. Geological Survey Techniques of Water Resources, Investigations, Technical Report. U.S. Geol. Survey, Resotn, VA Chapt. A 1.

McNab Jr., W.W., Rice, D.W. and Tuckfield, C. (2000), Evaluating chlorinated hydrocarbon plume behavior using historical case population analyses, *J. Bioremediat.*, 4:311–335.

Mercer, J.W. (1998), Transport and remediation of chemicals in groundwater. in R.A. Meyers, (Ed.), Encyclopedia of Environmental Analysis and Remediation, John Wiley & Sons, New York.

Meuzelaar, H.L.C. ed. (1998), *Field Analytical Chemistry and Technology*, Vol. 2, No. 2, John Wiley & Sons, New York.

Mulligan, C.N. (2002), *Environmental Biotreatment*, Government Institutes, Rockville, MD.

Mulligan, C.N., Yong, R.N. and Gibbs, B.F. (2001), The use of selective extraction procedures for soil remediation, *Proceeding International Symposium On Suction, Swelling, Permeability and Structure of Clays*, Balkema, Rotterdam, the Netherlands.

Němeček, J., Dolinová, I., Macháčkova, J., Špánek, R., Ševců, A., Lederer, T. and Černík, M. (2017), Stratification of chlorinated ethenes natural attenuation in an alluvial aquifer assessed by hydrochemical and biomolecular tools. *Chemosphere*, 184:1157–1167.

Newell, C.J., McLeod, R.K. and Gonzales, J. (1996), *BIOSCREEN Natural Attenuation Decision Support System*, EPA/600/R-96/087, August, https://www.epa.gov/water-research/bioscreen-natural-attenuation-decision-support-system

Newell, C.J., Rifai, H.S., Wilson, J.T., Connor, J.A., Aziz, J.A. and Suarez, M.P.(2002), *Calculation and Use of First-Order Rate Constants for Monitored Natural Attenuation*, Office of Research and Development, National Risk Management Research Laboratory, EPA/540/S-02/500, Cincinnati, OH.

New Jersey Department of Environmental Protection, (NJDEP). (2012), *Monitored Natural Attenuation Technical Guidance*. Site Remediation Program.

Nielsen, P.H., Bjarnadottir, H., Winter, P.L. and Christensen. T.H. (1995), In situ and laboratory studies on the fate of specific organic compounds in an anaerobic landfill leachate plume. 2. Fate of aromatic and chlorinated aliphatic compounds, *J. Contam. Hydrol.*, 20:51–66.

National Research Council (NRC). (1993), *In Situ Bioremediation: When Does It Work?* National Academic Press, Washington, DC.

National Research Council (NRC). (2000), *Natural Attenuation for Groundwater Remediation, Committee on Intrinsic Remediation*, Water Science and Technology Board and Board on Radioactive Waste Management, Commission on Geosciences, Environment and Resources, National Academy Press, Washington, DC.

Nzengung, V.A. and Ramaley, S. (2001), Coupling natural attenuation and phytoremediation to clean up a shallow chlorinated solvent plume at the former Naval Training Center in Orlando, Florida, *Proceedings of 2001 International Containment & Remediation Technology Conference and Exhibition* (10–13 June), Boca Raton, FL.

Odencrantz, J.E. (1998), Implications of MTBE for intrinsic remediation of underground fuel tank sites, *Remediation*, 9(3), 7–16.

Odencrantz, J.E., Varljen, M.D. and Vogel, R.A. (2002), Natural attenuation: Is dilution the solution?, *Lust Bulletin*, 8–12.

Pennington, J.C., Miyares, P.H., Ringelberg, D.B., Zakikhani, M., Reynolds, C.M., Felt, D., Coffin, R.B., Gunnison, D., Cifuentes, L., Fredrickson, H.L. and Jenkins, T.F. (1999), *Natural Attenuation of Explosives in Soil and Water Systems at Department of Defense Site*, Prepared for Strategic Environmental Research and Development Program, U.S. Army Corps of Engineers, Waterways Experiment Station, Technical Report SERDP-99-1, July 1999.

Rectanus, H., Darlington, R., Kucharzyk, K. and Moore, S. (2015), Attenuation pathways for munitions constituents in soils and groundwater, TR-NAVFAC EXWC-EV-1503. Navel Facilities Engineering Command (NAVFAC), Washington DC.

Renner, R. (2000), Natural attenuation's popularity outpaces scientific support, NRC finds, *Environ. Sci. Technol.*, 34: 203A–204A.

Rice, D.W., Dooher, B.P., Cullen, S.J., Everett, L.G., Kastenberg, W.E., Grose, R.D. and Marino, M.A. (1995), *Recommendation to Improve the Cleanup Process for California's Leaking Underground Fuel Tanks (LUFTs)*, report submitted to the California State Water Resources Control Board and the Senate Bill 1764 Leaking Underground Fuel Tank Advisory Committee, California Environmental Protection Department, Sacramento, CA.

Rifai, H.S., Bedient, P.B., Wilson, J.T., Miller, K.M. and Armstrong, J.M. (1988), Biodegradation modeling at aviation fuel spill site. *J. Environ. Eng.*, 114:1007–1029.

Rifai, H.S., Newell, C.J., Miller, R., Taffinder, S. and Rounsaville, M. (1995), Simulation of natural attenuation with multiple electron acceptors, *Bioremediation*, 3:53–58.

Rifai, H.S., Newell, C.J., Gonzales, J.R., Dendrou, S., Kennedy, L. and Wilson, J. (1997), *BIOPLUME III Natural Attenuation Decision Support System Version 1.0 User's Manual, prepared for the U.S. Air Force Center for Environmental Excellence*, Brooks Air Force Base, San Antonio, TX.

Rifai, H.S., Newell, C.J., Gonzales, J.R. and Wilson, J.T. (2000), Modeling natural attenuation of fuels with BIOPLUME III, *J. Environ. Eng.*, 126:428–438.

Rogers, S.W., Ong, S.K., Kjartanson, B.H., Golchin, J. and Stenback, G.A. (2002), Natural attenuation of polycyclic aromatic hydrocarbon-contaminated sites: Review, *Pract. Period. Hazard. Toxic, Radioact. Waste Manage.*, 6:141–155.

Roling, W.F.M. and van Verseveld, H.W. (2002), Natural attenuation: What does the subsurface have in store?, *Biodegradation* 13:53–64.

Scheutz, C., Broholm, M.M., Durant, N.D., Weeth, E.B., Jorgensen, T.H., Dennis, P., Jacobsen, C.S., Cox, E.E., Chambon, J.C. and Bierg, P.L. (2010), Field evaluation of biological enhanced reductive dechlorination of chloroethenes in clayey till, *Environ. Sci. Technol.*, 44:5134–5141.

Scheutz, C., Durant, N.D. and Broholm, M.M. (2014), Effects of bioaugmentation on enhanced reductive dechlorination of 1,1,1-dechlorination in groundwater: A comparison of three sites, *Biodegradation*, 25:459–478.

Schiefler, A.A, Tobler, D.J., Overheu, N.D. and Tuxen, N. (2018), Extent of natural attenuation of chlorinated ethenes at a contaminated site in Denmark, *Energy Procedia.*, 146:188–193.

Schurig, C., Melo, V.A., Miltner, A. and Kaestner, M. (2014), Characterization of microbial activity in the framework of natural attenuation without groundwater monitoring wells? A new Direct-Push probe, *Environ. Sci. Pollut. R.*, 21:9002–9015.

Seagren, E.A. and Becker, J.G. (2002), Review of natural attenuation of BTEX and MTBE in groundwater, *Pract. Period. Hazard. Toxic, Radioact. Waste Manage.*, 6:156–172.

Şimşir, B., Yan, J., Im, J., Graves, S. and Löffler, F.E. (2017), Natural attenuation in streambed sediment receiving chlorinated solvents from underlying fracture networks, *Environ. Sci. Technol.*, 51:4821–4830.

Sinke, A. (2001), Monitored natural attenuation; moving forward to consensus, *Land Contamination Reclamation*, 9(1):111–118.

Srivinsan, P. and Mercer, J.W. (1988), Simulation of biodegradation and sorption processes in groundwater, *Ground Water*, 26(4):475–487.

Stapleton, R.D., Ripp, S., Jimenez, L., Cheol-koh, S., Fleming, J.T., Gregory I.R. and Sayler, G.S. (1998), Nucleic acid approaches in bioremediation: Site assessment and characterization, *J. Microbiol. Meth.*, 32:165–178.

Stollenwek, K.G. (1994), Geochemical interactions between constituents in acidic groundwater and all uvum in an aquifer near Globe, Arizona, *Appl. Geochem.*, 9:353–369.

Strzelecki, D. (2002), Low-tech remedies save millions at DOE site, *Pollution Eng.*, 34(3):41.

Suarez, M.P. and Rifai, H. (1999), Biodegradation rates for fuel hydrocarbons and chlorinated solvents in groundwater, *J. Biophysical.*, 3:337–362.

Suarez, M.P. and Rifai, H.S. (2002), Evaluation of BTEX remediation by natural attenuation at a coastal facility, *GWMR.*, 22:62–77.

Sun, Y., Petersen, J.N., Clement, T.P. and Hooker, B.S. (1996), A monitoring computer model for simulating natural attenuation of chlorinated organics in saturated ground-water aquifers. In *Proceedings of the Symposium on Natural Attenuation of Chlorinated Organic in Ground Water*, Dallas, TX, Sept. 11–13, EPA/540/R-96/509, US EPA, Washington, DC.

Tartre, A. (2001), Plume delineation and monitoring of natural attenuation processes via in situ flux measurement. Presented at *2001 International Containment & Remediation Technology Conference and Exhibition*, 10–13 June, Boca Raton, FL.

Truex, M.J. and Carroll, K.C. (2013), Remedy Evaluation Framework for Inorganic, Non-Volatile Contaminants in the Vadose Zone, PNNL-21815; RPT-DVZ-AFRI-004. Pacific Northwest National Laboratory, Richmond, WA.

Truex, M., Brady, P., Newell, C., Rysz, M., Denham, M. and Vangelas, K. (2011), The Scenarios Approach to Attenuation-Based Remedies for Inorganic and Radionuclide Contaminants, SRNL-STI-2011-00459. Pacific Northwest National Laboratory, Richmond, WA.

USEPA, Office of Solid Waste and Emergency Response (OSWER). (1996), *Use of Risk-based Decision-Making in UST Corrective Action Programmes*, Directive 9610.17.

USEPA (1998), *Technical Protocol for Evaluating Natural Attenuation of Chlorinated Solvent in Groundwater*, EPA/600/R-98/128, National Risk Management Laboratory, Ada, OK.

USEPA Office of Solid Waste and Emergency Response (OSWER) (1999), *Use of Monitored Natural at Superfund, RCRA Corrective Action and Underground Storage Tank Sites*, Directive Number 9200.4-17P, April 21, 1999, 32pp.

USEPA Science Advisory Board (2001), *Monitored Natural Attenuation*, USEPA Research Program: An EPA Science Advisory Board Review, Science Advisory Board (1400A), Washington, DC.

USEPA Office of Solid Waste and Emergency Response (OSWER). (1999), *Use of Monitored Natural Attenuation at Superfund, RCRA Corrective Action and Underground Storage Tank Sites* Directive Number 9200.4-17P, April 21.

USEPA Science Advisory Board. (2001), Monitored Natural Attenuation; USEPA Research Programme: An EPA Science Advisory Board Review, Science Advisory Board (1400A) Washington, DC.

USEPA. (2007a) Monitored Natural Attenuation of Inorganic Contaminants in Ground Water, Volume 1: Technical Basis for Assessment, EPA 600-R-07-139, 2007.

USEPA. (2007b) Monitored Natural Attenuation of Inorganic Contaminants in Ground Water, Volume 2: Assessment for Non-Radionuclides, Including Arsenic, Cadmium, Chromium, Copper, Lead, Nickel, Nitrate, Perchlorate, and Selenium, EPA 600-R-07-140, 2007.

USEPA. (2010), Monitored Natural Attenuation of Inorganic Contaminants in Ground Water, Volume 3: Assessment for Radionuclides Including Tritium, Radon, Strontium, Technetium, Uranium, Iodine, Radium, Thorium, Cesium, and Plutonium-Americium. EPA 600-R-10-093, 2010.

USEPA. (2014), *Technical Resource Document on Monitored Natural Recovery*, US EPA, National Risk Management Research Laboratory, Cincinnati, OH. EPA 600-R-14-083.

USEPA. (2015), *Use of Monitored Natural Attenuation for Inorganic Contaminants in Groudwater at Superfund Site*, U.S. Environmental Protection Agency, Office of Solid Waste and Emergency Response, Directive 9283.1–36.

Vogel, T.M., Criddle, C.S., and McCarty, P.L. (1987), ES&T critical reviews: Transformations of halogenated aliphatic compounds, *Environ. Sci. Technol.*, 21:722–736.

Waters, R.D., Brady, P.V., and Borns, D.J. (1998), Natural attenuation of metals and radionuclides—An overview of the Sandia/DOE Approach, Presented at *Waste Management* '98, Tucson, AZ.

Weatherill, J.J., Atashgahi, S., Schneidewind, U., Krause, S., Ullah, S., Cassidy, N and Rivett, M.O. (2018), Natural attenuation of chlorinated ethenes in hyporheic zones: A review of key biogeochemical processes and in-situ transformation potential, *Water Research*, 128:362–282.

Weidemeier, T., Wilson, J.T., Kampbell, D.H., Miller, R.N. and Hansen, J.E. (1995), Technical protocol for implementing intrinsic natural remediation with long-term monitoring for natural attenuation of fuel contamination dissolved in groundwater (Volume I and II) U.S. Air Force Center for Environmental Excellence, San Antonio, TX.

Weidemeier, T.H., Rifai, H.S., Newell, C.J., and Wilson, J.T. (1999), *Natural Attenuation of Fuels and Chlorinated Solvents in the Subsurfaces*, John Wiley & sons, New York.

Weidemeier, T., Swanson, M.A., Moutoux, D.E., Gordon, E.K., Wilson, J.T., Wilson, B.H., Kampbell, D.H., Haas, P.E., Miller, R.N., Hansen, J.E. and Chapelle, F.H. (1998), *Technical Protocol for Evaluating Natural Attenuation of Chlorinated Solvents in Ground Water*, EPA/600//R-98/128, September, EPA Office of Research and Development, Washington, DC.

Weidemeier, T.H. and Haas, P.E. (1999), *Designing Monitoring Programmes to Effectively Evaluate the Performance of Natural Attenuation*, U.S. Air Force Center for Environmental Excellence, San Antonio, TX.

Weidemeier, T.H., Rifai, H.S., Newell, C.J. and Wilson, J.T. (1999) *Natural attenuation of Fuels and Chlorinated Solvents in the Subsurfaces*, John Wiley and sons, New York.

Weidemeier, T.H., and Haas, P.E. (2002), Designing monitoring programs to effectively evaluate the performance of natural attenuation, *Groundwater Monit. R.*, 22(3):124–135.

Wilson, J.T., Kaiser, P.M. and Adair, C. (2005), *Monitored Natural Attenuation of MTBE as a Risk Management Option at Leaking Underground Storage Tank Sites*. U.S. EPA, National Risk Management Research Laboratory, Ada, OK. EPA 600-R-04-179.

Wu, L., Verma, D., Bondgaard, M., Melvej, A., Vogt, C., Subudhi, S and Richnow, H.H. (2018), Carbon and hydrogen isotope analysis of parathion for characterizing its natural attenuation by hydrolysis at a contaminated site, *Water Research*, 143:146–154.

Zhao, J., Al, T., Chapman, S.W., Parker, B.L., Mishkin, K.R., Cutt, D. and Wilkin, R.T. (2017), Determination of Cr (III) solids formed by reduction of Cr(VI) in a contaminated fractured bedrock aquifer: Evidence for natural attenuation of Cr(VI). *Chemical Geology.*, 474:1–8.

Zwolinski, M.D., Harris, R.F. and Hickey W.J. (2000), Microbial consortia involved in the anaerobic degradation of hydrocarbons. *Biodegradation*, 11:141–158.

8 Application of Monitored Natural Attenuation

8.1 INTRODUCTION

As we have pointed out in the beginning chapters of this book, application of monitored natural attenuation (MNA) as a passive contaminated site or contaminant remediation tool requires proper knowledge of the contaminant mitigating processes required to achieve the necessary reduction in concentration and toxicity of the contaminants—the result of which is the reduction or elimination of health and environmental threats to biotic receptors. There are at least two different kinds of ab initio site contaminant remediation scenarios for which application of MNA as a remediation tool can be considered or implemented. These include: (a) remediation of an existing contaminated site, and (b) management of the contaminant plume in a site with a recent contaminant source such as a recent spill or discharge of hazardous substances. It is emphasized that the MNA technique is a knowledge-based remediation technique and the level of success in application of this technique is highly dependent on: (a) knowledge of contaminant-reduction and toxicity-reduction processes, (b) continuous and adequate monitoring of the ongoing in situ processes at key locations, and (c) defining multiple lines of evidence testifying to adequacy of contaminated site characterization, contaminant and toxicity characterization, tracking at the key locations, and evidence of success.

Most regulatory agencies now recommend the above three items as requirements or sets of protocols for application of MNA as a remediation tool for contaminated lands—see for example US EPA (2015). In essence, the three requirements form a three-legged platform for successful application of MNA. The first of the three legs—knowledge of contaminant and toxicity reduction—has been discussed in the previous chapters. The other two legs of the platform will form the basis for discussions in this chapter. For the sake of clarity, the term *natural attenuation* is italicized when it is used in the context of a material property to distinguish it from natural attenuation of contaminants as a process derived from the various mechanisms of contaminant assimilation and associated biological activities.

From the previous chapters, we have seen that the many variables in the interactions in the soil-water system between the various contaminants, aqueous phase, and soil fractions, can produce anywhere from minimal to significant contaminant attenuation in transport through the subsurface soil. Since the evidence of success or failure of the soil-water system to attenuate contaminants is not readily perceptible, monitoring, sampling, and testing of the various elements in the region of suspected contaminant plume presence and advance are necessary pieces of action. We need to determine the necessary sets of protocols that would provide us with the assurance that the processes that occur in the ground are indeed producing the desired sets of results. This is perhaps one of the more significant requirements for future application of natural attenuation as a process and as a tool in contaminant plume management. In other words, the necessary sets of proof that will testify to the attenuation performance of the subsurface soil need to be properly developed. While these sets of proofs would be site and problem specific, there are some basic elements of physical, chemical, and biological pieces of evidence that form the underpinning of the proofs. We need to be fully aware of these if we wish to exploit the natural attenuation property of soils.

8.2 MONITORING

Monitoring for application of MNA remediation means keeping track of "what is happening" in the contaminant zone (plume) at all times—the intent of which is to assess whether attenuation of the contaminants in the affected zone is occurring. Note that application of monitoring tools outside the contaminant zone is also required to ensure complete tracking of contaminant plume. Monitoring is an important management tool for MNA application and is used to provide information on pre-existing conditions, site characteristics and properties, ongoing nature and distribution of contaminants in the affected site, and changing conditions in the affected (contaminated) zone.

8.2.1 Site Characterization

As has been pointed out in the previous section, monitoring of the progress of attenuation processes in the contaminated site is one of the three legs of the platform upon supporting MNA application. What is required is information in the sub-regions of the contaminated site pertaining to: (a) nature of contaminants at various depths at selected spatial locations in the aqueous phase and sorbed onto soil solids at various time intervals, and (b) concentration and toxicity of the detected contaminants. However, before this can be implemented, information pertaining to the nature of the contaminated site and of the site itself is needed. This means gathering information on the site characteristics and properties. These should include the hydrogeological setting and hydrogeochemistry. The former is of utmost importance since this establishes the primary direction of transport of the contaminants and their ultimate fate. Contaminants from the contaminant source will be transported in the ground by advective and diffusive means. In other words, the main contaminant transport "vehicle" required is the presence of an aqueous phase (groundwater and porewater).

Figure 8.1 is a schematic of the bare essentials of initial site characterization and contaminated site problem delineation. It is essential to obtain intact soil samples from uncontaminated contiguous areas for compositional analyses, determination of porewater chemistry and contaminant assimilative capacities of the intact soil. It is also important to obtain groundwater samples—as distinct from porewater—from these same areas to determine their chemistry. Why? Because it is necessary to determine if there are geogenic contaminants and if there were, one needs to establish whether the geogenic contaminants have been partitioned—assimilated by the soil solids—and by what mechanisms or processes. It is important to have knowledge of the subsurface aqueous regime and underlying soil attributes to enable

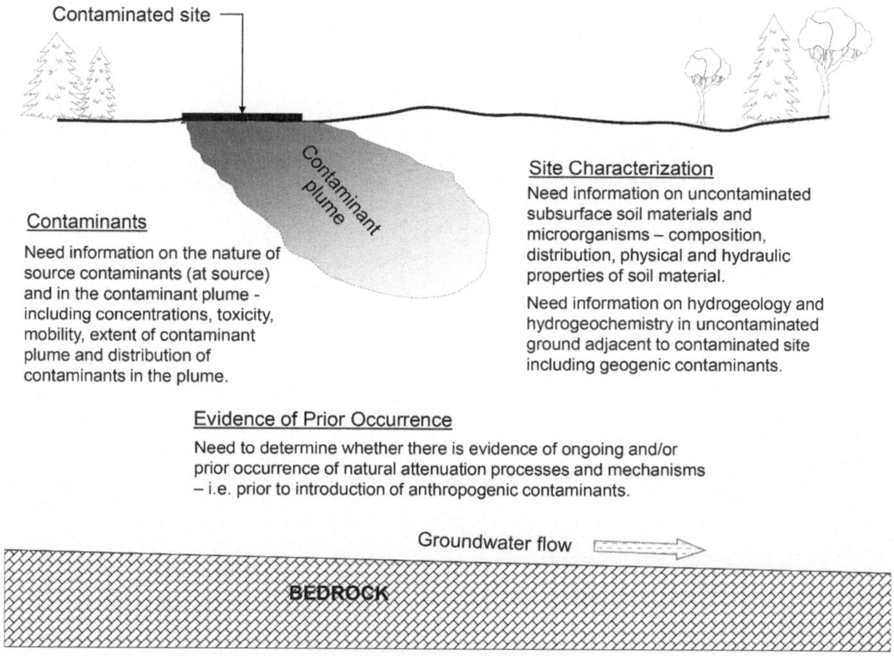

FIGURE 8.1 Schematic showing required site characterization requirements and determination of site contamination.

one to: (a) distinguish between geogenic and anthropogenic contaminants, (b) assess the potential or capability of the underlying soil to attenuate contaminant concentration and toxicity, (c) determine the flow or transport pattern of the contaminants (contaminant plume), (d) devise the monitoring requirements and strategy that will ensure competent capture of samples for analyses and studies, and (e) establish the investigative program required to ensure that the transport and fate of the contaminants are well monitored and managed.

8.2.2 Monitoring Essentials

The term *monitoring*, used in the context of monitoring of a contaminated site, is all encompassing—meaning that it includes initial site characterization requirements, development and implementation of the strategy for complete tracking of the transport and fate of the contaminants emanating from source and in the ground. What sets this apart from regular monitoring programs is the need for continuous monitoring—to provide information and evidence that contaminant attenuation is (or is not) occurring (see Figure 8.2).

The required companion set of activities that will provide support for the monitoring program is called *lines of evidence*. Monitoring during the NA process is used to determine whether the expected events associated with attenuation processes have occurred in the contaminated site, and whether contamination has spread beyond the expected contaminated plume boundaries. As Figure 8.2 shows, it is necessary to

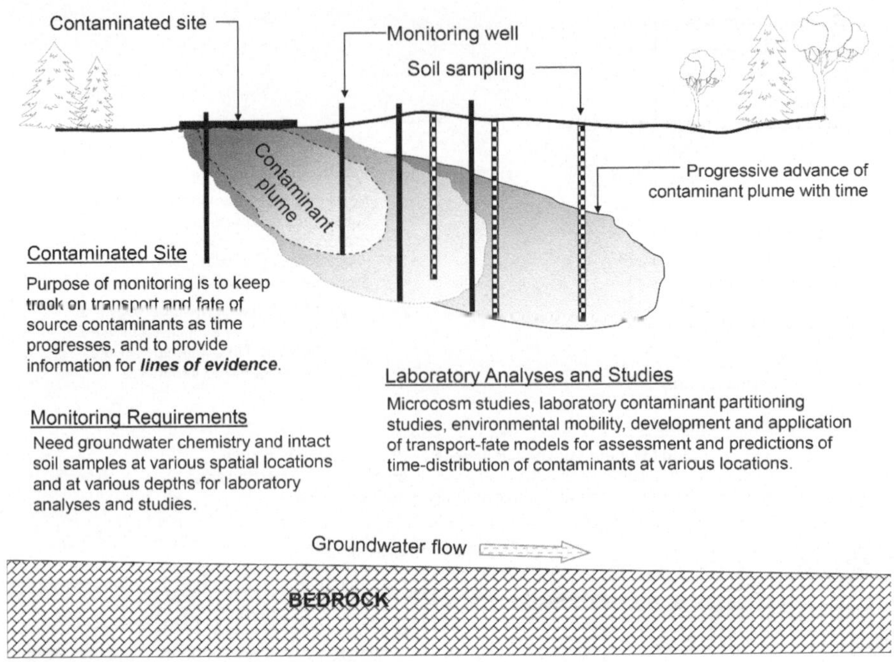

FIGURE 8.2 Schematic showing purpose and requirements for monitoring in application of monitored natural attenuation (MNA) remediation techniques.

gather all pertinent pieces of information providing evidence that those events have or have not occurred within the management zone shown in Figure 8.3.

The pieces of information required for determining the effectiveness of natural attenuation processes require extraction of intact soil samples and groundwater samples to allow for laboratory analyses. It is necessary to obtain information on the nature, concentrations, toxicity, characteristics, and properties of the contaminants in the attenuation zone. The contaminants reside in both the porewater and groundwater and on the surfaces of the soil solids. Residence associated with the soil solids can take the form of sorbates and co-precipitates. In turn, the sorbates can be complexed with the soil solids and will remain totally fixed within the structure of the soil solids. However, the sorbates can also be held by ionic forces that can be disrupted, thus releasing the sorbates.

What the preceding discussion of residence status of the contaminants tells us is that, we need to monitor and sample not only the porewater and groundwater but also the soil fractions in the contaminant attenuation zone. Two types of monitoring-sampling systems are needed. For porewater or groundwater, monitoring wells are generally used. These wells are necessary to provide access to groundwater at various locations (vertically and spatially) in a chosen location. The choice of monitoring well types and distribution or location of wells will depend on the purpose for the wells. In respect to determination of whether natural attenuation can be used as a treatment process, there are at least three separate and distinct monitoring schemes that need to be considered. These range from the initial

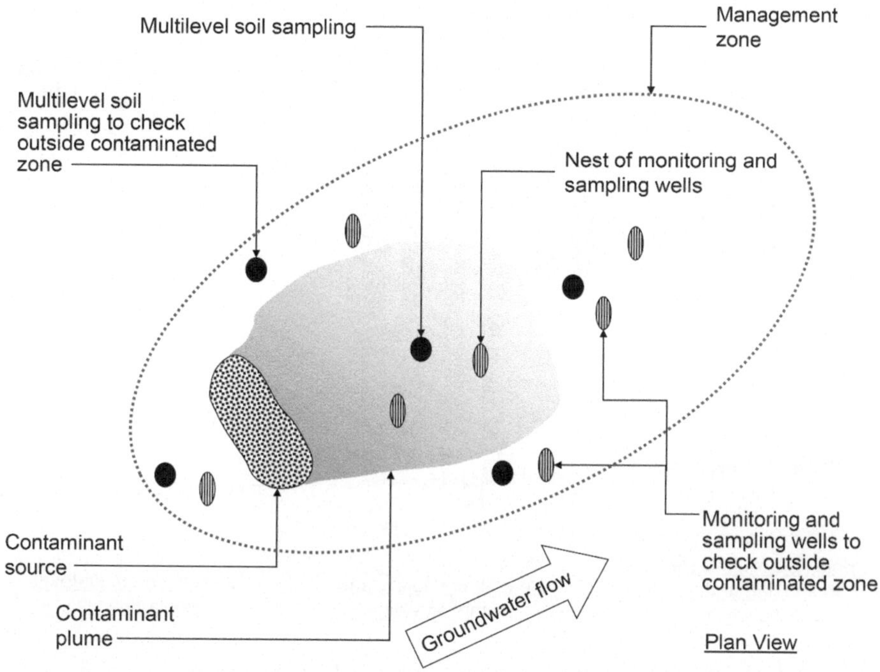

FIGURE 8.3 Plan view showing the management zone surrounding the contaminant source and the distribution of monitoring wells and soil sampling boreholes for verification and long-term monitoring.

site characterization studies to verification monitoring and long-term conformance monitoring. The term *monitoring scheme* is used deliberately to indicate the use of monitoring and sampling devices to obtain both soil and water samples. Figure 8.4 shows some typical devices used as monitoring wells to permit monitoring groundwater at various levels. In the left-hand group, we see individual monitoring wells with sampling ports located at different depths, but grouped together in a shared borehole. This is generally identified as a single bore-hole multilevel monitoring well system. The middle drawing shows a nest of single monitoring wells in their own separate boreholes, and the right-hand drawing shows a single tube system with monitoring portholes located at the desired depths. With present technological capabilities, monitoring wells and the manner of operation have reached levels of sophistication where many kinds of downhole groundwater analyses can be performed in situ. There are sophisticated in-hole devices on the market that can inform one of the porewater and groundwater chemistry, and even conduct various kinds of sensing tests that one usually conducts in the laboratory. Technology has improved to such an extent that in situ technical information gathering can obviate much of the associated laboratory test requirements.

Site characterization is necessary to provide information on the hydrogeology of the site. It is necessary to properly characterize subsurface flow to fully delineate or anticipate the transport direction and extent of the contaminant plume. Determining the

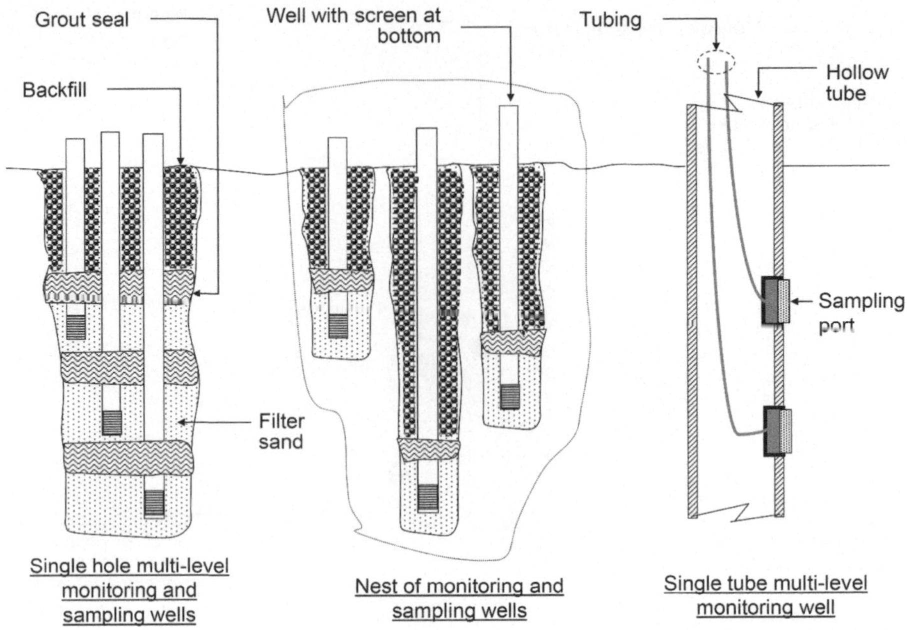

FIGURE 8.4 Some typical groundwater monitoring and sampling wells.

direction and magnitude of groundwater flow is most important. Obviously, this means a judicious distribution of monitoring wells upgradient and downgradient. A proper siting of the monitoring wells and analysis of the results should provide one with knowledge of the source of the contaminants and the characteristics of the contaminant plume.

Verification monitoring requires placement of monitoring wells and soil sampling devices within the heart of the contaminant plume and also at positions beyond the plume. Figures 8.9 and 8.10 show the vertical and plane views of how the wells and sampling stations might be distributed. It is a truism to state that the more monitoring and sampling devices there are, the better one is able to properly characterize the nature of the contaminant plume—assuming that the monitoring wells and sampling devices are properly located. The monitoring wells and sampling devices placed outside the contaminant plume, shown in Figure 8.10 will also serve as monitoring wells and sampling devices for long term conformance assessment.

8.2.3 Tests and Studies on Retrieved Samples

The tests required of samples retrieved from monitoring wells are designed to determine the nature of the contaminants in the porewater or groundwater. With the proper laboratory analyses, the results will inform one of the concentrations, composition, and toxicity of the target contaminants. For prediction of further or continued attenuation of the target contaminants, the partition coefficients and solubilities of the various contaminants are needed as input to transport and fate models. If biotransformation of the target contaminants has occurred, supporting laboratory research would be needed to determine the likely fate of the transformed or intermediate products.

Tests on retrieved soil samples at various locations and depths in the contaminant attenuation zone are required as part of the monitoring program. These tests are designed to determine the ongoing transport and fate of the contaminants in the contaminated site—i.e., to provide evidence of whether contaminant attenuation is (or is not) occurring or progressing. What one is looking for is the environmental mobility of the contaminants and also the nature and concentration of contaminants sorbed onto the soil particles (soil solids). With regard to attachment of contaminants to the soil solids, it is important to recognize that the contaminants or sorbates can be lightly bonded or strongly bonded to the soil solids. This means that one needs to determine if changes in the porewater chemistry, redox changes, pH changes etc. would dislodge the sorbates. Detailed discussions of many of the bonding mechanisms, their sensitivity to changes in the immediate environment, and the methods for determining partitioning of the contaminants have been discussed in Chapters 3 through 5. Determination of environmental mobility of contaminants sorbed onto soil solids is generally conducted using leaching columns. In such tests, the contaminated soil samples retrieved from the site are subject to leaching by deionized water as a calibration or standard case, and by various fluids selected to mimic the chemistry of the groundwater in the site of interest. An example of such a calibration desorption test is shown in Figure 8.5 for a laboratory study on arsenic release from arsenic-contaminated soils.

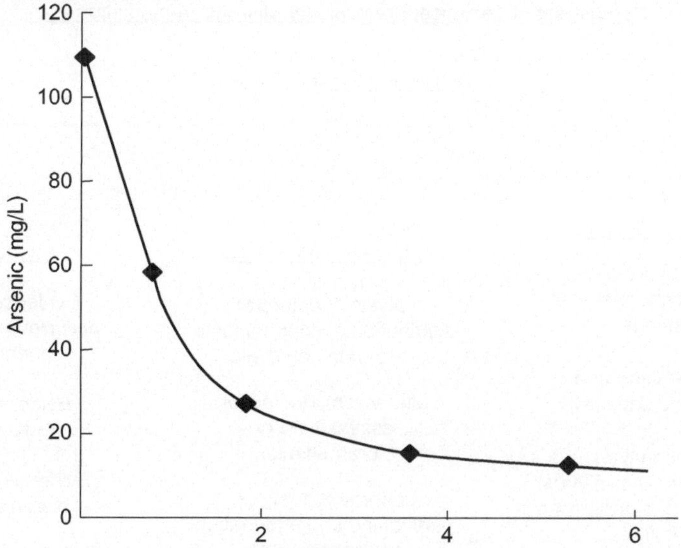

FIGURE 8.5 Desorption of arsenic from an arsenic-contaminated soil with deionized water. Measurements of concentration of arsenic are in the effluent. (Data from Tan, B.K., Update on arsenic leaching tests, unpublished report submitted to Geoenvironmental Research Centre, Cardiff University, Cardiff, UK, 2003.)

8.3 EVIDENCE OF CONTAMINANT ATTENUATION

From the discussions in the previous chapters, we can define natural attenuation broadly as a process consisting of "*biodegradation, dispersion, dilution, sorption, contaminant volatilization, chemical and biochemical reactions and transformations of the contaminants, and reduction of contaminant toxicity, volume, mass, and concentrations to levels considered as non-threatening to biotic receptors and the environment.*" It is important to recognize that forming conclusions solely on evidence-based events without relevant science-based foundation can lead to some grossly incorrect and faulty conclusions. The protocols associated with *lines of evidence* expect one to assess the various mechanisms and processes that combine to produce assimilation of contaminants in soils—in addition to the abiotic and biotic processes and reactions occurring in the soil. By doing so, one can is able to determine if the evidence-based results under scrutiny are reliable or perhaps spurious. The previous chapters have discussed many of these as specific processes involving interactions between contaminants and soil fractions. The sets of parameters that constitute the basic elements of the markers are shown assembled into three groups in Figure 8.6. The groups are distinct in that each group encompasses a set of tests and investigations that are needed to satisfy the technical protocols in the framework defined by the lines of evidence. For proper consideration of monitored natural attenuation (MNA) as a remediation option, the markers and information requirements specified must be addressed.

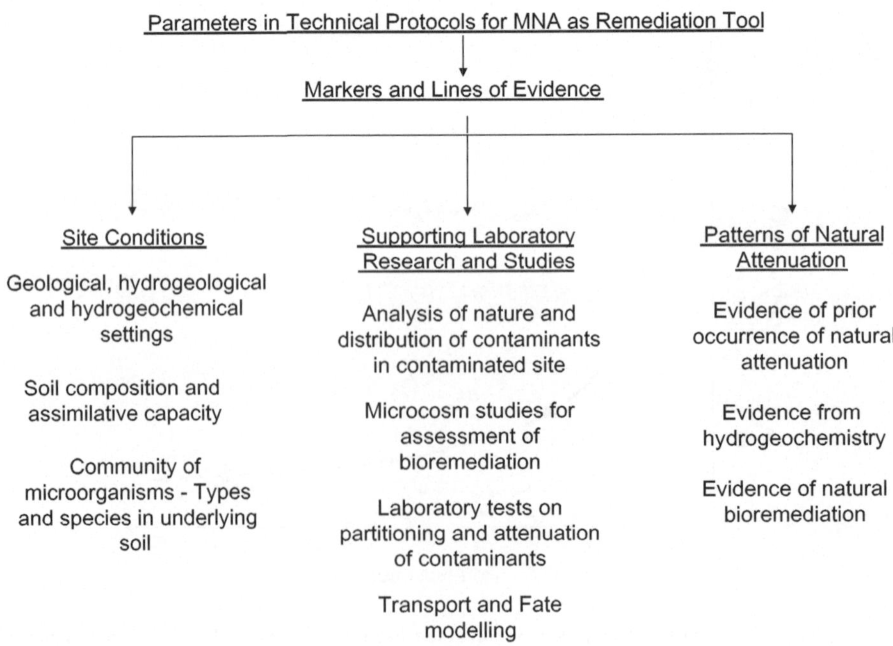

FIGURE 8.6 Parameters and markers for technical protocols in consideration of monitored natural attenuation (MNA) as a remediation tool.

Because of the importance of relevant science-based facts testifying to evidence-based information required in *lines of evidence* requirements, the discussions in previous chapters dealing with specific interactions between inorganic and organic chemical contaminants will now be used as part of the contaminant attenuation tracking routine. Once again, it is worth repeating that while monitoring information provides one with evidence-based accounting, it is necessary to obtain science-based information that will support the gathered evidence from monitored sites. We take into account the role of the soil fractions in the attenuation processes and consider that a proper account of these is needed in the program of scrutiny required to establish the success of MNA.

8.3.1 Tracking Organic Chemical Contaminants

To understand attenuation of organic chemical contaminants by soils, we need to go back to the previous chapters and recall some of the mechanisms of interactions between the organic chemicals and clay-type soils. As we have seen in the previous chapters, the chemical functional groups that make up the surface properties of the soil fractions contribute appreciably to the acidity of the soil particles. This is a significant property of the soil since surface acidity is very important in the adsorption of ionizable organic molecules of clays. Surface acidity plays a major role in clay adsorption of amines, *s*-triazines, amides, and substituted urea due to protonation on the carbonyl group. A good example of this is the hydroxyl groups in organic chemical compounds that consist of two broad classes of compounds, alcohols (ethyl, methyl, isopropyl, etc.) and phenols (monohydric and polyhydric), and the two types of compound functional groups—i.e., those having a *C-O* bond (carboxyl, carbonyl, methoxyl, etc.) and the nitrogen-bonding group (amine and nitrile). Amine, alcohol, and other organic chemicals possessing dominant carbonyl groups that are positively charged by protonation can be readily sorbed by clays. In amines, for example, the NH_2 functional group of amines can protonate in soil, thereby replacing inorganic cations from the clay complex by ion exchange. As we have seen in Chapter 5, the extent of sorption of these kinds of organic molecules depends on:

- The cation exchange capacity (CEC) of the clay minerals,
- The composition of the clay soil (soil organics and amorphous materials present in the soil),
- The number of reactive surfaces, and
- The molecular weight of the organic cations.

Large organic cations are adsorbed more strongly than inorganic cations because they are longer and have higher molecular weights, and polymeric hydroxyl cations are adsorbed in preference to monomeric species. This is because of the lower hydration energies and higher positive charges and stronger interactive electrostatic forces.

Because of the unsymmetrically shared electrons in the double bond, carbonyl compounds possess dipole moments. This permits sorption onto clay minerals by hydrogen bonding between the OH group of the adsorbent and the carbonyl group of the ketone or through a water bridge. For the carbonyl group of organic acids such as benzoic and acetic acids, sorption onto clays occurs directly with the interlayer of cation or by formation of hydrogen bonds with the water molecules (water bridging) coordinated to the exchangeable cation of the clay complex.

We will use an example of organic chemical contaminants in a contaminated site to demonstrate the importance of the protocols that attend the serious consideration of MNA as a remediation option. Organic chemicals that find their way into the land environment have origins in various chemical industrial processes and as commercial substances for use in various forms. These are generally classified as xenobiotic compounds. Chemical products such as organic solvents, paints, pesticides, oils, gasoline, creosotes, and greases, etc. are responsible for many of the chemicals found in contaminated sites. Because of the many thousands of organic chemical compounds in existence, it is not possible to categorize them all in respect to how they would interact in a soil-water system. The more common organic chemicals found in contaminated sites can be grouped into three convenient groups as follows:

- Hydrocarbons—these include the PHCs (petroleum hydrocarbons), the various alkanes and alkenes, and aromatic hydrocarbons such as benzene, MAHs (multicyclic aromatic hydrocarbons), e.g., naphthalene; and PAHs (polycyclic aromatic hydrocarbons), e.g., benzo-pyrene;
- Organohalide compounds—of which the chlorinated hydrocarbons are perhaps the best known. These include: TCE (trichloroethylene), carbon tetrachloride, vinyl chloride, hexachlorobutadiene, PCBs (polychlorinated biphenyls) and PBBs (polybrominated biphenyls);
- Miscellaneous compounds—including oxygen-containing organic compounds such as phenol and methanol, and nitrogen-containing organic compounds such as TNT (trinitrotoluene).

The density of these compounds in comparison to that of water is of particular interest since this will control the transport characteristics of the organic chemical. We classify non-aqueous phase liquids (NAPL) into two groups. These are: (a) light NAPLs identified as LNAPLs and (b) dense ones called DNAPLS, as illustrated in Figure 5.5 in Chapter 5. Since LNAPLs are lighter than water and the DNAPLs are heavier than water, one would expect the NAPL to stay above the water table, and the DNAPL, which is denser than water to sink through the water table and come to rest at an impermeable bottom (bedrock or impermeable substrate layer). Light NAPLs include gasoline, heating oil, kerosene, and aviation gas. Dense NAPLs include the organohalide and oxygen-containing organic compounds such as 1,1,1-trichloroethane, chlorinated solvents such as tetrachloroethylene (PCE), trichloroethylene (TCE) and carbon tetrachloride (CT), PCBs, PCPs (pentachlorophenols) and TCPs (tetrachlorophenols).

The various results of transformations and biodegradation of organic chemicals have been discussed in various forms in Chapter 6. The significant outcome of the application of MNA as a remediation process is the evidence of occurrence of biodegradation and transformation of the target organic chemicals in the MNA process—a key requirement in *lines of evidence*. In the case of abiotic transformation processes, these occur without the mediation of microorganisms. These processes include chemical reactions such as hydrolysis, and oxidation-reduction. The markers used to satisfy *lines of evidence* requirements involve determination of both the decrease in concentration of the contaminants and the transformation of the organic chemical contaminants. This requires one to determine the nature and composition of the transformed products of the original organic chemical contaminants. It is important to distinguish between the products obtained via abiotic and biotic processes. As also noted in Chapter 6, *biotic* transformation processes are biologically mediated transformation reactions—with associated chemical reactions arising from microbial activities. The major difference between the transformation products from abiotic and biotic processes is that abiotic transformation products are generally other kinds of organic chemical compounds, whereas transformation products resulting from biotic processes are mostly seen as stages (intermediate products) towards mineralization of organic chemical compounds.

To interpret the information, it is necessary to understand the evidence of occurrence of natural bioremediation. We recall from the previous chapters that biologically mediated transformation processes are the only types of processes which can lead to mineralization of the subject organic chemical compound. Complete conversion to CO_2 and H_2O (i.e., mineralization) does not always occur. However, intermediate products can be formed during the mineralization. Note that intermediate products obtained from abiotic and biotic transformation processes can themselves become greater environmental threats, as for example in the transformation of perchloroethylene (PCE, C_2Cl_4) to trichloroethylene (TCE, C_2HCl_3), to 1,2-dichloroethylene (DCE, $C_2H_2Cl_2$) to vinyl chloride (VC, C_2H_3Cl) and to ethane (CH_3CH_3). Beginning with PCE, where the log k_{oc} value indicates good partitioning to the soil fractions, degradation of the PCE to TCE and onward to VC, shows that the log k_{oc} values diminish considerably to a very low value for the vinyl chloride. As the PCE continues to degrade, more of the chemical substance is released into the aqueous phase (porewater). This is particularly true for VC, where the low values of log k_{oc} and high-water solubility values suggest that this chemical can be environmentally mobile.

Because of the many varied nature of organic chemicals found in the groundwater, and because degradation of the organic chemical contaminant may be extremely slow, tracking of degradation of the offending chemical requires one to firstly identify the chemical compounds in question, and secondly, to determine how one should or could identify degradation occurrence. A case in point is the study reported by Hatzinger, et al. (2018) on the use of stable isotope analysis to track the biodegradation of hexahydro-1,3,5-trinitro-1,3,5-triazine (RDX) using a technique that they report as compound-specific stable isotopic analysis (CSIA). In their report in the use of this technique, they indicate that "biological contaminant destruction can be documented as changes in the ratio of stable isotopes of specific elements in a molecule" and for the particular case of RDX this meant ratios of $^{15}N/^{14}$ and $^{13}C/^{12}$.

8.3.2 Tracking Heavy Metal Contaminants

For sites contaminated by heavy metals (HMs), we note from Chapter 4 that environmental mobility of these metals is dependent upon whether they are in the porewater as free and complexed ions or sorbed onto the soil particles. Mobility of free ions and complexed ions in the porewater will be governed by advection and diffusion mechanisms. So long as the full assimilative potential of the soil for HM is not reached, attenuation of the HMs will continue. The mechanisms for HM assimilation discussed in Chapter 4 show for example that the metals that are sorbed onto the soil particles are held by different sets of forces—determined to a large extent by the soil fractions and the pH of the soil-water system. The selective sequential extraction (SSE) procedure and test results shown previously in Figure 4.24 (Chapter 4) for Pb sorbed by an illite soil, and in Figures 4.25 and 4.26 demonstrated that the various types of soils and their different soil fractions have different sorption capacities, dependent on the nature, and distribution of the HMs and pH of the system.

Although precipitation of HMs is not, strictly speaking, a sorption phenomenon, precipitation of HMs as hydroxides, sulfides, and carbonates generally classify as part of the assimilative mechanism of soils because the precipitates form distinct solid material species. They classify under the category of "removal of solutes from the porewater" and are thus most often considered as part of the attenuation process. Either as attached to soil particles or as void pluggers, precipitates of HMs can contribute significantly to attenuation of HMs in contaminant plumes. Hydroxide precipitation is favored in alkaline conditions as for example when $Ca(OH)_2$ is in the groundwater in abundance. With available sulfur and in reducing conditions, sulfide precipitates can be obtained. Sulfide precipitates can also be obtained as a result of microbial activity—except that this will not be a direct route. Sulfate reduction by anaerobic bacteria will produce H_2S and HCO_3^- thus producing the conditions for formation of metal sulfides.

8.4 CONSIDERING MONITORED NATURAL ATTENUATION APPLICATION

There are several points of consideration before one adopts MNA as a remediation tool. Many of these require some considerable site investigative work and laboratory analyses. Obviously, one cannot "look" at a contaminated site and immediately decide that MNA would work as a remediation tool. Figure 8.7 shows the general protocol for considering MNA as a remediation tool for treating contaminated ground—using Figures 8.1 through 8.3 as reference guides. The data and information inputs are shown on the left-hand side of the diagram. These tell us what is required to satisfy site specific conditions, and whether the markers that point towards evidence of natural bioremediation and contaminant attenuation are sufficient to proceed with further examination to satisfy that the MNA is a viable treatment option. A "No" response from the first two decision steps will automatically trigger a technological and/or engineered solution to the remediation problem. The laboratory research input in the third step refers to the microcosm studies and the various laboratory procedures for determination of contaminant partitioning. These, together with the transport and fate predictions from the models developed for the specific situation, constitute the "supporting studies" shown in the third decision box. Analyses of the results from the supporting studies (laboratory research and predictions) should inform one about the ability of the site materials and conditions

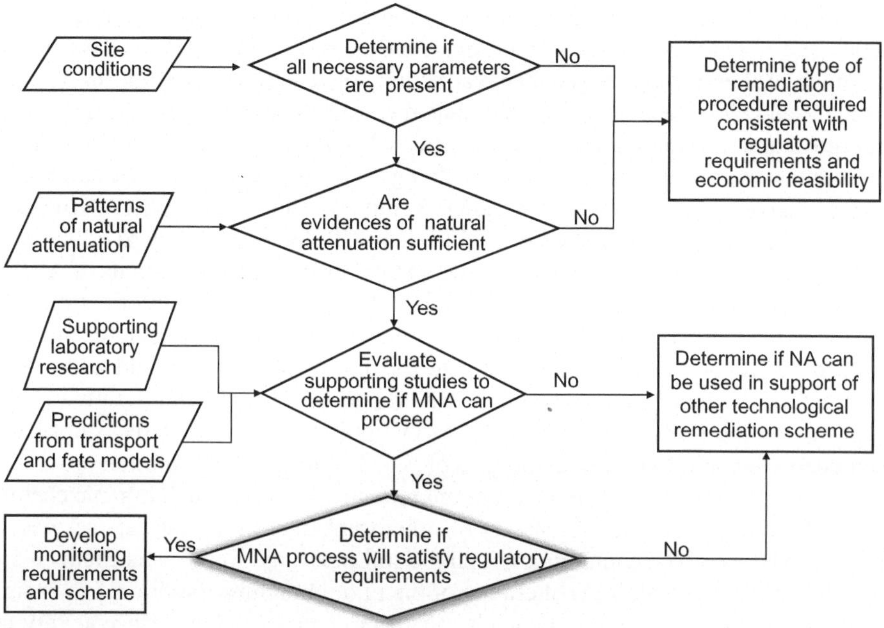

FIGURE 8.7 General protocol for considering monitored natural attenuation (MNA) as a remediation tool.

to attenuate the contaminants. A "No" response from the third decision step will allow one to incorporate natural attenuation processes as part of a technological remediation solution. This will be discussed in a later section.

The most critical pieces of information required in considering the general protocols shown in the flow chart of Figure 8.7 are the criteria or goals set in the central diamond-shaped boxes. These could be threshold values, indicators, trends, etc. Decisions need to be made as to "what is desired" or "what must be obtained" in order to achieve the "Yes" sign to proceed to the next level of consideration.

8.4.1 Contaminant Release and Augmentation

Environmental mobility of contaminants is an issue that requires careful examination since subsurface groundwater flow can dislodge already sorbed (partitioned) contaminants. Monitoring information can only provide "snapshots" of what is held by the soil particles and what is in the porewater at the time of sampling. It is up to the investigators and decision-makers to determine whether some previously held contaminants have been released and have become mobile; therefore, having added to the contaminant load (augmentation of contaminants) arriving at the point of sampling. This can occur when the bonds holding the contaminants to the soil's solids are disrupted or ruptured, thereby causing the contaminants to be released into the porewater. The released contaminants will add to the concentration of the incoming leachate thereby resulting in an increase in the concentration of the contaminants.

Dislocation and removal (desorption) of inorganic contaminants such as heavy metals and organic chemicals sorbed or bound to soil solids arise because of the dynamics of the various biogeochemical actions or reactions occurring in the soil water system. We need to remember that the various soil mineral particles can undergo changes in the surface characteristics when the chemistry of the immediate environment (porewater, for example) is changed. If the chemistry of the soil-water system changes such that low pH values result, solubility of exposed structural Al(III) on the clay particle surfaces will occur. The amount of Al(III) dissolved from the surface will be a function of time and pH level. Greater amounts of Al(III) will be dissolved with time and lower pH values (Wieland and Stumm, 1992). If perchance the pH is raised to a high level, solubility of the Si(IV) could occur. The released Al(III) and Si(IV) could compete for the metal ions or could add to the contaminants in the contaminant plume. The report by Wieland and Stumm (1992) suggests that the preferential release of Si(IV) obtained in their study could have been due to the simultaneous re-sorption of Al(III) during the dissolution process.

For disruption of bonds between the contaminants (inorganic and organic chemicals) and clay soil solids, it is necessary to either weaken or overcome the energies of interaction between the contaminants and the soil fractions. Another mechanism for bond disruption or breakage is alteration of the initial bonding condition. For example, de-ionization of ionized organic chemicals or changes to the surface acidity of clay mineral particles could trigger release of sorbed organics. Introduction of ligands could set up competition for sorption of inorganic contaminants. Biotransformation of sorbed organic chemicals could also alter the bonding relationships previously established between the organic chemical and the respective soil fraction, resulting, therefore, in the release of the organic chemical from the soil solids. It is useful to remember that the soil-water or soil-aqueous system is a dynamic system subject to biogeochemical processes at all times. The results of such processes cannot be easily predicted. We need to be fully aware of the dynamics of the system and to anticipate that both augmentation and attenuation mechanisms exist since these are in essence two sides of the same coin.

8.4.2 Contaminant Release Studies

Studies designed to establish the potential for contaminant release from the soil fractions are necessary to determine if contaminant augmentation could: (a) be responsible for the non-conforming or unexpected results from measurements of contaminant concentrations and performance characteristics, (b) pose a serious health and environmental threat downstream, and (c) be a significant problem in itself. Since the biogeochemical environment in a soil-aqueous system is never constant, we should investigate the possibilities for bond disruption and contaminant release because of the changing environment.

8.4.2.1 Batch Equilibrium Studies

The simplest procedure to determine contaminant release in soils contaminated with inorganic contaminants is to conduct batch equilibrium tests (see Chapter 4). At least two types of soil suspension studies can be conducted: (a) determination of desorption characteristics of the candidate contaminated soil, and (b) soil washing-type studies designed to weaken the bonds. In the case of soils contaminated with heavy metals, desorption-extraction tests using extractants at various pH levels can inform one about the "extractability" of the metals in the metal-contaminated soil. A typical set of desorption test results is shown in Figure 8.8 for a Pb-contaminated illite soil. The results show that the number of "washings" needed to achieve metal release from the soil is also dependent on the initial state of the soil.

Neutral salts are generally used as the reagent solution of interest in the evaluation of desorption of heavy metals. The use of reagents in the desorption characterization studies are principally designed to seek extraction of the contaminants attached to the soil fractions. Selective sequential extraction (SSE) studies such as those described in Chapter 4 are very useful for evaluation of the ease of removal of heavy metal contaminants in soil samples. Figure 8.9 shows the sets of typical reagents used as extractants to remove heavy metal contaminants. Mulligan et al. (2001) have used this procedure to evaluate the effectiveness of various kinds of biosurfactants in the removal of sorbed heavy metals in a soil contaminated with both organic chemicals and heavy metals.

Bonding between Pb and the different soil fractions (minerals, natural organic matter, oxides/hydroxides, carbonates), as shown for example in Figure 8.9, for the illitic soil is determined by mechanisms which range from ion-exchange

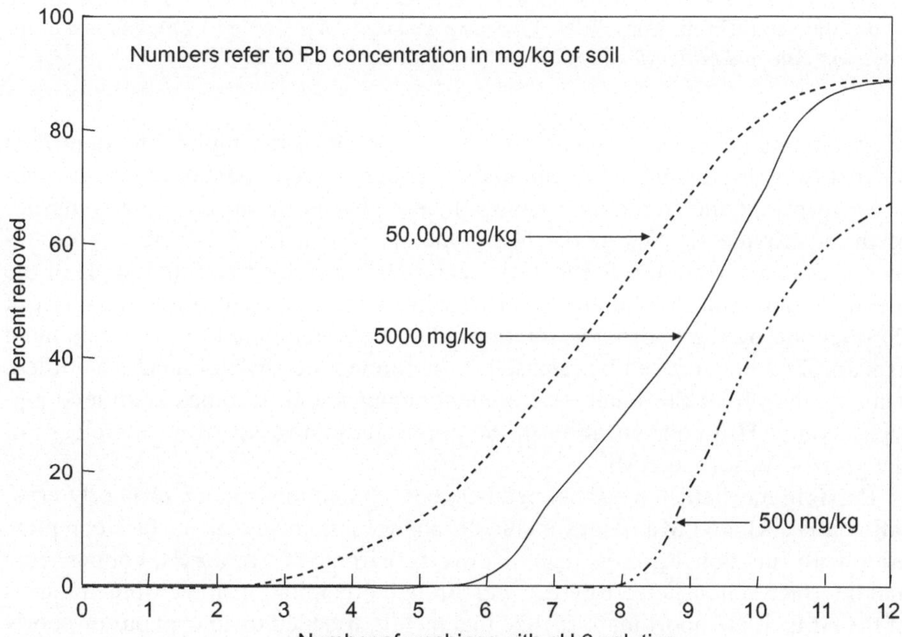

FIGURE 8.8 Desorption batch equilibrium test results for Pb contaminated illite soil using HNO_3 as the "washing solution." Numbers refer to initial Pb concentration in the soil. (Adapted from Yong, R.N., *Geoenvironmental Engineering: Contaminated Soils, Pollutant Fate and Mitigation*, CRC Press, Boca Raton, FL, 307, 2000.)

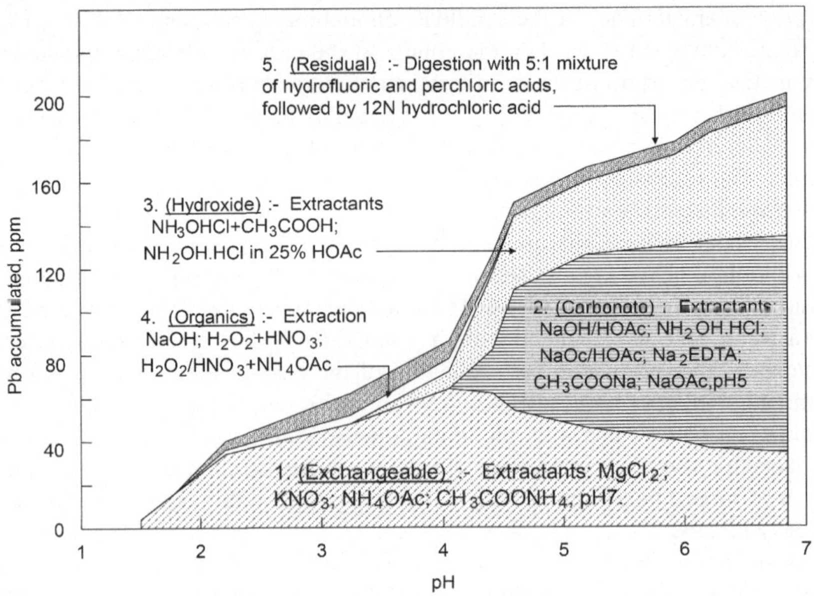

FIGURE 8.9 Some representative reagents that will extract heavy metals sorbed by the various soil fractions shown in the diagram. Results shown in the diagram are for Pb sorbed by an illitic soil. (From Yong, R.N., *Geoenvironmental Engineering: Contaminated Soils, Pollutant Fate and Mitigation*, CRC Press, Boca Raton, FL, 307, 2000.)

to precipitation and/or co-precipitation. As shown in the figure, the dominant mechanisms responsible for accumulation of heavy metal contaminants are sensitive to pH of the immediate environment. This is because of the solubility of the hydroxide species of the heavy metals. When the pH in the porewater increases to a certain level (generally seen to be near the precipitation pH of the metal contaminant), Pb begins to form hydroxy species, resulting in the onset of Pb retention by the hydroxide fractions. The Pb precipitated or co-precipitated as natural carbonates can be released if the immediate environment is acidified. Thus, if the pH of the local environment changes and becomes acidified, augmentation of HM concentration in the porewater can result from the release of the previously sorbed HMs.

Bonding mechanisms established between Pb and amorphous or poorly crystallized Fe, Al, and Mn oxides include exchangeable forms via surface complexation with functional groups (e.g., hydroxyls, carbonyls, carboxyls, amines, etc.) and interface solutes (electrolytes), and through precipitation and co-precipitation of the HMs. It is important to realize that metals attached to amorphous or poorly crystallized Fe, Al, and Mn oxides can be removed by redox gradients. pH sensitivity in HM bonding to soil solids is a very serious issue as can be seen in the results in Figure 8.9. When low pH values exist in the soil-water system, neutral salts such as $MgCl_2$, $CaCl_2$, and $NaNO_3$ become effective extractants. These can promote release of ions physically bound by electrostatic attraction to the negatively charged sites on the soil particle surfaces. Because of the affinity of group II and II cations (valence of 2 or 3) for most surface sites, the cations in these extractant solutions have to be present in larger concentrations than the HMs bound by electrostatic attraction.

8.4.2.2 Leaching Column Tests

Leaching columns are used not only to determine sorption of leachates in soil samples, but also to determine leachability or environmental mobility of sorb contaminants in contaminated soils. To determine leachability, the contaminated samples are placed in the leaching cells or columns, and the fluids used as extractants are those that are expected to be present in the site. Common practice is to leach initially with water as a reference base. Most often, little evidence of leaching of sorbed contaminants will be seen. This is to be expected since the SSE experiments will confirm that even with contaminants sorbed by exchange mechanisms, leaching with water will not result in significant exchange of the sorbed cations. Since the porewater in the soil will contain dissolved solutes, leaching with representative solutions is required to determine leachability and environmental mobility of the sorbed contaminants. The results of two leaching tests are shown in Figures 8.10 and 8.11. In Figure 8.10, the results show leaching of arsenic from an arsenic-contaminated soil in relation to the number of pore volumes of water used as the leaching fluid.

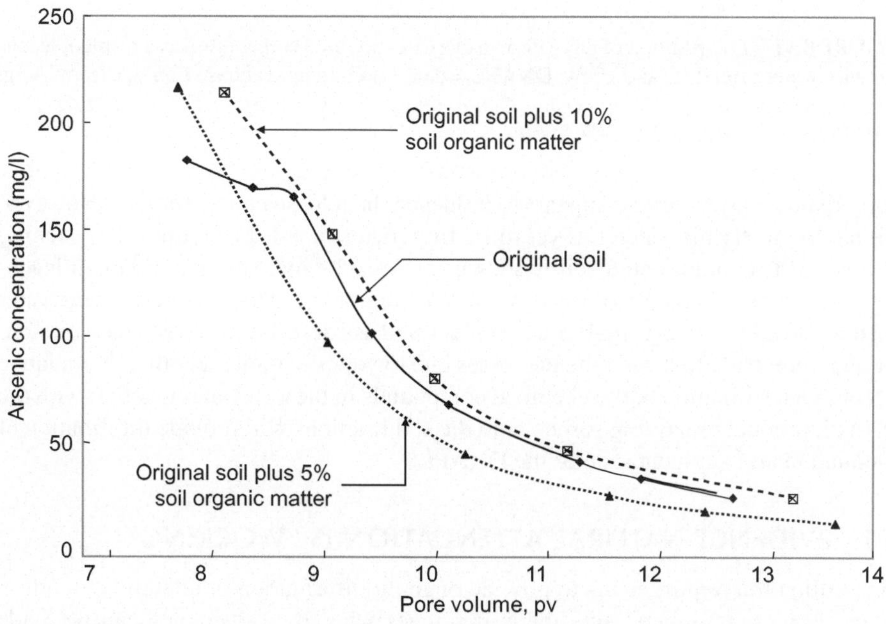

FIGURE 8.10 Leaching of arsenic from arsenic contaminated soil in relation to number of pore volumes of water used as leaching solution. Values in the ordinate refer to concentration of As in the effluent. (Data from Tan, B.K., Update on arsenic leaching tests, unpublished report submitted to Geoenvironmental Research Centre, Cardiff University, Cardiff, UK, 2003.)

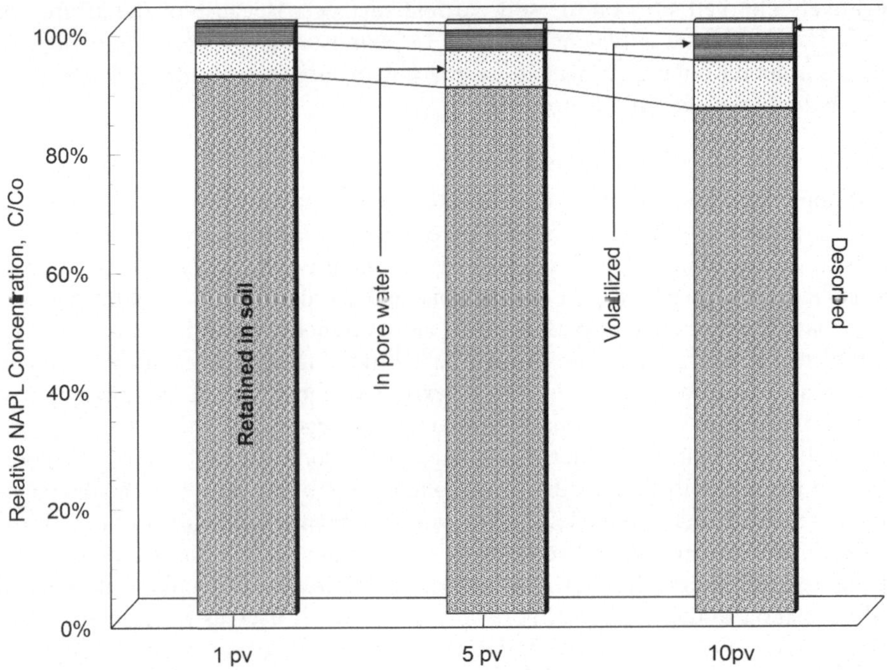

FIGURE 8.11 Distribution of DNAPL in a DNAPL-polluted soil column as a result of leaching with water after 1, 5, and 10 pv. DNAPL = dense non-aqueous phase liquid. (From Yong, R.N., *Geoenvironmental Engineering: Contaminated Soils, Pollutant Fate and Mitigation*, CRC Press, Boca Raton, FL, 307, 2000.)

Soil organic matter content appears to influence the release concentrations. However, the nature of the influence has yet to be fully determined. In Figure 8.11, leaching of a DNAPL-contaminated soil is shown. The results indicate that although leaching with water is not a major issue, some leaching of the DNAPL into the porewater occurs—in addition to a small fraction identified as "desorbed." Mass balance calculations were made to determine the losses due to volatilization and other degradative effects. Determination of the chemical compounds in the porewater together with the DNAPL product remaining sorbed onto the soil fractions will provide information of attenuation and augmentation of the DNAPL.

8.5 EVIDENCE NATURAL ATTENUATION IS "WORKING"

In fulfilling the requirements to provide ongoing information of contaminant attenuation in a contaminated site, there are stages whereby judgements can be made regarding whether the attenuation process is working well, or there needs to be changes made to the treatment process. The criteria or indicators used to arrive at any of the decisions will be varied depending on the issues presented. The protocols for *evidence of NA feasibility*, shown in Figure 8.12 are in many ways similar to those developed for lines of evidence (Figure 8.5).

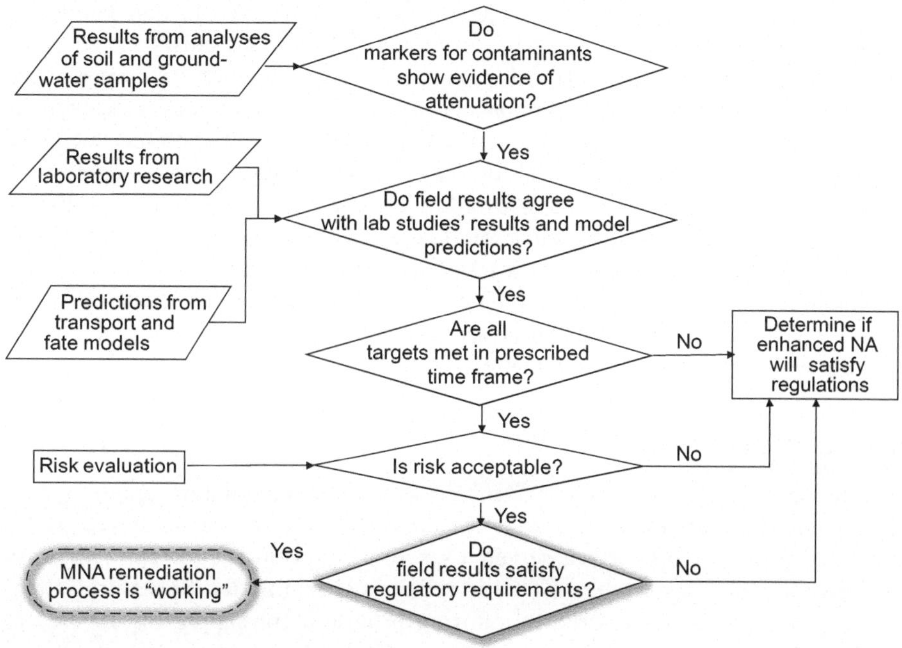

FIGURE 8.12 Example of protocols for evidence of natural attenuation (NA) feasibility decision flowchart using information from multiple *lines of evidence* to determine if monitored natural attenuation (MNA) remediation is "working."

Determination of evidence of NA feasibility begins with requirements for proper siting of the monitoring wells and sampling devices as shown in the top decision box. The lower decision boxes which deal with the verification requirements are the essence of the feasibility testimony for MNA. Without positive verification, it is not likely that application of MNA as a remediation process will satisfy regulatory requirements. The key elements that are essential to the verification monitoring scheme portion of the evidence of NA feasibility come from the supporting studies and model predictions. The supporting studies and model predictions are the same as those shown in Figures 8.5 and 8.8. The importance of verification of the markers for attenuation of contaminants using comparative leaching column studies and microcosm studies cannot be overstated. Exact or close corroboration is not expected between field and laboratory values because of size and time-scaling problems. One looks for patterns and similarities in partitioning performance and degradation, and pathways. Further corroboration is sought between field performance and predictions from analytical-computer models of system performance. While it would be desirable to obtain good accord between predicted and observed field performance, the more realistic view is to look for trends in field performance to closely match model predictions. If the trends are in the right direction, and if the degree of attenuation predicted by the fate and transport model accords with the observed field performance, we can be reassured that MNA can work as a remediation option.

One of the biggest challenges in evaluating results from a verification monitoring program is to determine an acceptable level of risk that would satisfy the decision-makers associated with the verification process. Risk evaluation, risk assessment, and risk analysis are necessary components of the evaluation scheme—as seen in the fourth decision box. In essence, we can describe MNA as a deliberate and calculated risk management scheme, with levels of risk carefully managed by strict application of multiple *lines of evidence* and assessed by evidence of NA feasibility protocols. It is useful to note that if all attenuation markers are not totally satisfied, it is perhaps possible to use NA as part of a set of structured technological solutions such as enhancements to natural attenuation capability of the subsoil—as shown in the right-hand side of the diagram. These aspects will be discussed in the next chapter together with other uses for natural attenuation.

8.6 CONCLUDING REMARKS

Monitored natural attenuation (MNA) as a passive remediation and/or intrinsic remediation can work well when evidence-based and science-based pieces of information are properly assessed. A prime requirement for application of the natural processes in the soil to remediate contamination by natural attenuation processes is: (a) a strict monitoring program, (b) supporting laboratory tests, (c) studies on partitioning mechanisms, (d) development of mechanistic and analytic transport and fate models, and (e) protocols and acceptance criteria for the multiple *lines of evidence*. The major pieces of information and knowledge needed include the following:

- Proper site characterization to establish: (a) that the proper hydrogeology exists to promote transport of the contaminants in the soil-water system that possesses the qualities that will promote attenuation of the contaminants; (b) the nature of the groundwater geochemistry and, in particular, the redox condition; and (c) that the essential site elements that can promote and sustain *natural attenuation* processes are present, e.g., composition and assimilative capacity of the soils, CEC and SSA of the soils, soil organic matter, pH and Eh, microorganisms, electron acceptors and donors.
- Determination of evidence of prior occurrence of natural attenuation at the site. The question that needs to be answered is: "Can we find evidence of the effects or results of natural attenuation of contaminants previously present in the site?" The pieces of evidence needed include determination of: (a) occurrence of natural bioremediation, (b) occurrence of partitioning of contaminants, (c) occurrence of transition products from abiotic reactions, (d) biotransformation and biodegradation products, (e) reduced and oxidized compounds—e.g., sulfate and iron reduction, (f) speciation, and (g) groundwater geochemistry and hydrogeology.
- Establish that the source of contaminants can be controlled, limited, or even removed. Ideally, one would want to remove or contain the source of the contaminants so that the natural attenuation process will be sufficient

to achieve effective complete remediation of the contaminant plume, i.e., contaminant plume will be completely attenuated.

- Supporting laboratory research—including microcosm studies, sorption characterization, and tests for assimilative capacity of soils. These include: (a) microcosm studies for evaluation of the degradation and/or transformation of the target organic chemical contaminants under specified conditions should be structured to provide information on the various factors and conditions that inhibit or promote mineralization and biotransformation of the target organic chemicals; (b) partitioning tests for both inorganic and organic chemical contaminants should include tests for determination of k_{oc}, k_d, k_{ow}, adsorption isotherms, leaching column tests, and desorption tests; and (c) modelling of transport and fate of contaminants—to provide one with the capability to track and predict transport and fate of the contaminants in the contaminated site. Uncertainties in prediction of fate of the various contaminants can often be traced to: (1) improper and incomplete accounting or understanding of the complex processes contributing to attenuation of the contaminants, and (2) poor quality information on the various parameters, site conditions, and boundary conditions.
- For proper prediction, the following items need to be considered: (a) accurate site-specific data and proper quantitative descriptions of the various processes controlling mass transport and chemical mass transfer such as redox reactions, hydrolysis, precipitation, complexation, and biological processes. and (b) changes in reaction rates and constants because of transformation, speciation and complexation;
- Examination and evaluation of the potential advantages and disadvantages of application of MNA. The balance between length of time needed to obtain acceptable attenuation of contaminants and costs needs to be examined in conjunction with long term monitoring requirements and health and safety assessments. The list of some of the major items for scrutiny include: (a) less site disturbance and less chances of exposure of above-surface biotic receptors to contaminants, (b) the overall costs for application of MNA as a remediation process should be much less than other conventional methods, (c) the possibility that transformation and intermediate products resulting from natural attenuation processes may be more toxic than original contaminants, and (d) that the toxicity, persistence, and mobility of the intermediary metabolites that result from incomplete biodegradation of the parent organic chemical compound can also occur.
- It should be noted that: (a) requirements for long term monitoring and sampling, together with associated tests to ensure attenuation occurs at the rate predicted by the transport models, can be a distinct disadvantage, and (b) changes in the hydrogeology of the site due to circumstances occurring outside the boundaries of the contaminated site over the treatment life could have negative impacts on the effectiveness of the MNA process.

REFERENCES

Hatzinger, P., Fuller, M., Sturchio, N.C., and Bohike, J.K. (2018), Validation of stable isotope ratio analysis to document the biodegradation and natural attenuation of RDX, Environmental Security Technology Certification Program (ESTCP) Project Report, Project Number ER-201208.

Mulligan, C.N. (2001), *Environmental Biotreatment: Technologies for Air, Water, Soil, and Wastes*, Government Institutes, Rockland, MA, 395p.

Tan, B.K. (2003), Update on arsenic leaching tests, unpublished report submitted to Geoenvironmental Research Centre, Cardiff University, Cardiff, UK.

USEPA (2015), Use of monitored natural attenuation for inorganic contaminants in groundwater at superfund sites, OSWER Directive 9283.1-36.

Wieland, E., and Stumm, W. (1992), Dissolution kinetics of kaolinite in acidic aqueous solution at 25°C, *Geochim. Cosmochim. Acta*, 56:3339–3355.

Yong, R.N. (2000), *Geoenvironmental Engineering: Contaminated Soils, Pollutant Fate and Mitigation*, CRC Press, Boca Raton, FL, 307p.

9 Monitored Natural Attenuation Enhancements

9.1 INTRODUCTION

There are several reasons why one would want to seek enhancement of natural attenuation (NA) processes in application of NA as a remediation tool. Two main reasons include: (a) shortening the time required for NA processes to reach remediated standards or criteria, and (b) controlling and directing—i.e., physically managing—the contaminant plume to optimize NA effectiveness. All the enhancement options falling within these two main reasons or categories should result in the improvement of NA as a passive or semi-passive contaminated site remediation tool.

The flow chart in Figure 8.12 (in the previous chapter) shows that if the assessment of NA is not working as expeditiously as expected as a remediation tool in the contaminated site under treatment, one should investigate the use of *enhanced natural attenuation* (ENA) treatments. Adherence to all the protocols shown in the figure is important because the results of these will provide the necessary information to determine the kinds of enhancement tools that would be appropriate. The two avenues for enhancing natural attenuation processes shown in Figure 9.1 are: (a) introduction of "aids" to the indigenous microorganisms to obtain greater effectiveness from these microorganisms, and (b) improving the substrate soil to obtain more effective partitioning of the contaminants. While the terminology ENA has been used in the literature, it must be noted that we are talking about improvement of the assimilative capacity of the soil and/or the biological activities therein. These will be discussed in the next section.

9.2 ENHANCED NATURAL ATTENUATION

9.2.1 Bioremediation Enhancement

Bioremediation of soils contaminated by organic chemical contaminants occurs when soil micro-organisms metabolize the contaminating organic chemical compounds. For soils that contain natural organic matter, it would not be surprising to find a consortium of indigenous micro-organisms that are well adapted to the specific habitat. Since many natural soil organics (soil organic matter, SOM) are closely similar to synthetic organic chemical compounds, bioremediation of the contaminating organic chemicals can be expected. Natural aromatic soil organics such as vanillin, lignin, and tannin are closely similar to the synthetic aromatic organic compounds represented by benzene, toluene, polycyclic aromatic hydrocarbons

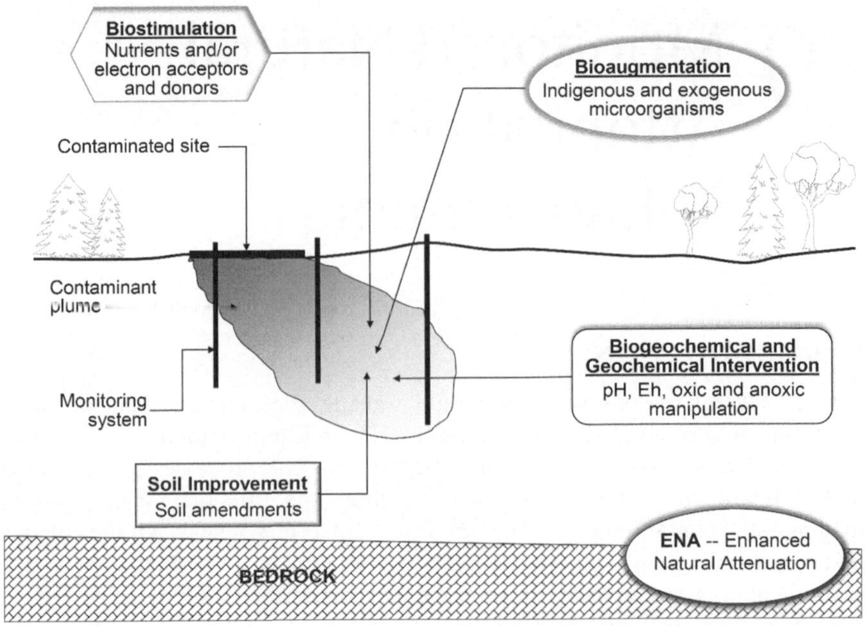

FIGURE 9.1 Schematic showing the two "avenues" of enhancement techniques that can be used to obtain more effective natural attenuation (NA) treatment of a contaminated site.

(PAHs), etc. The naturally occurring consortia of micro-organisms, ranging from bacteria and fungi to viruses should be able to successfully address the synthetic organic chemicals since they are well adapted to the specific habitat. The available energy sources and all the other micro-environmental factors such as pH, temperature, water content, etc. will produce the suites of biomass that have adapted to the micro-environment.

9.2.1.1 Biostimulation

Probably the simplest procedure for improving the intrinsic bioremediation capability of a soil is to provide a stimulus to the microorganisms that already exist in the site. This procedure is called *biostimulation*—i.e., adding nutrients and other growth substrates, together with electron donors and acceptors. The intent of biostimulation is to promote increased microbial activity with the set of stimuli to better degrade the organic chemical contaminants in the soil. With the addition of nitrates, Fe(III) oxides, Mn(IV) oxides, sulfates, and CO_2 for example, anaerobic degradation can proceed. This technique is used for sites contaminated with organic chemical contaminants, and while it is one of the least intrusive of the methods of enhancement of natural attenuation, delivery of the nutrients, growth substrates, and electron acceptors and donors can be challenging. Delivery through natural fractures or through boreholes can be effectively managed by taking advantage of the groundwater flow pattern.

9.2.1.2 Bioaugmentation

The other method of enhancement which falls in the same class of "less-intrusive" enhancement procedure is bioaugmentation. If the native or indigenous microbial population is not capable of degrading the organic chemicals in the soil—for whatever reason, e.g., concentrations, inappropriate consortia, etc.—other micro-organisms can be introduced into the soil. These are called exogenous micro-organisms. Their function is to augment the indigenous microbial population such that effective degradative capability can be obtained. If need be, biostimulation can also be added to the bioaugmentation to further increase the likelihood of effective degradative capability. We need to be conscious of the risks that arise when unknown results are obtained from interactions between the genetically engineered micro-organisms and the various chemicals in the contaminated ground. The use of micro-organisms grown in uncharacterized consortia which include bacteria, fungi, and viruses can produce toxic metabolites (Strauss, 1991). In addition, the interaction of chemicals with microorganisms may result in mutations in the microorganisms themselves, and/or microbial adaptations.

9.2.2 Geochemical and Biogeochemical Intervention

When the assimilative capacity of a soil is not capable of providing: (a) complete contaminant attenuation, or (b) the desired level of attenuation of the contaminants in the contaminant plume, or (c) effective contaminant attenuation within a specified time frame, there are several ways in which we can increase the assimilative capacity of the soil. While some may consider the procedures to be slightly intrusive, they are nevertheless useful procedures since they are relatively inexpensive and simple to implement. The techniques for introducing the nutrients, growth substrate, and exogenous microorganisms to implement biostimulation and bioaugmentation can also be used to provide geochemical intervention. By and large, the simplest forms of geochemical and/or biogeochemical interventions are pH and pE or Eh manipulation. These types of intervention are also well suited for treatment of inorganic contaminants or contaminants. Changes in toxicity for some inorganics (e.g., Cr and As) can result because of changes in their oxidation state, resulting either in a decrease or increase in the toxicity level of the contaminant. We can cite chromium (Cr) as an example. Cr(III) is an essential nutrient that helps the body use sugar, protein and fat, while Cr(VI) on the other hand has been determined by the World Health Organization (WHO) to be a human carcinogen. Cr(III) can be oxidized to Cr(VI) by dissolved oxygen and quite possibly with manganese dioxides. Since this is not a desirable situation, geochemical and/or biogeochemical means can be used to create a reducing environment in the subsurface. One useful method is to deplete the oxygen in the subsurface to create a reduced condition in the soil. In the case of As, As(III) is more toxic than As(V). What is needed, therefore, is to provide an oxygen source to ensure oxidation of the As(III).

Manipulation of pH is quite common in agricultural practice. When applied to assist the assimilative capacity soils, pH manipulation, amongst other things, addresses the precipitation of heavy metals in solution (porewater) or dissolution of precipitated heavy metals. Changes in pH will also result in changes in the sign of surface electrostatic charges for those materials that have pH-dependency charge characterization. When this happens, bonding of metals or release of heavy metals from disruption of bond disruptions will occur. Both pH and Eh changes will have considerable influence also on the various acid-base reactions and on abiotic and biotic electron transfer mechanisms. Transformations and degradation of organic chemical contaminants resulting from acid-base and oxidation-reduction reactions are less significant than those obtained via biotic processes. Reaction kinetics in relation to such processes, and those initiated by the catalytic action of soils resulting in abiotic transformation, are considered to be relatively slow.

9.2.3 Soil Improvement

The procedures for soil improvement generally include addition of soil amendments to increase the sorption or assimilative capacity of the soil. While soil improvement is perhaps the most intrusive of the enhancement procedures, it is not uncommon to find soil amendments used quite widely. We consider the problem of a site contaminated by heavy metals as an example. In the past, lime has been used as the additive of choice because of its ability to raise the pH of the soil-water system. This would precipitate the heavy metal contaminants and thus make them less bioavailable. However, we must recognize that this is not a permanent solution because if the pH of the system is subsequently reduced by environmental forces or external events, the same heavy metals will become mobile again. To obtain a more permanent solution, we should be thinking in terms of not only a pH change, but also of amendments that would bind the metals to the soil. This can take the form of addition of those soil fractions that have the greatest sorption capability (see Chapters 2 and 3), such as clay minerals, soil organic matter, and amorphous materials. The aim is to increase the cation exchange capacity (CEC) and the specific surface area (SSA) of the soil. These, in turn, would increase the partition coefficient of the soil. Figure 9.2 shows the breakthrough curves for arsenic in relation to a soil in its original state and with additions of soil organic matter. It would appear that changes in soil structure with the SOM have significant influence on the initial sorption of arsenic. Including the addition of beneficial soil fractions, one could introduce inorganic and organic ligands in the porewater, to promote speciation and formation of various compounds.

9.2.4 Treatment Zone

Figure 9.3 shows an example of the enhanced natural attenuation technique where a *treatment zone* is established with the addition of any or all of the enhancements discussed in the previous sub-sections, and monitored to determine the efficacy of the enhanced treatment techniques. Adjustments to enhancement treatments can be made when indicated by the monitoring results. It is essential to establish the treatment zone downgradient from the contaminated site to permit the ENA to function as planned.

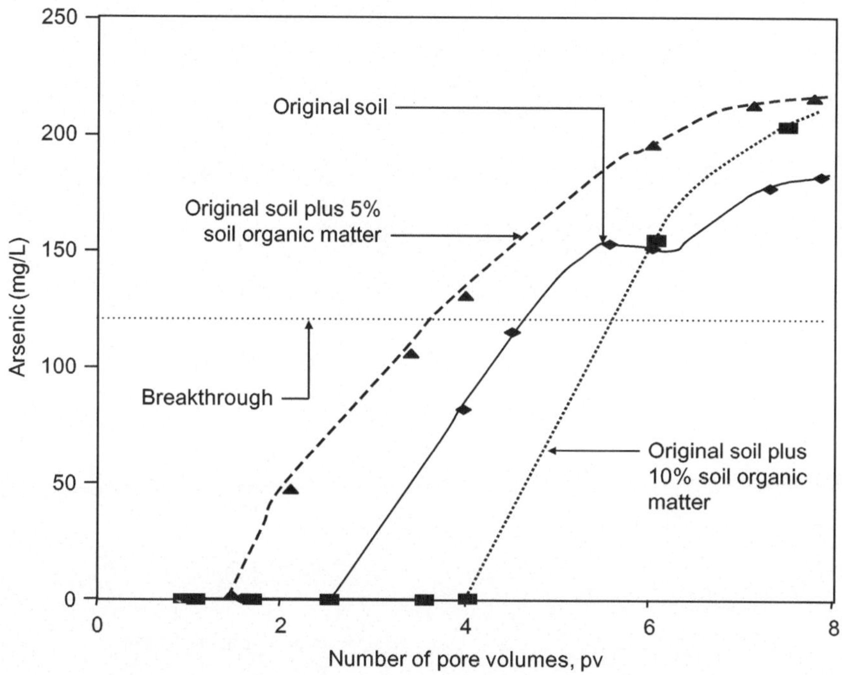

FIGURE 9.2 Sorption of arsenic showing breakthrough performance and influence of soil organic matter content on performance. (Data from Tan, B.K. Update on arsenic leaching tests, unpublished report submitted to Geoenvironmental Research Centre, Cardiff University, Cardiff, UK, 2003.)

9.3 ENGINEERED ENHANCEMENTS

9.3.1 Permeable Reactive Barrier

In the event that further remedial treatment procedures—additional to the ENA procedure adopted—are required to assist in remediation of a contaminated site, engineered treatment aids can be used. Of the many options are available, the placement of a permeable reactive barrier (PRB) in the *treatment zone* ahead of the contaminant plume is perhaps the simplest system to adopt—as shown in Figures 9.4 and 9.5. The intent of a PRB is to provide treatment as a remediation procedure to a contaminant plume as it is transported through the PRB so that the plume no longer poses a threat to biotic receptors when it exits the PRB. The PRB needs to be strategically located downgradient to intercept the contaminant, and if needed, the contaminant plume can be channeled to flow through the PRB—as will be discussed in the next sub-section.

Permeable reactive barriers (PRB) are also known as treatment walls. The soil materials in these walls or barriers can include a range of oxidants and reductants, chelating agents, catalysts, micro-organisms, zero-valent metals, zeolite, reactive clay soils, nanocomposites such as biochar, ferrous hydroxides, carbonates and

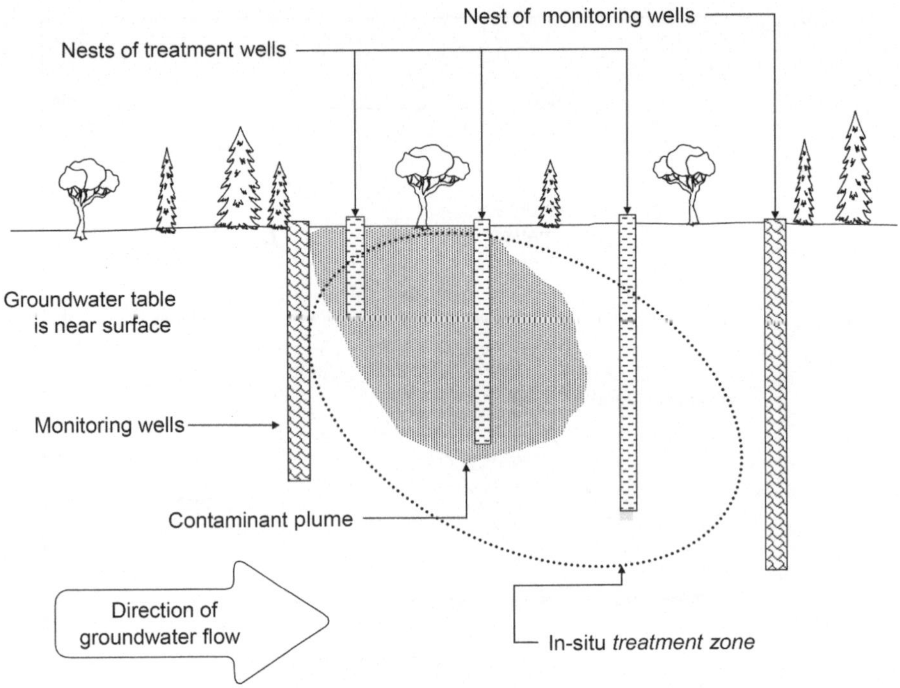

FIGURE 9.3 Enhancement of natural attenuation using treatment wells. Treatments for enhancement can be any or all of the following: geochemical intervention, biostimulation, and bioaugmentation. Treatment occurs in the contaminant plume and downgradient from the plume.

sulfates, ferric oxides and oxyhydroxides, activated carbon and alumina, nutrients, phosphates, and soil organic materials. The choice of any of these treatment materials is made on the basis of site-specific knowledge of the interaction processes between the target contaminants and material in the PRB. Laboratory tests and treatability studies are essential elements of the design procedure for the treatment walls (PRBs). When designed properly, a PRB provides the capability for assimilation of the contaminants in the contaminant plume as it migrates through the barrier.

The region immediately in front of the permeable reactive barrier serves not only to function as an ENA vehicle, but also as a pre-treatment or preconditioning aid for the PRB. The successful use of PRBs as engineered enhancements of the natural attenuation remediation process relies upon the movement of the contaminated groundwater into and through the permeable barrier. By doing so, the PRB will function as an active contaminant attenuating tool. Locating the treatment barrier to intercept the contaminant plume is of utmost importance. Determination of the proper sets of reactive materials in the barrier requires laboratory assessment of the kinds of contaminants retrieved from the contaminated site and their interaction with candidate reactive agents for placement in the PRB. The choice of reagents and

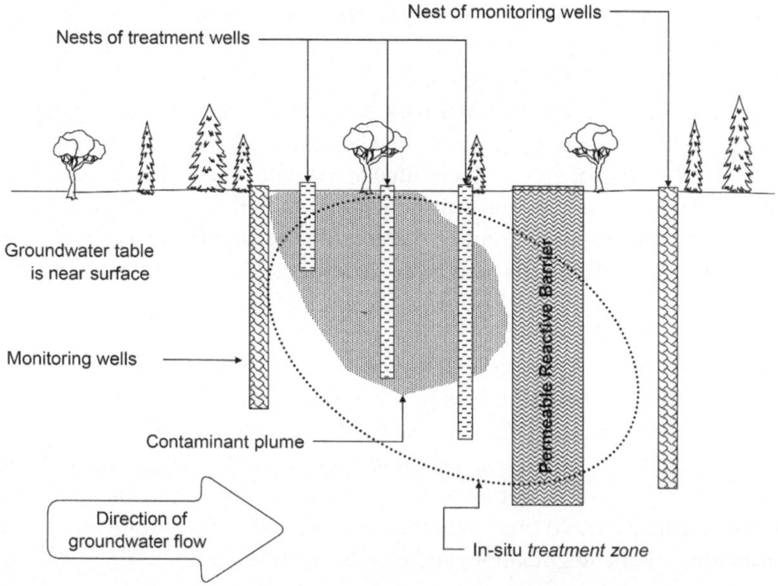

FIGURE 9.4 Placement of permeable reactive barrier (PRB) ahead of advancing contaminant plume to aid in final partitioning (removal) of contaminants in contaminant plume.

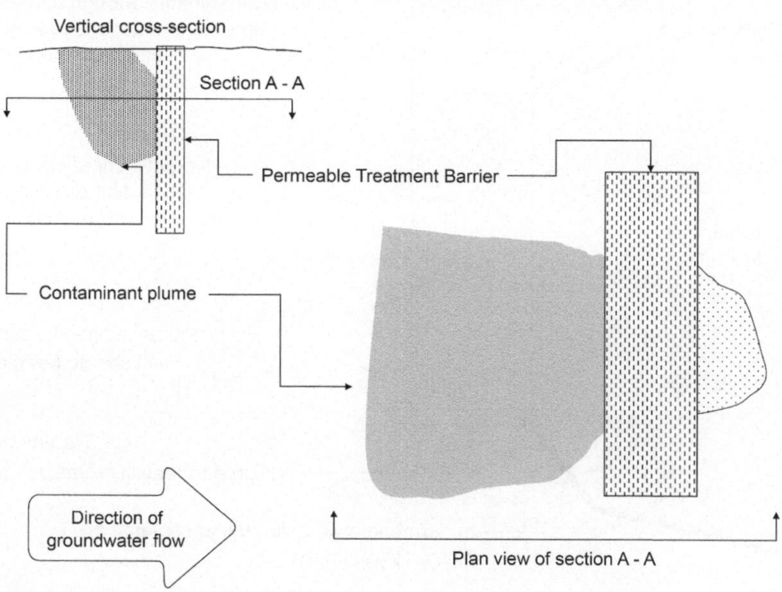

FIGURE 9.5 Cross-section and plan view of permeable reactive barrier (PRB) shows leachate plume entering the PRB with contaminants and leaving the PRB without contaminants.

compounds, and the manipulation of the pH-pE micro-environment in the PRBs will need to be made on the basis of site-specific knowledge of the interaction processes between contaminants and soil fractions.

Another key factor in the use of PRBs is the residence time. By this we mean the time spent in the PRB by a prescribed volume of contaminated plume. What is needed is the proper sets of circumstances for the reaction kinetics—i.e., contaminant-reactants interaction—to function efficiently (see Figure 9.6). This means obtaining the proper hydraulic conductivity through the PRB to achieve equilibrium or close-to-equilibrium conditions in the interaction processes. Laboratory permeability tests of candidate materials for the PRB are required to determine the bounding values of hydraulic conductivity. In general, these would be Darcy permeability tests with results expressed in terms of the Darcy coefficient of permeability k. The soil materials and their additives in the PRB should be "designed" to take into account the major attenuating (contaminant-removal and immobilization) processes. For inorganic contaminants, some of these would include sorption, precipitation, substitution, transformation, complexation, oxidation, and reduction. In the case of organic contaminants, these processes include sorption, abiotic transformation, biotransformation, abiotic degradation, and biodegradation.

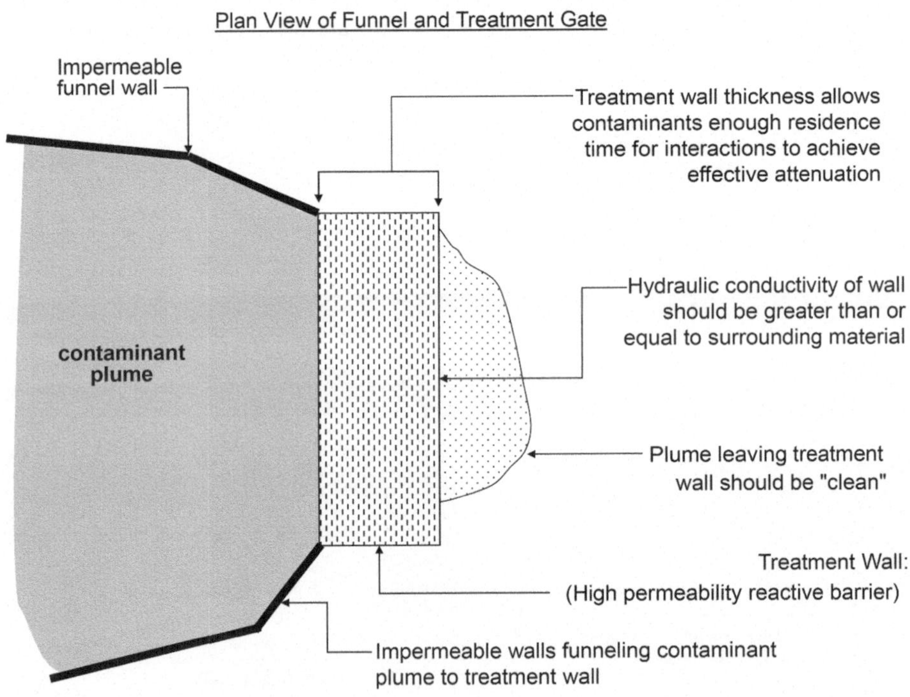

FIGURE 9.6 Funnel and gate arrangement of permeable reactive barrier (PRB) treatment of contaminant plume. Funnel effect is provided by the impermeable walls that channel contaminant plume transport to the PRB gate.

Other treatment procedures can also be used in place of the PRB. The presence of heavy metals in combination with organic chemicals in the contaminant plume is not an uncommon occurrence. One could, for example, envisage using the treatment zone as a "bioremediation pre-treatment" procedure for subsequent PRB procedures designed to fix or remove the metals. This is generally part of a multiple-treatment process. A good example of this is the use of pump-and-treat as the first phase of the remediation program. The efficiency of clean-up using pump-and-treat methods rapidly decreases as greater contaminant extraction from the groundwater or porewater is required. It is not unusual to remove some large proportion of the contaminants from the groundwater or porewater, and to leave the remaining proportion to be removed via natural attenuation processes in an in situ reactive region.

The "funnel-gate" technique shown in Figure 9.6 is one of the more common techniques used to channel the contaminant plume to flow through the reactive barrier. This funnel which is constructed or placed in the contaminated ground is composed basically of confining boundaries of impermeable material (e.g., sheet pile walls), which narrow towards the funnel mouth where the reactive wall is located. Other variations of the funnel-gate technique exist—obviously in accord with site geometry and site specificities.

9.4 ENHANCED NATURAL ATTENUATION AS ENGINEERED BARRIER

The use of *enhanced natural attenuation* (ENA) as an engineered barrier against the progression of contaminant plumes is an extension of the practice of using natural subsurface soils as a planned barrier to assimilate contaminants in their transport through the soils. The soil attenuation layer of an engineered barrier system is often composed of soil materials that are chosen for their contaminant assimilative capability—a designed soil-water system that is generally called an *engineered clay barrier*. Figure 9.7 shows a general view of an engineered barrier system used for containment of a waste pile. The details of the filter, membrane, and leachate collection system and the nature and dimensions of the contaminant attenuation layer are specified by regulatory command and control requirements or by performance requirements.

The intent in the design details of the engineered natural or enhanced natural attenuation barriers shown in the figure is that if leachates inadvertently leak through the high-density polyethylene membrane (HDPE) and are not captured by the leachate collection system, the contaminants in the leachate plumes will be attenuated by the engineered clay barrier. The engineered clay barrier serves as the second line of defense or containment. There are several options for the foundation base seen in the diagram. Since a fully compacted foundation base is a standard requirement—to provide support for the material contained above—one has the option of working with the native material if it has the proper assimilative potential, or with enhanced natural attenuation techniques described in the previous sections. Once again, the purpose of the foundation base is to provide attenuation of contaminants as a third line of defense against contaminant transport into the subsurface soils.

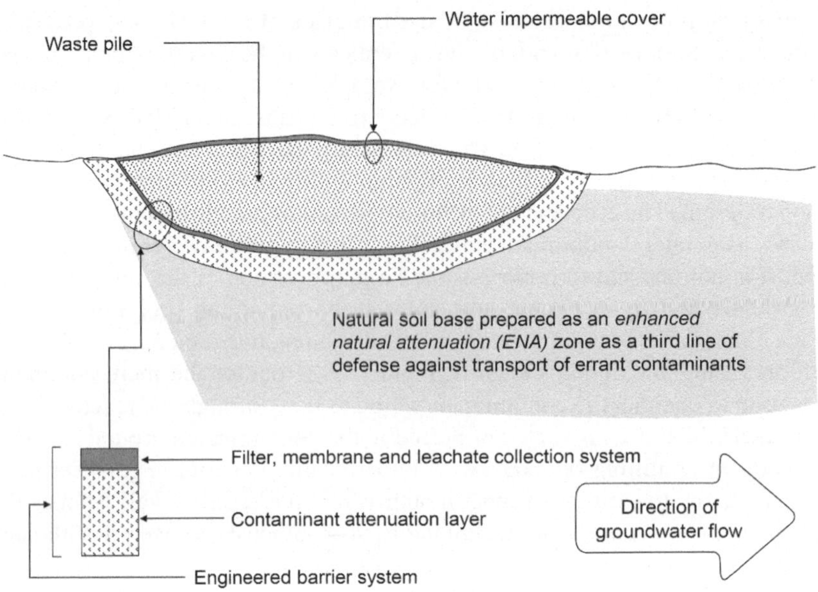

FIGURE 9.7 Contaminant attenuation layer constructed as part of an engineered barrier system. The dimensions of the attenuation layer and the specification of the various elements that constitute the "filter, membrane and leachate collection system" are generally determined by regulations or by performance criteria.

For assurance that the engineered barrier material, or that the foundation base material and preparation technique will provide for a capable barrier or foundation, it is necessary to obtain evidence of capability of the material to function according to specifications. At the present time, there are few mandatory specifications. If specifications are written, these are generally site and material specific and may not include tests that would provide evidence of the necessary assimilative capacity of the soil material used for the barrier. Prudent practice requires that one obtains evidence of capability of the material to function as required, and that the placement technique provides the proper barrier performance capability.

Evidence of capability of the engineered natural attenuation barrier to provide effective attenuation of contaminants requires attention to a set of tests and analyses as outlined in the diagram shown in Figure 9.8. The supporting laboratory studies are the same supporting laboratory studies shown in Figure 8.5 in Chapter 8. As in the case of the evidence of success protocols, the procedures for determining positive evidence that the engineering application of natural and/or enhanced natural attenuation is "working" require almost the same sets of procedures used in determining whether MNA is "working" (Figure 8.13 in Chapter 8).

One of the significant benefits in applying the protocols for determination of whether the engineering application of NA or ENA is "working" is the determination of the required thickness of the contaminant attenuating layer shown in Figure 9.7.

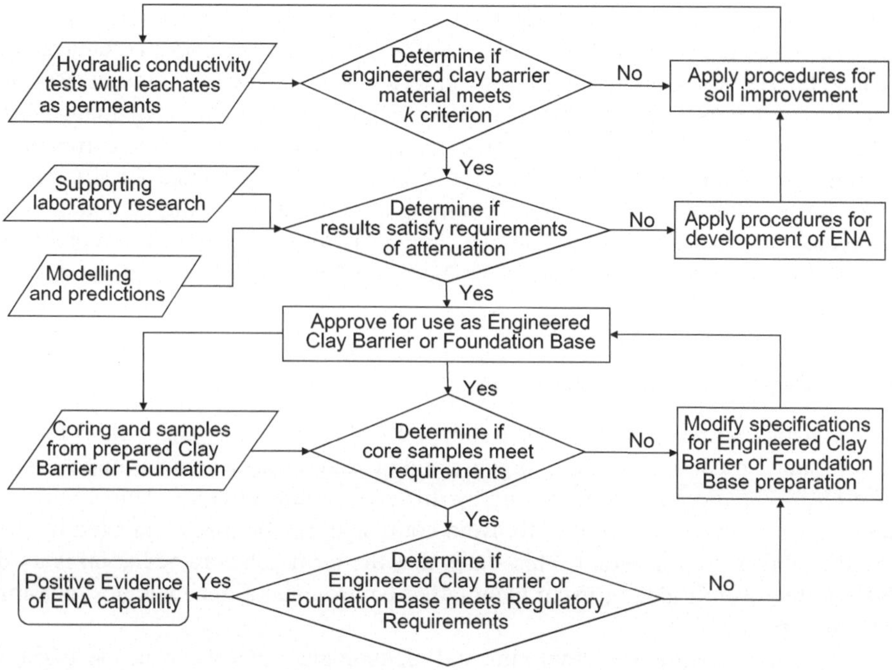

FIGURE 9.8 Protocol for determination of positive evidence that the engineering application of natural or enhanced natural attenuation is "working." ENA = enhanced natural attenuation.

As in the case of PRBs, the residence time in the engineered NA or ENA barrier is critical. The first set of decision diamonds shown in Figure 9.8 highlights this issue in terms of the *k* value required to meet residence time objectives.

9.5 PLANT-BASED NATURAL ATTENUATION ENHANCEMENT—PHYTOREMEDIATION

There are some that might argue that phytoremediation should not be included in the group of processes contributing to the natural attenuation of contaminants in soils because they are not totally associated with actions by the soil-water system. However, upon a closer look at how plants require nutrients that classify as contaminants in the discussions contained in all the previous chapters of this book, it becomes evident that the use of plant-based technology—i.e., phytoremediation—as an enhancement to natural attenuation remediation processes is not misplaced.

Plants require micro- and macronutrients to grow and to thrive. The distinction between micro- and macronutrients lies in the amounts needed for the plants to grow. We should bear in mind that different kinds of plants and different edaphic settings or conditions will mean that there will be differences in the distinction between micro and macro classification. In general, trace amounts of micronutrients are needed in comparison to somewhat larger amounts of macronutrients. Included in the list of micronutrients are chloride (Cl), the metalloid boron (B), and some heavy metals such as copper (Cu), iron (Fe), zinc (Zn), nickel (Ni), manganese (Mn), and molybdenum (Mo). The macronutrients include calcium (Ca), carbon (C), hydrogen (H), magnesium (Mg), nitrogen (N), oxygen (O), phosphorous (P), potassium (K), and sulfur (S). From this accounting of nutrients, one can see that the use of specially cultivated plants to "uptake" and even degrade contaminants can be a potentially attractive remediation tool. The "trick" is to be able to cultivate such types of plants.

9.5.1 Phytoremediation Processes

Research and various field trials have shown that there are more capabilities for the use of plants in control of both inorganic and organic chemical contaminants. The term *phytoremediation* has been used by different groups of researchers and practitioners to mean several things. To avoid confusion, the definition adopted by the USEPA (2000) is used herein. This states that the term "phytoremediation is used to mean the overall idea of using plant-based environmental technologies, not any specific application."

The various plant-based environmental technologies referred to in the USEPA document can be considered as separate and different kinds of remediation processes—all of which are dependent on edaphic conditions allowing for cultured plants to grow and to perform their function. The various phytoremediation processes classify as follows (USEPA, 2000; Ali et al., 2013; Pinto et al. 2015):

- Phytoextraction—uptake of contaminants in soil by plant root system. Specifically, cultured hyperaccumulator plants can uptake more than the normal quantities of heavy metal micronutrients. According to Ali et al. (2013), phytoextraction is most useful in removal of heavy metals and metalloids from contaminated soils. The heavy metals considered to be amenable for phytoextraction (USEPA, 2000) include Ag, Cd, Co, Cr, Cu, Hg, Mn, Mo, Ni, Pb, and Zn.
- Phytostabilization—the use of plants to immobilize and/or impede the mobility of contaminants (mainly heavy metals) by binding them to the soil mass in the rhizosphere. through assimilative processes discussed previously in Chapter 4—thus reducing or even eliminating their bioavailability.
- Phytovolatilization—according to USEPA (2000) is the uptake and transpiration of contaminants by plants, with release of contaminants (or modified form) into the atmosphere. Processes involved include plant metabolism of uptake contaminants and plant transpiration.
- Phytostimulation (Rhizodegradation)—activity directed towards organic chemical contaminants consisting of plant root systems involved in biodegradation and bioremediation in the rhizosphere. In essence, the processes involved are akin to the earlier discussion on biostimulation except in this case the plant root system is the stimulating agent.

- Rhizofiltration—pertains more to the contaminated soil-water within the root system. Sorption and precipitation mechanisms resulting from both biotic and abiotic processes are involved.
- Phytodegradation—uptake and breakdown of organic chemical compounds by plants' metabolic processes. Organic compounds found to have been amenable for phytodegradation include chlorinated solvents (McCutcheon, 1996) herbicides (Burken and Schoor, 1997), and insecticides (McCutcheon, 1996).

Ongoing research has shown that uptake of heavy metals can also be accomplished by some species of plants under processes that classify as *phytoremediation*. Many plants have the ability to extract and concentrate certain kinds of elements in the soil. Their root systems absorb and accumulate the necessary nutrients (and water) to sustain their growth. While metal-tolerant plants have some tolerance for toxic metal ions uptake, by and large, their tolerance level for such metals is very low. However, hyperaccumulating plants have higher levels of tolerance for toxic metal ions and can take HM ions up to several percent of their dry weight. Genetic engineering or selective breeding of the plants can lead to faster uptake and higher yields of contaminants. Plant species that accumulate metals include *Thlaspi calaminare* (zinc accumulation), *Alyssum* sp. (nickel accumulation), *Brassica juncea* (lead accumulation) (PRC Environmental, 1997) and *Populus* (TCE remediation) (Chappell, 1997).

Continued research into this area shows considerable promise in the use of specific plants for phytoextraction or phytodegradation. Mulligan et al. (2001) explains phytodegradation as the metabolism of contaminants in the leaves, shoots, and roots, resulting in the release of enzymes and other components for stimulation of bacterial activity or biochemical conversion. Examples cited from various sources by Mulligan (2001) show removal of trichloroethylene (TCE), perchloroethylene (PCE), 1,2-dichloroethylene (DCE), and xylene from groundwater by hybrid poplars.

The use of plants for remediation of contaminated lands is still in its infancy. As more and more research and field trials are conducted, there will be more benefits discovered. One of the biggest challenges is to determine how to extend the treatment zone (depth) beyond the rhizosphere and vadose zone. Another challenge is to continue work on development on hyperaccumulator plants.

9.6 ROLE OF SOILS IN NATURAL ATTENUATION PROCESSES

Other than the many assimilative, partitioning, and bioremediation processes offered by soils in respect to interactions with inorganic and organic chemicals, there are other beneficial NA processes offered by soils in their natural state. Two prominent and important functional roles fulfilled by soils are described in the next two sub-sections.

9.6.1 Soil Catalysis

The use of clay soils as contaminant attenuating material has by and large been considered in terms of physical interactions between sorbent and sorbate. Chapters 3 through 5 have elaborated on the interactions between contaminants and soils in respect to reactive forces and mechanisms associated with the surfaces of the soil solids and their respective functional groups. The catalytic role of clay soils, and mineral soils in particular, has not been well exploited. In part, this is due to the fact that soil catalysis is not a well appreciated phenomenon, and also in part because soil improvement methods for contaminant assimilation have been confined to the previously stated sorbent-sorbate interacting relationship.

In Section 3.4.1 we briefly discussed hydrolysis as an acid-base reaction. The importance of hydrolysis or hydrolytic reactions in respect to contaminants in a soil-water system cannot be overstated. Section 4.4.1 has addressed hydrolysis reactions of metal ions and has indicated that these reactions are influenced by the pH of the active system, temperature, redox, and type and concentration of the metal ions. The hydrolysis reactions of metals (MX) can be expressed as:

$$MX + H_2O \rightarrow MOH + X^- + H^+ \tag{9.1}$$

With respect to organic chemical contaminants, hydrolysis involves reactions between an organic molecule and water, resulting in the formation of a new covalent bond with OH. Designating the organic molecule as RX, cleavage of the covalent bond with X occurs. The net reaction is given by Mill and Mabey (1988) for displacement of X by OH^- as:

$$RX + H_2O \rightarrow ROH + X^- + H^+ \tag{9.2}$$

Metal ion catalysis occurring at interfaces between two phases (porewater and mineral soil particle surface) is referred to as heterogeneous catalysis. Clay particle surfaces exhibit surface acidity which can be from 2 to 4 units lower than the acidity of the bulk porewater. Soil-catalyzed hydrolysis reactions associated with the surface acidity of clay minerals can be significant because they can affect the hydrolysis half-lives of the reacting organic chemicals,—i.e., they affect the kinetics of hydrolysis. Figure 9.9 shows the surface acidity of kaolinite in relation to the moisture content of the soil. This surface acidity which is derived from the surface hydroxyls on the octahedral layer of the mineral particles is reduced significantly as the moisture content of the soil is increased. Surface acidity in the case of montmorillonites is due to isomorphous substitution and to interlamellar cations. The charge and nature of the cations affect the degree of catalytic activity since these (cations) impact directly on the polarizing power and the degree of dissociation of the water in the inner Helmholtz plane (adsorbed water). In addition, the surface acidity (of montmorillonite) increases as we increase the valency of the exchangeable cations,—e.g., for Na^-, Mg^-, and Al-montmorillonites, surface acidity increases in the order given as Na, Mg, and Al.

For sites contaminated with heavy metals, metal-ion catalysis of hydrolysis occurs through the heavy metals sorbed by the soil fractions. Direct polarization results from coordination of hydrolysable functional groups by the metals. According to Buckingham (1977), direct polarization mechanisms can accelerate hydrolysis rates

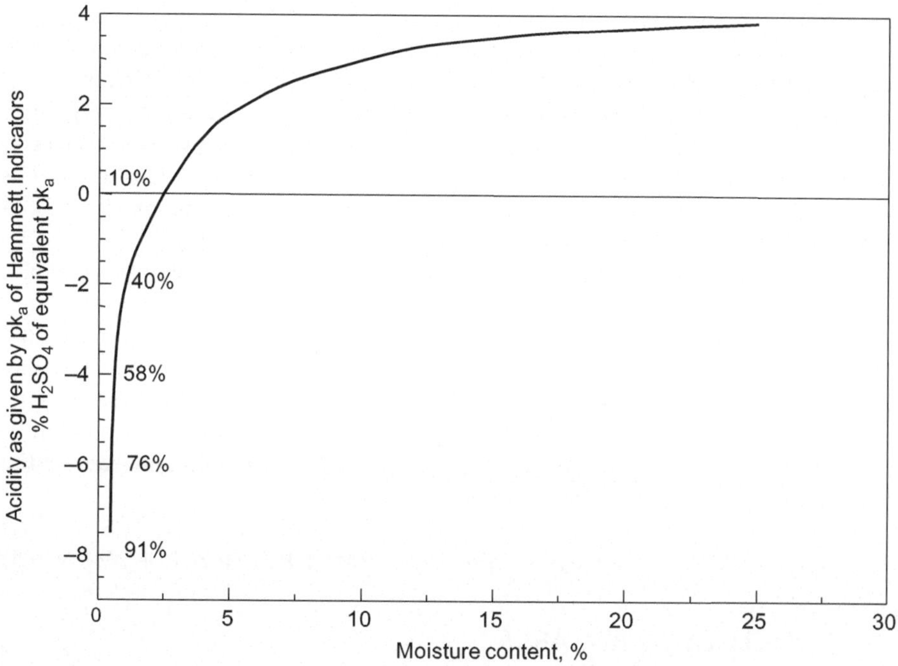

FIGURE 9.9 Surface acidity of kaolinite expressed as pk_a of Hammett indicators and also as percent of H_2SO_4 of equivalent pk_a. (Adapted from Solomon, D.H. and Murray, H.H., *Clays and Clay Minerals*, 20, 135–141, 1972.)

by factors of ten thousand or more. It is also possible for metal-ion catalyzed hydrolysis to form a metal-coordinated nucleophile that is more reactive than a comparable free nucleophile (Plastourgou and Hoffman, 1984).

$$M(H_2O)_m{}^{n+} \leftrightarrow M(H_2O)_{m-1}OH^{(n+1)+} + H^+ \quad (9.3)$$

$$M(H_2O)_{m-1}OH^{(n+1)+} + RCOX \rightarrow (H_2O)_m M\text{-}O\text{-}CR^{n+} + X: \quad (9.4)$$

The increase in acidity of the water molecules, in situ generation of OH^- results from metal coordination.

9.6.2 Promotion of Oxidation-Reduction Reactions

The promotion of oxidation-reduction reactions by soils is generally called: soil-catalyzed oxidation and soil-catalyzed reduction. In essence, what we have is soil-mediated electron transfer. Soil-catalyzed oxidation depends on several factors, the most important of which are the presence of redox-active metals such as iron, manganese, copper, aluminum and trace metals. Free radical oxidation (homolytic oxidation) in soils is dependent on lower solubility limits and chemical structure of the organic chemicals

(Dragun, 1988). The structural groups are classified according to the characteristics of electron withdrawal and donation, as for example: aromatic compounds with only electron-donating substituents and aromatic compounds with electron-withdrawing with: (a) weak electron-donating fragments, and (b) strong electron-donating groups. In the example of the oxidation of phenols by montmorillonite given in Chapter 5, we see from the work reported by Yong et al. (1997), the greater oxidizing capability of the Fe(II)-clay in transforming the monomer 2,6-dimethylphenol of mass 122 to a 2,6-dimethylphenol dimer of mass 242—in comparison to other clays.

As noted in Chapter 5, the major groups of organic chemicals that undergo reductive transformation in reducing environments are the halogenated aliphatic and aromatic compounds. For abiotic processes, we look towards electron-mediated reductions to accomplish reductive transformations. Soil organic matter and clay minerals provide electron-mediated reductions through iron and sulfide components in the mineral structure. Kriegman-King (1991) reported on the transformation of hexachloroethane (HCA) to tetrachloroethylene (PCE) by biotite and vermiculite according to the following mechanism:

$$2[Fe^{2+}, nM^{+}]_{mineral} + HCA \rightarrow 2[Fe^{3+}, (n-1)M^{+}] + PCE + 2Cl^{-} + 2M^{+} \quad (9.5)$$

9.7 CONCLUDING REMARKS

The existence of natural bioremediation assures us that, given an allowed time factor, exploitation of natural attenuation for remediation of organic chemical contaminants in the ground can be successful either by itself, or more likely in terms of an added element in the total treatment of a contaminated site. By this, we mean that we can exploit the assimilative potential of soils by using this potential in concert with other technologies for remediation of contaminated sites. This can be enhanced with biostimulation and/or bioaugmentation, plus geochemical and biogeochemical manipulation. For leachate streams containing heavy metals, natural attenuation works in the form of partitioning, precipitation, and speciation of the metals in the soil. Natural bioremediation of soils contaminated with heavy metals occurs through processes such as bioaccumulation and biological oxidation and reduction. It is important to bear in mind that there are both advantages and disadvantages to the use of natural attenuation in contaminant plume management. Many of these have been discussed in the previous chapter. That being said, we have every expectation that exploitation of the various processes that contribute to the natural attenuation of contaminants will flourish and expand. A good example of this is the more recent use of phytoremediation.

It seems fairly clear that we will want to continue exploiting the natural assimilative and bioremediative potential of soils—with or without enhancements. It also seems clear that this property of natural attenuation will be a useful tool in a multiple treatment process—e.g., treatment train. The protocols set forth in the multiple *lines of evidence* in this chapter and the previous chapter rely on the execution of several mandatory tests and analyses which we consider as being of utmost importance and significance. The diagram in Figure 9.10 provides a visual summary of the use of enhancements that can be applied to aid in the efficacy of natural attenuation processes.

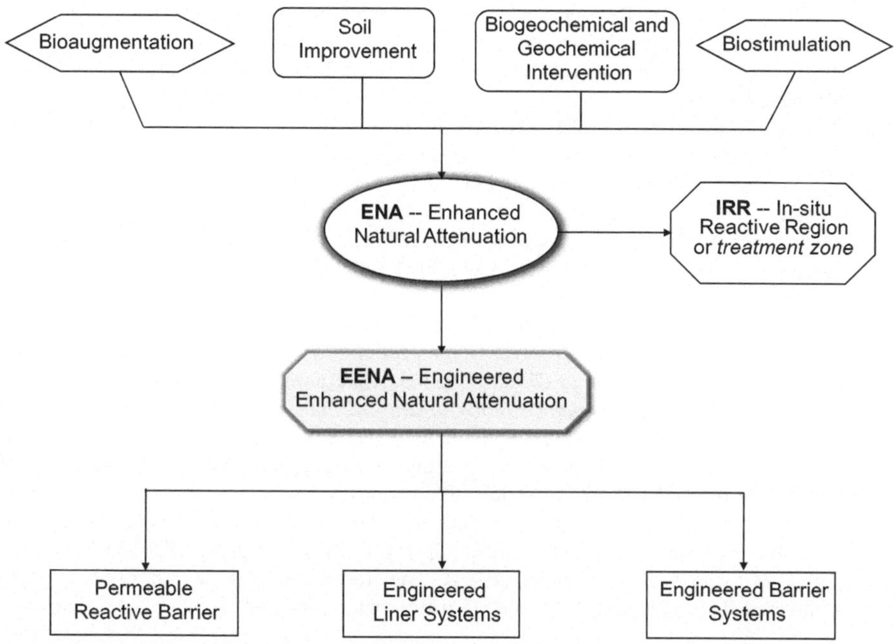

FIGURE 9.10 Enhanced techniques for natural attenuation of soils and engineering application of Natural attenuation (NA) and enhanced natural attenuation (ENA).

Whilst we recognize and value the importance of monitoring and analyzing groundwater and porewater samples, we cannot over-emphasize the importance of analyses of soil samples from contaminated sites and from monitoring stations. We need to be fully aware of the nature of the contaminants sorbed onto the soil solids, and to be conscious of the possibility that these can in themselves pose health threats to biotic receptors, and at some point in time they may become mobile. The material presented in this book has attempted to present the total picture of attenuation from the soils' point of view, in the hope that the reader will begin to better appreciate the need for more attention to the role of soils in the determination of contaminant attenuation.

REFERENCES

Ali, H., Khan, E., and Sajad, M.A. (2013), Phytoremediation of heavy metals—concepts and applications, *Chemosphere*, 91:869–881.

Buckingham, D.A. (1977), Metal-OH and its ability to hydrolyze (or hydrate) substrates of biological interest, in A.W. Addison, W.R. Cullen, D. Dolphin, and B.R. James (Eds.), *Biological Aspects of Inorganic Chemistry*, John Wiley & Sons, New York.

Burken, J.G., and Schnoor, J.L. (1997), Uptake and metabolism of atrazine by poplar trees, *Environ. Sci. Technol.*, 31:1399–1406.

Chappell, J. (1997), Phytoremediation of TCE using *Populus*, Report prepared for U.S. EPA Technology Innovation Office, August.

Dragun, J. (1988), *The Soil Chemistry of Hazardous Materials*, Hazardous Materials Control Research Institute, Silver Spring, MD, 458p.

Kriegman-King, M.R., and Reinhard, M. (1991), Reduction of hexachloroethane and carbon tetrachloride at surfaces of biotite, vermiculite, pyrite, and marcasite, in R.A. Baker (Ed.), *Organic Substances and Sediments in Water, Vol. 2*, Lewis Publishers, Chelsea, MI, pp. 349–364.

Mill, T., and Mabey, W. (1988), Hydrolysis of organic chemicals, in O. Hutzinger (Ed.), *Handbook of Environmental Chemistry, Vol. 2D: Reactions and Processes*, Springer-Verlag, New York, pp. 71–111.

McCutcheon, S.C. (1996), Phytoremediation of organic compounds: Science validation and field testing, in W.W. Kovalick and R. Olexsey (Eds.), *Workshop on Phytoremediation of Organic Wastes*, December 17–19, 1996, Ft. Worth, TX. An EPA unpublished meeting summary (http://www.rtdf.org).

Mulligan, C.N., Yong, R.N., and Gibbs, B.F. (2001), The use of selective extraction procedures for soil remediation, *Proceeding International Symposium on Suction, Swelling, Permeability and Structure of Clays*, Balkema, Rotterdam, the Netherlands.

Mulligan, C.N. (2001), *Environmental Biotreatment: Technologies for Air, Water, Soil, and Wastes*, Government Institutes, Rockland, MA, 395p.

Pinto, A.P. , de Varennes, A., Fonseca, R., Martins Texeira, D. (2015), Phytoremediation of Soils Contaminated with Heavy Metals : Techniques and Strategies, In Phytoremediation Management of Environmental Contaminants, Volume 1, Editors: Ansari, A.A., Gill, S.S., Gill, R., Lanza, G.R., Newman, L., Springer, Switzerland, pp 133–155

Plastourgou, M., and Hoffman, M.R. (1984), Transformation and fate of organic esters in layered-flow systems: The role of trace metal catalysis, *Environ. Sci. Technol.*, 18:756–764.

PRC Environmental Management. (1997), Recent developments for in situ treatment of metal contaminated soils, *Prepared for U.S. Environmental Protection Agency, Office of Solid Waste and Emergency Response*, Technology Innovation Office, Washington, DC.

Solomon, D.H., and Murray, H.H. (1972), Acid-base interactions and properties of kaolinite in non-aqueous media, *Clays Clay Miner.*, 20:135–141.

Strauss, H. (1991), Final report: An overview of potential health concerns of bioremediation, Environmental Health Directorate, Health Canada, Ottawa, Canada.

Tan, B.K. (2003), Update on arsenic leaching tests, unpublished report submitted to Geoenvironmental Research Centre, Cardiff University, Cardiff, UK.

USEPA. (2000), Introduction to phytoremediation, EPA/600/R-99/107.

Yong, R.N., Desjardins, S., Farant, J.P., and Simon, P. (1997), Influence of pH and exchangeable cation on oxidation of methylphenols by a montmorillonite clay, *Appl. Clay Sci.*, 12:93–110.

Index

Note: Page numbers in italic and bold refer to figures and tables, respectively.